Essential Gas Safety
Eighth Edition

Technical Editor: Blue Flame Associates

Previous contributors:
Chris Long
Frank Williamson
Graham Elkins
Mark Rolfe
Jamie Cooper
Gary Carter
Tony Lloyd
David J Smith
David Smith
Fiona Talbot (TQI)

Notice

CORGI*direct* has used its best efforts to prepare this manual, but makes no warranty about the content of the manual and will not be liable under any circumstances for any direct or indirect damages resulting from any use of this manual.

Acknowledgement

CORGI*direct* gratefully acknowledges the use in this manual of reference material published by the British Standards Institution (BSi), the Institution of Gas Engineers and Managers (IGEM) and Uklpg.

CORGI*direct* also gratefully acknowledges the assistance provided by OmegaFlex Ltd in supplying information and photographs used within this manual.

CORGI*direct*
Essential products and services delivered direct to your door

Published by

CORGI*direct*

CORGI*direct*

Telephone: 0800 915 0490
Website: www.corgi-direct.com

First Edition:	May 1999
Second Edition:	June 2001
Third Edition:	November 2002
Third Edition – Revised:	October 2003
Fourth Edition:	May 2006
Reprinted:	2007
Fifth Edition:	June 2008
Reprinted:	December 2008
Reprinted:	March 2009
Fifth Edition – Revised:	March 2010
Fifth Edition – Second Revised:	April 2011
Sixth Edition:	March 2012
Seventh Edition:	August 2013
Eighth Edition:	August 2016

ISBN: 978-1-907723-30-8

Contents

Each section includes its own detailed contents list

Gas safety legislation – 1

1 – Gas safety legislation

Gas safety legislation – 1

The competence you need to carry out work

- you must not carry out any work (see **Part 19 – Definitions**) in relation to gas appliances and other gas fittings or gas storage vessels covered by this manual unless you are competent to do so
- when any work is carried out, gas businesses must be registered with Gas Safe Register and their gas operatives (engineers) must hold valid competency certification for each work activity they wish to undertake.
- the competency certification must be issued under one of Gas Safe Registers accepted routes to registration (see **What competency certification is valid?** in this Part)
- no employer, member of the public or other responsible person should knowingly employ any gas operative who cannot comply with these requirements
- you must install all appliances and other gas fittings in accordance with the Gas Safety (Installation and Use) Regulations, British Standards, Building Regulations, Regulations for Electrical Installations or those Regulations appropriate to the geographical region in which they are to be installed
- you must also install appliances to comply both with the Water Supply (Water Fittings) Regulations – and manufacturers' installation instructions

How is your competence as a gas operative assessed?

Gas Safe's rules for registration require individual gas operatives to be competent to carry out gas work.

Competence in safe gas work, in relation to gas fittings, requires:

- knowledge;
- understanding;
- the practical skills;

to carry out the work in hand in such a way as to prevent danger to life and property.

Which competency certification is valid?

Currently, Gas Safe Register accepts that gas operatives are competent if they hold valid competency certification to cover the areas of gas work that they intend to undertake.

Subject to conditions that apply, Gas Safe Register accepts competency certification issued under the following schemes:

1. Nationally Accredited Certification Scheme for Individual Gas Fitting Operatives (ACS).
2. National/Scottish Vocational Qualifications (N/SVQ) in Gas Services that have been aligned to ACS in matters of gas safety.
3. Qualification and Credit Framework (QCF) Qualifications in Gas Utilisation and Plumbing that have been aligned to ACS in matters of gas safety (a number of these qualifications may still have the designation N/SVQ).
4. Group Competence Scheme (GCS).

What do the certifications involve?

The Nationally Accredited Certification Scheme (ACS) for Individual Gas Fitting Operatives

ACS is the gas industry's recognised scheme under which individual gas operatives demonstrate competence, to comply with the Gas Safety (Installation and Use) Regulations (ACoP and Guidance) and Gas Safe Register's rules for registration.

ACS has significantly changed the assessment of competence by introducing:

- national standards of assessment
- separation of assessment from training
- accreditation by The United Kingdom Accreditation Service (UKAS) to BS EN ISO/IEC 17024: 2012 'Conformity Assessment. General requirements for bodies operating certification of persons'

Prior to undertaking any assessments, candidates must ensure they have all the necessary information to enable them to prepare for and fully understand the assessment process.

They can do this by obtaining information from certification bodies and assessment centres.

All assessments require candidates to:

- complete a practical assessment
- complete a written assessment
- where necessary, answer oral questions posed and recorded by an assessor

The results of all assessments are downloaded directly to Gas Safe Register by one of the UKAS Accredited Gas Certification Bodies and only these results and associated certificates will be accepted by Gas Safe Register (see **Part 21 – References**).

Practical assessments cover safety in practice

They are designed to enable the candidate to demonstrate competence using a range of appropriate equipment. The practical assessments are straightforward, based upon the work carried out by gas operatives and should not present any undue difficulties.

Written assessments cover competence in knowledge and understanding

To enable the candidate to demonstrate this type of competence, different types of questions are used. These may be multiple choice, 'true or false', or short response.

The majority of questions are of the 'open book' type. Candidates are given suitable reference material to enable them to resource the answers if neccessary.

Qualifications that have been aligned to ACS in matters of gas safety

The ACS is designed to assess the competent performance of experienced gas operatives in the areas of safe gas work they undertake

The N/SVQ assessment process is designed to assess inexperienced individuals who are developing broad work competencies using a number of assessment methods

So, at a glance, the N/SVQ process is designed to assess operatives entering the gas industry:

- for the first time, or
- changing career direction, or
- inexperienced individuals

There are a large number of N/SVQ and QCF Qualifications in the Gas Services / Utilisation and Plumbing Sectors that have been aligned to ACS in matters of gas safety.

The recognised N/SVQ Qualifications in the Gas Services Sector that have been aligned to ACS in matters of gas safety are issued by the Awarding Organisation (AO) City & Guilds. All the qualifications have the Scheme Code 6012.

The recognised QCF Qualifications in the Gas Utilisation and Plumbing Sector that have been aligned to ACS in matters of gas safety are issued by the AO's City & Guilds, EAL, BPEC and Logic Certification. All these qualifications have been issued with a code by Energy and Utility Skills with a 'EUSGU' prefix.

The completion of any of these qualifications means that persons who are successful in being awarded the N/SVQ's or QCF's can apply for registration without the need to undertake further assessment.

The successful completion of these qualifications are downloaded directly to Gas Safe Register by one of the AO's listed and only these results and associated certificates will be accepted by Gas Safe Register.

A detailed list of which ACS Categories each qualification has equivalence to is available from the respective AO's.

Unlike the recognised N/SVQ and QCF Qualifications in the Gas Services / Utilisation and Plumbing Sectors mentioned there are many other qualifications in the Gas, Plumbing & Heating, Refrigeration & Air Conditioning and Heating & Ventilation Sectors that have not been aligned with ACS and are not recognised for registration purposes.

If you have completed one of these qualifications you will be required to undertake additional ACS assessments to be able to register with Gas Safe Register. You may also need to undertake further on the job and off the job training to gain additional knowledge and work experience before being eligible to undertake ACS assessments.

A detailed list of those qualifications that are and are not accepted is available from the Energy and Utility Skills website:

www.euskills.co.uk

Group Competence Scheme (GCS)

This is a relatively new process for companies that have a detailed Competence Management System that proves the competence of the workforce by a range of means, including workplace assessment and audits, such as quality assurance processes. It involves the approval of the Employer to run their Competence Management System rather than the certification of an individual person.

The scheme is still aligned to ACS in matters of gas safety, but these can be demonstrated in the workplace rather than an assessment centre. The Certification Bodies that offer this Scheme will be accredited by UKAS against the standard BS EN ISO/IEC 17021-1:2015 Conformity Assessment. Requirements for bodies providing audit and certification of management systems.

The verification of a gas operative's ongoing competency is downloaded directly to Gas Safe Register by a UKAS Accredited Gas Certification Body. No certificate is issued to the gas operative by the Certification Body, but a certificate may be issued by an employer.

The current Certification Bodies that are offering approval against the GCS are:

- Blue Flame Certification
- NICEIC Certification
- Stroma Certification

1 – Gas safety legislation

The Gas Safety (Installation and Use) Regulations (GSIUR) – what do they cover?

GSIUR deal with the safe installation, maintenance and use of gas installations and appliances mainly in domestic and commercial premises such as offices, shops, public buildings and similar places, with certain exceptions (see also **Certificate of Exemption No. 1** in this Part).

The current regulations update, consolidate and replace any previous regulations or amendments and apply to Manufactured gas, Natural gas and Liquefied Petroleum Gas (LPG) installations.

Responsibilities are placed on a wide range of people including those:

- installing
- servicing
- maintaining, or
- repairing gas appliances and other gas fittings as well as:
  - suppliers and users of gas
  - certain landlords

Certificate of Exemption No. 1

There is in place a Certificate of Exemption No. 1 in relation to the Gas Safety (Installation and Use) Regulations (GSIUR).

Background to exemption certificate

It is a requirement of the GSIUR and in particular Regulation 26 Gas appliances – safety precautions clause (9) "that where a person performs work on a gas appliance he shall immediately thereafter examine –

1. The effectiveness of any flue;
2. The supply of combustion air;
3. Its operating pressure or heat input or, where necessary, both;
4. Its operation so as to ensure its safe functioning"

As can be seen, the Regulation asks for the appliance operating pressure or heat input to be measured and checked as part of a range of tests and checks to be applied in all cases where "work" is undertaken on a gas appliance. It is recognised that this is not always possible or practicable to carry out these particular checks to confirm the safety of the appliance.

In the cases where the requirements of Regulation 26(9)(c) cannot be met, (measuring the gas rate, and/or the operating pressure), another method of checking the appliance for correct operation must be employed (see Note).

The 'Certificate of Exemption' permits appliances in certain circumstances to be checked for safe operation using a suitably calibrated electronic portable combustion gas analyser (ECGA), verifying that the combustion performance of the appliance meets that specified by the individual appliance manufacturer.

Alternately, in situations where manufacturer's information is not available that the combustion performance ratio (CO/CO_2) of the appliance is within the defined action levels specified in the relevant part of BS 7967: 2015 'Guide for the use of electronic portable combustion gas analysers for the measurement of carbon Monoxide in dwellings and the combustion performance of domestic gas-fired appliances'

Note: This is only applicable to appliances which incorporate air/gas ratio valve technology where no gas meter is installed.

Who enforces these regulations?

The Health and Safety Executive (HSE) and Local Authorities enforce them (as determined by the Health and Safety (Enforcing Authority) Regulations).

How the Approved Code of Practice (ACoP) helps you comply

The ACoP supports the regulations by giving practical advice. Follow this and you should be doing enough to comply with the law relating to the matters it covers. It is true that you may use alternative methods to those set out in the code and still achieve equivalent standards that comply with the law.

However, the code does have a special legal status. If gas operatives are prosecuted for a breach of health and safety law and it is proved that they did not follow the relevant provisions of the code, they will need to show that they have complied with the law in some other way. Otherwise a court may find them at fault.

The regulations and the ACoP also come with guidance notes that do not actually form part of the code. The guidance is not compulsory and you are free to take other action. But if you do follow it, you will normally be doing enough to comply with the law.

The HSE seek to secure compliance with the law and may refer to the guidance as highlighting good practice.

How important are a manufacturer's installation instructions?

The GSIUR highlight the fact that they are very important. The manufacturer's instructions supplied with a gas appliance will normally specify it must be installed in accordance with the British Standards – or other published guidance applicable at the time of type testing of the appliance (see **Gas Appliance (Safety) Regulations (GASR)** and **CE Marking** in this Part).

However, the manufacturer's instructions may recommend special requirements specific to the appliance type and model. Where they do, you should follow these.

How you assess if an existing appliance is installed correctly

1. First consult the manufacturer's instructions for the appliance.
2. Note any special requirements (where the manufacturer's instructions for the appliance set special requirements for its installation, assess the appliance against these requirements).
3. Where the manufacturer's instructions for the appliance simply call up the British Standards or other published guidance, assess the installation against the requirements of the current versions of these standards.

What if the manufacturer's instructions are not available?

Assess the installation against the requirements of the current and relevant British Standards or other published guidance.

1 – Gas safety legislation

The Gas Act 1995 updated the law – as competition is now permitted in the gas industry

The Gas Act 1995 updated the Gas Act 1986, including new licensing arrangements for Gas Transporters (GTs) as the Government now allows competition in the domestic gas market/supply industry.

From 1st April 1996, The Gas Act allows gas users, including domestic users, to purchase gas from whichever supplier they wish to use. It also includes provision for safety regulations to be made in the following areas:

- The Gas Safety (Installation and Use) Regulations
- The Gas Safety (Rights Of Entry) Regulations

You can obtain a copy of The Gas Act 1995 from The Stationary Office (TSO) – (see **Part 21 – References – Organisations**).

The Health and Safety at Work etc. Act (HSWA) – the wide ranging duties explained

This act places duties on everyone concerned with work activities including employers, the self-employed and employees. This includes manufacturers, designers, suppliers and importers of equipment/material for use at work and people in control of premises. It also includes provisions to protect members of the public. The duties apply to:

- individual people
- corporations
- companies
- partnerships
- local authorities
- nationalised industries, etc.

The duties are expressed in general terms, applying to a wide range of work activities and work situations. Every employer has a duty to ensure, so far as is reasonably practicable, the health, safety and welfare at work of their employees.

The principles of safety responsibility and safe working are expressed in sections 2 – 9 of the HSWA.

Both employers and the self-employed are required to carry out their work so as to ensure, as far as is reasonably practicable, that they do not expose people who are not their employees to health and safety risks (see section 3(1) and 3(2) of HSWA).

In some areas the general duties have been supplemented by specific requirements in regulations made under the Act and such regulations will continue to be made e.g. GSIUR.

Specific legal requirements are also included in earlier legislation which is still in force.

What happens if the act is not complied with?

Failure to comply with the general requirements of the act, or with the specific requirements found elsewhere, may result in legal proceedings.

Although some of the duties imposed by the act and related legislation are absolute, many are qualified by the words 'so far as is reasonably practicable' or 'so far as is practicable'.

What does 'practicable or reasonably practical' mean?

If someone is prosecuted for failing to comply with a duty which is qualified by these words, it is up to the accused to show to the court that it was not reasonably practicable for them to do more to comply with the duty.

As a rule of thumb, the judgement of what is 'reasonably practicable' involves weighing up the seriousness of the risk against the difficulty and cost of reducing/eliminating it.

It can mean that where the difficulty and cost of reducing the risk are high and the benefits to be gained are low, action may not be necessary. However, such decisions are based on a range of factors and are dependent upon the individual circumstances.

The following sections of HSWA specify particular requirements:

- Section 15 Regulations to be made
- Sections 16 and 17 Approved Codes of Practice and their use in legal proceedings
- Sections 21 and 23 Improvement and prohibition notices
- Section 33 Offences and penalties

The Gas Safety (Management) Regulations (GSMR) – setting out a safety framework that transporters and suppliers must meet

The Health and Safety Executive (HSE), the enforcing body for gas safety matters, uses this legislation which covers the differing requirements of gas transporters and suppliers.

1. Transporters must:
    - submit a 'safety case' laying out the systems and procedures that will be adopted and applied, to ensure that gas is conveyed safely. The HSE will review the safety case and approval to convey gas will be granted subject to its ratification
    - operate a full gas emergency service where the service is provided either by the gas transporters employees or by a sub-contractor
    - operate a central telephone emergency number for gas users to report gas escapes
2. Suppliers must provide a gas incident investigation service which reports on incidents e.g. Carbon Monoxide (CO) poisoning.

A copy of the GSMR can be obtained from TSO. See **Part 21 – References – Organisations**.

The Gas Safety (Rights of Entry) Regulations – invoked where threat to life or property

A Gas Transporter (GT) has Rights of Entry to make safe in a dangerous situation.

When and how do you contact this help?

Under these regulations, in England, Scotland and Wales (see Note), a GT has 'Rights Of Entry' to a property to deal with a dangerous situation to make that situation safe.

Remember that in such circumstances, the GT could only respond to a situation where life or property is threatened.

Note: Similar regulations apply in Northern Ireland and the Isle of Man.

1. If you find an actual or suspected escape of gas, or the products of incomplete combustion:

The Gas Emergency Contact Centre receives calls, provides safety advice and passes information to the Emergency Service Provider (ESP).

Tel: 0800 111 999 reporting line.

2. If you find an Immediately Dangerous situation and the responsible person for the premises refuses to allow the appropriate action, then you must refer the matter on immediately, as follows:
   - in England, Scotland and Wales, contact the Gas Emergency Contact Centre, on telephone number 0800 111 999. In the case of Liquefied Petroleum Gas (LPG), contact the gas supplier, whose details can be found on the bulk storage vessel or cylinder (if no label can be found, information can be found under GAS in the local telephone directory)
   - in Northern Ireland contact Phoenix Natural Gas on 0800 002 001, or the LPG supplier
   - in the Isle of Man contact Manx Gas Ltd for all areas (including LPG) on 0808 1624 444
   - in the Channel Islands contact either Guernsey Gas Ltd on 01481 749000 or Jersey Gas Company Ltd. on 01534 755555, as appropriate

What happens next?

The ESP will then take the necessary action through their GT.

In exceptional circumstances, where they are refused access to the equipment, they may obtain a warrant to exercise their Rights of Entry and powers to disconnect.

Note: Where you work in a customer's premises, you do so by invitation of the customer. Actions you may wish to take must be with the user/responsible person's permission.

See also Part 2 – Gas emergency actions and procedures – Gas Safety (Rights of Entry) Regulations – for further guidance.

Gas Regulations that affect Northern Ireland and the Isle of Man

1. Northern Ireland

The Health and Safety Executive for Northern Ireland (HSENI) is the enforcing authority for the gas safety regulations there. The regulations that apply in Northern Ireland are:

1. The Gas Safety (Installation and Use) Regulations (Northern Ireland) 2004.
2. The Gas (Northern Ireland) Order 1996. Schedule 5 deals with Rights of Entry.
3. The Gas Safety (Management) Regulations 1996.

2. Isle of Man

The enforcing body for the Isle of Man is the Health and Safety at Work Inspectorate, part of the Department of Local Government and Environment (DoLGE) and the Isle of Man regulations are:

1. The Gas Safety (Installation and Use) Regulations 1994 as amended and applied by the Gas Safety (Application) Order 1996.
2. The Gas Safety (Rights of Entry) Regulations 1983 as applied by the Gas Safety (Application) Order 1996.

Building Regulations/Standards – requirements you must follow for any building work

When you undertake any building work you must follow the requirements of Building Regulations in England, Wales and Northern Ireland and Building Standards in Scotland. The Regulations/Standards are supplied in the form of 'Approved Documents' that are sectioned into different parts.

The following documents detail requirements for gas installations in domestic premises. They give guidance on how to meet some of the specifications required to comply with the GSIUR.

England and Wales – Building Regulations

- Approved Document B – Fire safety. Volume 1 - Dwellinghouses
- Approved Document F – Ventilation
- Approved Document G – Sanitation, hot water safety and water efficiency
- Approved Document J – Combustion appliances and fuel storage systems
- Approved Document L1A – Conservation of fuel and power in new dwellings
- Approved Document L1B – Conservation of fuel and power in existing dwellings
- Approved Document P – Electrical Safety – Dwellings

Northern Ireland – Building Regulations

- Technical Booklet E – Fire safety
- Technical Booklet F1 – Conservation of fuel and power in dwellings
- Technical Booklet K – Ventilation
- Technical Booklet L – Combustion appliances and fuel storage systems
- Technical Booklet P – Sanitary appliances, unvented hot water storage systems and reducing the risk of scalding

Scotland – Building Standards (Scotland) Regulations

Technical Handbook – Domestic:

- 0 – General
- 1 – Structure
- 2 – Fire
- 3 – Environment
- 4 – Safety
- 5 – Noise
- 6 – Energy
- 7 – Sustainability

Building Regulations – definition of competence – how does this affect you?

To meet these Regulations, anyone who installs a gas appliance or fitting, including chimney systems, must be competent.

Under the GSIUR, if you are a registered installer with current certificates in the relevant ACS modules, you are deemed competent. This same definition is used to define competence for the purposes of the Building Regulations.

A registered gas operative is therefore deemed competent for some other types of building work necessary for the installation of the appliance, such as the installation of a chimney system.

Note: Competence does not extend to the provision of a masonry chimney.

Under the regulations:

- you must notify building work to the appropriate local Building Control Body (BCB)

Where the installation of a gas appliance is concerned, the Building Regulations stipulates that where you are competent, as defined in the regulations:

- you do not need to notify BCB prior to carrying out the work
- you are, however, required to complete a compliance report

Approved Document J (ADJ) requirements

Notification of building work – how it affects competent operatives

ADJ, amended in October 2010, extended the definition of building work to be notified to BCBs. A previous amendment to the regulations – called the Building (Amendment) Regulations 2002 – defined 'competency' and introduced an exemption from the need to notify.

Building work under ADJ – what you must do

Chimney systems are defined as 'controlled services' under the Building Regulations. This means that, in addition to new chimney systems having to meet the requirements of ADJ, if you are to carry out work on existing chimney systems then these must also meet the requirements of ADJ.

If the work involves lining a brick/masonry chimney (either by means of introducing a new or replacement liner) then this is 'building work' according to the regulations.

The liner, which may be of rigid, flexible or prefabricated components or may be cast in situ, could alter the internal flue dimensions. It is, therefore, important that you test its performance as with a new chimney system.

In addition, if a brick/masonry chimney:

- is to be used for a different type of appliance
- is to be used for an appliance with a different output, or
- is being brought back into use

you must test it to ensure it is compliant.

Chimneys in voids

New guidance is included in ADJ on access for visual inspection of chimneys concealed within voids, i.e. those typically found in multi-occupancy dwellings (flats) where room-sealed chimneys are run within ceiling voids.

See **Part 13 – Chimney standards – Chimneys in voids** – for further guidance.

Meeting compliance requirements of ADJ – who is responsible?

The person carrying out the work is responsible for meeting the requirements of ADJ.

To demonstrate this, when the building work has been completed e.g. installation of a flexible metallic flue liner or chimney system, that person must notify the local authority of the work undertaken.

Registered Installers are permitted to self-certify that their work meets the requirements of ADJ.

Government-led changes stipulate that it is a legal requirement in England and Wales for the relevant BCB to be notified of the installation or exchange of any heat producing appliance (Building Regulations, Part J) and associated fittings or services served by the appliance (Building Regulations, Part L), within a residential dwelling.

For particular Building Regulations requirements relating to ADJ on the installation of ventilation and chimney systems for gas burning appliances, see **Part 4 – Ventilation** and **Part – 13 Chimney standards** in this manual.

Approved Document L (ADL1A and ADL1B) requirements

Notification of building work – how it affects competent operatives

ADL 'Conservation of fuel and power', amended in October 2010 (ADL1A (new dwellings) and ADL1B (existing dwellings)), extended the definition of building work requiring notification to BCBs. A previous amendment to the regulations called the Building (Amendment) Regulations 2002, defined 'competency' and introduced an exemption from the need to notify.

Building work under ADL – what you must do

The person who carries out the work is responsible for meeting the requirements of ADL1A or ADL1B. To demonstrate this, when the building work is completed e.g. installation of a flue liner or chimney system, this person must notify the local authority of the work undertaken.

Registered Installers are permitted to self-certify that their work meets the requirements of ADL1A or ADL1B, as appropriate.

Government-led changes stipulate that it is now a legal requirement in England and Wales for the relevant BCB to be notified of the installation or exchange of any heat producing appliance (Building Regulations, Part J) and associated fittings or services served by the appliance (Building Regulations, Part L), within a residential dwelling.

Similar requirements have also been introduced into the Building Standards (Scotland). Reference should be made to Part J.

ADL about energy efficiency – to conserve fuel and power

The revised building regulations are closely linked with SAP (the Government's Standard Assessment Procedure for Energy Rating of Dwellings), SEDBUK (Seasonal Efficiency of Domestic Boilers in the UK), Code for Sustainable Homes and the Domestic Building Services Compliance Guide.

ADL affects the conservation of fuel and power in all types of buildings (ADL1A and ADL1B covers dwellings, while ADL2A and ADL2B covers buildings other than dwellings).

Requirements for dwellings and other buildings outline:

- construction techniques
- lighting
- insulation and heating

to promote the most economic use of fossil fuels.

Ways needed to achieve this are:

1. By reducing heat escaping from the fabric of the building.
2. By roofs, walls, windows, doors and floors having an adequate resistance to loss of heat.
3. By hot water pipes, hot air ducts and hot water vessels also limiting any heat loss.
4. Space heating and hot water systems being energy efficient and adequate controls being provided to control appliances, to avoid inefficient usage and waste.
5. By ensuring gas users are given the right information to help them operate and maintain heating and hot water systems economically.

A major revamp of the Building Regulations has been necessary due to:

- buildings contributing to a high proportion of the carbon dioxide (CO_2) emissions
- the government's targets of reducing CO_2 levels

Additional benefits for gas users

These factors may also bring the additional advantage of lowering users' heating bills.

Energy rating – helps show compliance in dwellings

For compliance in dwellings, various energy efficiency-rating methods are used to show reasonable provision has been made for the conservation of fuel.

Both new and modified dwellings that are within the regulations are subjected to an Energy rating, i.e. SAP (The Government's Standard Assessment Procedure for Energy Rating of Dwellings).

The SAP scale is from 1 to 100 where:

- 1 denotes a very poor standard of energy efficiency, and
- 100 is exceptionally high

Compliance with SAP may be achieved by combining a number of energy conservation methods. The regulations however also contain stringent provisions regarding gas central heating appliances.

The performance standards in ADL are significantly higher than those introduced in 1995 and this may have an impact on the selection and the installation of domestic and non-domestic appliances and plant.

Within the requirements, consideration is made for a reasonable provision for appliance efficiency. This may be established by utilising an appliance with a minimum SEDBUK rating.

Approved Document P (ADP) requirements

- All electrical work associated with the installation of a gas appliance and its controls needs to comply with manufacturer's instructions and BS 7671: 2008 (Incorporating Amendment No.3 2015) Requirements for electrical installations. IET Wiring Regulations Seventeenth Edition
- electrical work needs to be undertaken by suitably competent persons and, where to be self-certified as complying with the Building Regulations, by a member of a competent person scheme
- suitable documentation – electrical installation certificates or minor works certificates – needs to be completed upon completion of works and presented to the user of the installation

Gas Appliance (Safety) Regulations (GASR) – about conformity and safety in use

These regulations, which implement a European Community (EC) Directive on gas appliances, require certain appliances and fittings:

- to conform with specified essential requirements; and
- to be safe when normally used

Products must bear the CE marking and their safety must be underpinned by valid certification or a declaration of conformity, or their supply is prohibited. The regulations include detailed procedures for product conformity and approval by third-party notified bodies appointed by the Secretary of State.

Information you need to ensure is provided with new gas appliances

The Regulations also specify requirements for information to be provided with new gas appliances, e.g. instructions covering safe installation, operation and maintenance.

CE Marking – showing that all products for sale meet safety standards

CE marking requirements have been introduced for new gas appliances through the Gas Appliance Directive. This came into force on 1st January, 1996.

Retailers must not offer for sale any new gas appliance that is not CE marked.

CE marking of gas appliances is a requirement that opens the European-wide gas appliance market to all member states and ensures that all products offered for sale meet agreed safety standards.

In recent years, consumers have become accustomed to seeing CE marks on goods for sale. A good example is children's toys, which must display the CE mark.

All appliances placed on the EC market (which includes appliances manufactured within the EC or outside e.g. in Japan or the USA) have to meet the requirements contained in the Gas Appliance Directive (GAD).

The GAD requires all EC member states to introduce national legislation ensuring that retailers can only supply gas appliances that have been type tested and have had their production monitored by government-approved certification bodies.

There are a number of such UK bodies including the British Standards Institution (BSI). Appliances must also have a CE mark affixed to them.

In the UK, GAD was implemented through the Gas Appliances (Safety) Regulations 1995.

The Gas Safety (Installation and Use) Regulations (GSIUR) state:

- "No person shall install a gas appliance which does not comply with any enactment imposing a prohibition or restriction on the supply of such an appliance on grounds of safety"

Published guidelines for the HSE on the Regulation state:

- "So far as new gas appliances are concerned, these should conform to the Gas Appliances (Safety) Regulations"

How may you be liable as a gas operative?

The GSIUR require that only CE marked appliances should be installed. These arrangements are not only designed to make the gas appliance market safer for the gas user: they also help you, as they ensure that all appliances are tested and approved to an agreed standard and are deemed safe to use.

So any appliances without CE marking are likely to be in contravention of the regulations and may leave the gas operative liable to prosecution if installed.

Note: This prohibition does not extend to previously used appliances manufactured prior to the implementation of these requirements.

Who is responsible for enforcing these regulations?

Trading Standards Officers have this responsibility.

Reporting of Injuries, Diseases and Dangerous Occurrences Regulations 2013 (RIDDOR)

For many years, there has been a requirement under the 'Reporting of Injuries, Diseases and Dangerous Occurrences Regulations' (RIDDOR), for certain types of Dangerous Gas Fittings (DGFs) to be reported to the HSE.

The purpose has been to allow the HSE to investigate reports of actual gas incidents and those with the potential to be gas incidents.

Aside from deciding what action might be needed, this enables the HSE to gauge the scale and nature of gas safety problems and give appropriate publicity to them in the interests of public safety.

The following guidance gives additional advice to gas installation businesses that have a duty to report these details. It should be read in conjunction with the general guidance contained in the HSE guide to RIDDOR, leaflet INDG453 (rev 1) (available from HSE books see **Part 21 – References – Organisations**).

Making the installation safe – who is involved?

Though it is important that the HSE are given information to follow up matters such as the activities of unregistered and incompetent gas installers, it is essential that everyone involved in the reporting process should recognise that the safety of gas users is of paramount importance.

The GSIUR require that all gas operatives finding a dangerous gas fitting should:

1. Advise the responsible person, normally the occupier, of the danger.
2. Inform the landlord and/or the managing agent if the property is rented.
3. Seek permission to disconnect any appliances that would obviously pose an immediate threat to life.
4. If permission is refused, contact the following authorities:
    - in England, Scotland and Wales, contact the National Gas Emergency Service Call Centre, on telephone number 0800 111 999. In the case of Liquefied Petroleum Gas (LPG), contact the gas supplier, whose details can be found on the bulk storage vessel or cylinder (if no label can be found, information can be found under GAS in the local telephone directory)
    - in Northern Ireland contact Phoenix Natural Gas on 0800 002 001, or the LPG supplier

- in the Isle of Man contact Manx Gas Ltd for all areas (including LPG) on 0808 1624 444
- in the Channel Islands contact either Guernsey Gas Ltd on 01481 749000 or Jersey Gas Company Ltd. on 01534 755555 as appropriate

The ESP will then take the necessary action through their GT. In exceptional circumstances, where access to equipment is refused, they may obtain a warrant to exercise their Rights Of Entry and disconnection powers (see also **Part 8 – Gas Industry Unsafe Situations Procedure**).

5. Inform the HSE about what was originally found.

Guidelines on what you need to report

General principles

The following guidance applies to Great Britain (England, Scotland and Wales). Other geographical areas may have similar reporting requirements.

There is a requirement under the Reporting of Injuries, Diseases and Dangerous Occurrences Regulations 2013 (RIDDOR) for certain types of injuries and dangerous gas fittings to be reported to the HSE

What to report under RIDDOR – Gas incident

RIDDOR regulation 11(1) applies when someone has died, been unconscious or taken to hospital in connection with gas, as a result of carbon monoxide (CO) poisoning, exposure to un-burnt gas, fire and/or explosion incidents. Incidents where people have taken themselves to hospital, or have been taken to another medical facility are not reportable.

The duty to report under regulation 11(1) rests solely with the conveyor of natural gas or filler, importer or supplier (except retail supplier) of LPG.

What to report under RIDDOR – Dangerous Gas Fittings

RIDDOR regulation 11(2) requires registered gas businesses/engineers to report any gas fittings (including appliances and flues or ventilation used with appliances) which they consider to be dangerous (to such an extent people could die, made unconscious, or need to be taken to hospital) because the 'design, construction, manner of installation, modification, or incorrect servicing', could or has resulted in:

- An accidental escape of gas;
- Incomplete combustion of gas; or
- Inadequate removal of products of combustion of gas.

In practice, this means that registered engineers should generally report situations classed as being immediately dangerous (ID) under the GIUSP.

You must report within 14 days any dangerous gas fittings you find

By law (RIDDOR), you should send a report within 14 days of finding any DGF.

How you report

The HSE has moved to a full online reporting service as of September 2011, meaning the use of pre-printed RIDDOR forms – F2508G2 – and sending these forms on by post is no longer appropriate.

- report Flammable Gas Incidents and DGFs directly to the HSE on-line by accessing the web site – www.hse.gov.uk/riddor – and clicking on 'Report an Incident' and then by clicking on either 'Report of Flammable Gas Incident' or 'Report of a

Dangerous Gas Fitting'

- a telephone service is available for fatal and major injuries only – Incident Contact Centre on 0845 300 9923

The HSE website also contains all the information you may need on RIDDOR requirements – visit www.hse.gov.uk/riddor

Other matters of concern

Some gas fittings may not have been installed in accordance with the requirements of the gas safety legislation in force at the time the work was completed. Unless you have good reason to believe they are dangerous, you do not strictly need to report them to the HSE.

However, there are some types of 'illegal' installations that call into question the competence of the original gas operative. The HSE would prefer you to notify them by using the online service at www.hse.gov.uk/riddor

These include:

- all illegally installed open-flued appliances in bathrooms and shower rooms (those known to have been installed after 24th November 1984)
- appliances in rented accommodation that have not been safely maintained

You may find some gas installations that do not fully comply with current regulations and Standards.

Engineers can still inform the gas user / responsible person via thier jobs reports or 'best advice' but use their judgment around what advice to give and the likelihood of ot being feasible to implement (for further guidance, see **Part 8 – Gas Industry Unsafe Situations Procedure**).

The Office of Gas and Electricity Markets (Ofgem) – protecting and promoting the interests of all gas and electricity customers

The Office of the Gas and Electricity Markets (Ofgem) regulates the gas and electricity industries in Great Britain. Ofgem's overriding aim is to protect and promote the interests of all gas and electricity customers by promoting competition and regulating monopolies. Ofgem operates under the direction and governance of the Gas and Electricity Markets Authority which determines strategy and policy priorities.

Ofgem's powers and duties are provided for under the Gas Act 1986, the Electricity Act 1989, the Competition Act 1998 and the Utilities Act 2000 and other statutes.

Ofgem's website (www.ofgem.gov.uk) provides:

- information on changing supplier
- links to price comparisons and consumer information published by Energywatch, the independent gas and electricity consumer watchdog

Consumers who have a complaint about their energy supplier should:

- first contact that company

If they do not feel the response is satisfactory, they should:

- contact their local Citizens Advice Bureau

If after exhausting the energy suppliers complaint process you feel that there is still a case to answer, then additional assistance can be obtained from:

- Energy Ombudsman – 0330 440 1624 or by visiting www.ombudsman-services.org

Gas emergency actions and procedures – 2

Figures

2 – Gas emergency actions and procedures

Gas emergency actions and procedures – helpful guidance for you

The approach you take when dealing with a gas emergency can be crucial. You need to know:

- what to do
- who to contact
- how to deal with the different situations which can occur

You also need to recognise that there is a defined order (priority) when dealing with gas emergencies, particularly gas escapes and these follow a commonsense approach:

1. Safeguard life
2. Safeguard property
3. Find and secure the escape
4. Carry out a final examination of the installation

In extreme cases where the occupants' safety is at risk it may well be necessary to evacuate the premises (and even adjacent properties) until it is confirmed that the property is safe to re-enter.

How does the national gas emergency service deal with gas emergencies from the Natural gas network?

1. The Gas Emergency Contact Centre receives calls, provides safety advice and passes information to the Emergency Service Provider (ESP).
2. The ESP has the responsibility to operate an effective round-the-clock, 365-days-a-year emergency service for gas escapes and smells of "fumes". LPG suppliers provide a similar service.

Gas operatives employed by an ESP to carry out emergency work:

- have the necessary competence to deal with gas escapes
- can, if necessary, carry out property evacuation
- can also identify unsafe situations and make appliances safe (isolate from the gas supply) for other registered installers to repair

Where an ESP operative is called out to escapes of gas on installation pipework that cannot be rectified, they will make safe the situation by isolating and disconnecting the complete installation.

Where they suspect that products of combustion are entering the property from an appliance, they will disconnect and label it.

What happens if you are a registered operative not employed by an ESP?

Due to the specialist nature of emergency work and the statutory obligations set for ESP companies, a registered gas operative not employed by an ESP would not normally be expected to respond to a reported gas escape or "fumes" from a member of the public (except to give safety advice and to ensure the situation is properly reported).

However, if for any reason a gas operative is responding to a reported gas escape it is extremely important that when arriving at the site where a smell of gas has been reported, the gas operative does NOT risk the possibility of an explosion by pressing any electrically operated (mains or battery powered) door bell or voice intercom system.

The operative must announce his arrival with due consideration of the fact that operating a door bell may create sufficient spark to ignite any flammable gas mixture, therefore the operative will need to knock at the door or window for example to advise the occupant of their arrival.

Note: Always work safely and where in doubt contact the ESP for assistance.

2 – Gas emergency actions and procedures

Responsibilities of the gas user in an emergency

1. If the person responsible for any premises knows, or has reason to suspect, that gas or fumes are escaping into the premises, they must immediately take all reasonable steps:
   - to turn the gas supply off at the emergency control valve (ECV), to prevent further escape of gas or fumes
   - to ventilate the property
2. If gas continues to escape after the supply of gas has been shut off, or where a smell of gas persists, the person responsible must immediately report it as follows:
   - in England, Scotland and Wales, contact the National Gas Emergency Service Call Centre, on telephone number 0800 111 999

     In the case of Liquefied Petroleum Gas (LPG), contact the gas supplier, whose details can be found on the bulk storage vessel or cylinder (if no label can be found, information can be found under GAS in the local telephone directory)
   - in Northern Ireland contact Phoenix Natural Gas on 0800 002 001, or the LPG supplier
   - in the Isle of Man contact Manx Gas Ltd for all areas (including LPG) on 0808 1624 444
   - in the Channel Islands contact either Guernsey Gas Ltd on 01481 749000 or Jersey Gas Company Ltd. on 01534 755555, as appropriate

The gas supply must not be used until remedial action has been taken to correct the defect and the installation has been re-commissioned by a competent person.

What must you do when advised of a gas escape, when you are not on site?

You must instruct the gas user(s) as follows:

- they must turn off the gas supply immediately at the ECV
- they must extinguish all sources of ignition
- they must not smoke
- they must not operate electrical light or power switches (on or off)
- they must ventilate the building(s) by opening doors and windows
- they must ensure access to the premises can be made
- they must report the escape to the appropriate gas emergency service call centre listed under **Responsibilities of the gas user in an emergency** in this Part

Note: Also see Figure 2.1.

What must you do when you detect (or are advised of) a gas escape, when on site?

You must:

- turn off the ECV
- extinguish all naked flames and remove any sources of ignition
- ensure that no one operates electrical lights or power switches (on or off)
- ventilate the building(s) by opening doors and windows
- test the installation for gas tightness (see **Part 6 – Testing for gas tightness (Natural gas)** for further guidance)

Note: Where you are advised of a smell of gas or detect a gas escape, no leakage tolerance is permitted (see Figure 2.1 also).

Figure 2.1 Gas escape procedure

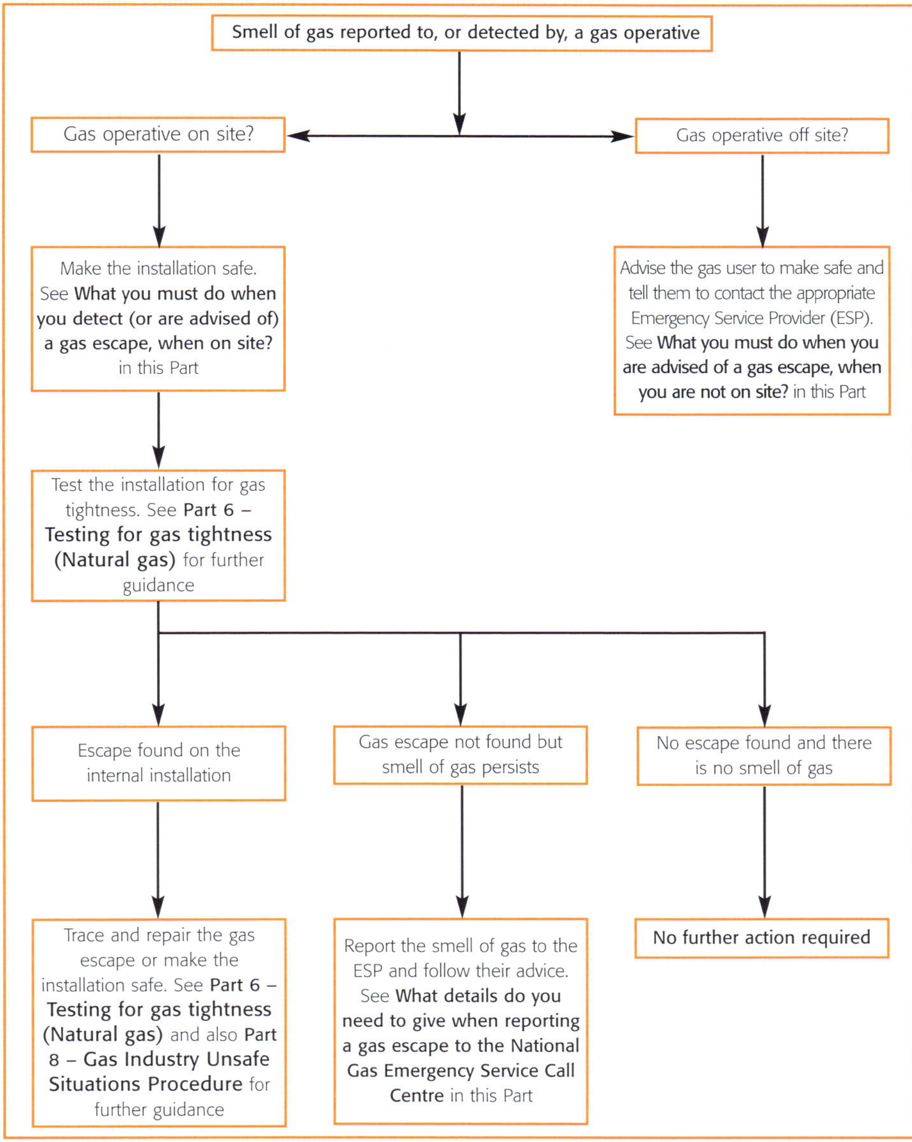

What details do you need to give when reporting a gas escape to the Gas Emergency Contact Centre?

Be prepared to give as many of the following details as possible:

- the address where the escape has been detected
- the name and telephone number of the gas user
- name and telephone number of the person reporting the suspected escape
- the place where the smell is most noticeable
- the time when the smell was first noticed
- whether the ECV is turned off
- if the ECV is turned off, whether there is still a smell of gas
- if it is not turned off, the reason why (ECV not accessible or broken etc.)
- whether there is a smell of gas outside the property
- any special circumstances (e.g. elderly or handicapped residents, access problems etc.)

You should obtain the following information from the Gas Emergency Service Call Centre and keep a record of:

- the job number
- the date and time of report
- the person to whom you reported the incident

Note: Also see Figure 2.1.

ESP Engineers attending reports of fumes

When an ESP engineer (or LPG Supplier's emergency response engineer) is called to a report of fumes, a visual inspection of the gas appliances in the property shall be carried out. Where no obvious unsafe situations are identified a Danger Do Not Use label shall be attached to all appliances and with the users permission all appliances shall be turned off.

ESP's have their own procedures for dealing with CO alarms activated due to reasons other than the presence of CO e.g. end of life or battery failure.

The Warning Notice shall state:

"Appliances have been visually inspected by an emergency service engineer who cannot confirm that they are safe to use. The appliances should not be used until they have been checked by a Gas Safe registered business".

Gas Safety (Rights of Entry) Regulations – advice on when you contact the Gas Emergency Service

When you carry out gas work on an appliance or installation, there will be occasions when you identify faults that require immediate attention to ensure that the gas installation is made safe, or can continue to be used safely.

What happens where you identify that an appliance/installation is Immediately Dangerous (ID)?

- you must encourage the gas user to have the remedial work carried out

What do you do if the gas user refuses?

- you need to inform them of the situation in writing (a suitable form is available from CORGI*direct*, order Ref: CP14 – see **Part 10 – Emergency notices, warning labels and forms**), warning that continued use of such equipment may constitute an offence
- you must also attach a suitable warning label to the appliance/installation (again suitable labels are available from CORGI*direct*, order Ref: WL1)
- don't forget you have no legal right to turn off, or disconnect an appliance/installation without the gas user's consent

What must you do next?

If the gas user refuses to allow you to take action, the current Gas Industry Unsafe Situations Procedure provides clear guidance as to what you must do next (a copy of which is available from CORGI*direct*, order Ref: USP1) or via Gas Safe's website at https://engineers.gassaferegister.co.uk.

Get help by contacting the Gas Emergency Contact Centre and requesting their assistance.

The Gas Transporter (GT) may then exercise their powers under the Gas Safety (Rights of Entry) Regulations.

The emergency telephone number for this purpose are:

- in England, Scotland and Wales, contact the Gas Emergency Contact Centre, on telephone number 0800 111 999

  In the case of Liquefied Petroleum Gas (LPG), contact the gas supplier, whose details can be found on the bulk storage vessel or cylinder (if no label can be found, information can be found under GAS in the local telephone directory)
- in Northern Ireland contact Phoenix Natural Gas on 0800 002 001, or the LPG supplier
- in the Isle of Man contact Manx Gas Ltd for all areas (including LPG) on 0808 1624 444
- in the Channel Islands contact either Guernsey Gas Ltd on 01481 749000 or Jersey Gas Company Ltd. on 01534 755555 as appropriate

GTs are not responsible for action in the case of LPG emergencies and the Gas Emergency Contact Centre will not generally be able to help in such circumstances. In such situations, seek advice from the LPG supplier.

With regard to identifying the 'gas supplier' for a given property:

- NG – contact the GT helpline on 0870 160 0229 and provide the call operator the first line of the address along with its postcode
- LPG – in most cases, it will be the LPG Company (i.e., Shell, Calor, etc).

Retailers or contract fillers of cylinders for instance are NOT regarded as suppliers for the purpose of the GSIUR (although a landlord can attract duties as a gas supplier in certain limited situations, e.g. rented caravan holiday homes/mobile homes).

The Gas Safety (Rights of Entry) Regulations: these only apply where life or property is threatened.

Under these regulations, in England, Scotland and Wales, a GT has 'Rights of Entry' to a property to deal with a dangerous situation to make that situation safe. Be aware that the GT could only respond to a situation where life or property is threatened (Immediately Dangerous ID).

Note: Northern Ireland and the Isle of Man have their own arrangements in place. See Part 1 – Gas Safety Legislation – Gas Regulations that affect Northern Ireland and the Isle of Man.

Under these regulations the GT will have no rights of entry where an appliance/installation is classified as 'At Risk' (AR)

For additional information on the classification of 'ID' and 'AR' situations and how to deal with these categories, see **Part 8 – Gas Industry Unsafe Situations Procedure.**

When informed by a gas operative regarding a dangerous gas appliance/installation, the Gas Safety (Rights of Entry) Regulations allow certain authorised officers of the ESP to apply for duly authenticated documentation.

They need to produce this to gain legal access to enter premises to make safe a gas installation. This includes suspected gas escapes and escapes of products of combustion.

What to do when you contact the Emergency Contact Centre:

You must ensure that all relevant information is passed to their Call Agent, giving full and precise details of the fault(s) found.

1. You should include the following information:

- the name of the person reporting and the registration number
- the address and post code at which the 'ID' situation exists
- the name of the person responsible for the property
- details of the 'ID' situation
- the type of appliance/installation
- its location within the property
- confirmation that it is an 'ID/Rights of Entry' disconnection request

2. During the call also:

- request a job reference number from the ESP Call Agent
- record the date and time of report
- note the persons name to whom you reported the incident

3. After the call:

- retain the job reference on file for future reference

When the GT has helped you make an appliance/installation safe:

Complete (where necessary) additional documentation concerning the dangerous situation found.

This requirement is known as RIDDOR (Reporting of Injuries, Diseases and Dangerous Occurrences Regulations 2013). It places a duty upon the gas operative to report certain types of dangerous gas fittings to the HSE.

Note: For further guidance on RIDDOR see Part 1 – Gas safety legislation – Reporting of Injuries, Diseases and Dangerous Occurrences Regulations (RIDDOR).

Action by the Gas Transporter (GT) to make safe

- when they have been informed about an unsafe situation, the GT will issue the job to one of their Authorised Officers who will visit the site address in order to make safe
- if the gas user will not allow the appliance/installation to be made safe, the GT will then take an appropriate course of action, which could lead to the gas supply being terminated

The procedures exist to help you succeed – and promote the safety record of the industry

When faced with a difficult situation where the safety of a gas appliance or installation is concerned, you are not facing the problem alone.

As long as you follow the correct procedures and deal with these situations in the appropriate manner, the safety record of the gas industry will continue to improve.

Note: Don't be tempted to contact the ESP to make safe, as an easy way out to avert a confrontation with the gas user. If you handle the situation sympathetically and explain it properly, you will normally be able to make the installation safe yourself.

Characteristics of combustion – 3

3 – Characteristics of combustion

Figures

Tables

What you need to know about gas and combustion

As you know, users depend on you to ensure the safe working of their gas appliances.

Gas appliances are a safe, convenient way to utilise the benefits of heat produced by combustion.

1. You do need to understand the reason why complete combustion (i.e. the correct mixture of the fuel gas and oxygen) is necessary to ensure gas safety.

The guidance in this part relates to:

- Methane (CH_4), generally referred to as Natural gas (NG)
- Propane (C_3H_8) and Butane (C_4H_{10}), which are types of Liquefied Petroleum Gas (LPG)

2. You also need to see The Gas Safety (Installation and Use) Regulations (GSIUR). These outline compulsory duties to ensure correct combustion (affecting you and users alike).

Combustion – understanding the process

It's a chemical reaction requiring 3 essential elements: FUEL + OXYGEN + IGNITION reacting to create heat. The by-products created are the products of combustion (POC).

Complete combustion is the safe outcome

The ideal mixture of a fuel gas and air is called the 'stoichiometric' mixture (a mixture of gas and air in the proportions determined by the theoretical air requirement).

To ensure complete combustion, 1 volume of Methane must react fully with 2 volumes of Oxygen. Similarly 1 volume of Propane must react fully with 5 volumes of Oxygen and 1 volume of Butane must react fully with 6.5 volumes of Oxygen.

The resulting products of combustion would mainly consist of Carbon Dioxide (CO_2) and Water Vapour (H_2O).

The basic chemical reaction for complete combustion of Methane, Propane and Butane can be shown as follows:

Methane

$CH_4 + 2O_2 + \text{IGNITION} = \text{HEAT} + CO_2 + 2H_2O$

Methane + Oxygen + Ignition =

Heat + Carbon Dioxide + Water Vapour

Propane

$C_3H_8 + 5O_2 + \text{IGNITION} = \text{HEAT} + 3CO_2 + 4H_2O$

Propane + Oxygen + Ignition =

Heat + Carbon Dioxide + Water vapour

Butane

$2C_4H_{10} + 13O_2 + \text{IGNITION} =$

$\text{HEAT} + 8CO_2 + 10H_2O$

Butane + Oxygen + Ignition =

Heat + Carbon Dioxide + Water Vapour

Figure 3.1 Combustion process

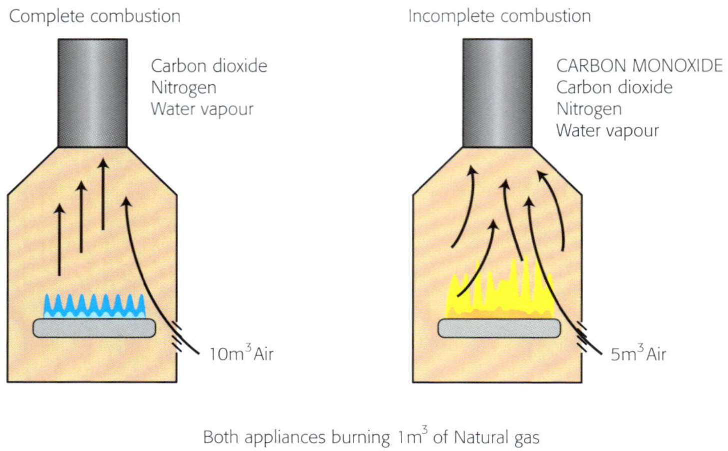

The air (atmosphere) we breathe consists basically of 20.9% Oxygen and 79.1% Nitrogen (Nitrogen (N_2) is an inert gas and takes no part in the combustion process).

As an example, the complete combustion equation of one cubic metre (m^3) of Methane is:

$1m^3\ CH_4 + 2m^3\ O_2 + 8m^3\ N_2 =$

$1m^3\ CO_2 + 2m^3\ H_2O + 8m^3\ N_2$

Methane + Oxygen + Nitrogen =

Carbon Dioxide + Water Vapour + Nitrogen

You'll see from this equation that $1m^3$ of Methane plus $10m^3$ of air gives a total of $11m^3$ of products of combustion (see Figure 3.1).

Incorrect combustion can lead to an explosive mix of gases

As Methane, Propane and Butane are carbon-based gases, Carbon Monoxide (CO) can be produced when they are burned. Mixtures of these gases with Oxygen (O_2) are also highly explosive if not controlled during combustion.

Incomplete combustion – the dangers and how to avoid it

An incorrect fuel/oxygen mix may cause incomplete combustion. See **Part 4 – Ventilation** and **Part 11 – Checking/setting appliance burner pressures and gas rates**. When this occurs, CO is produced.

As CO is a highly toxic gas, it is essential you prevent this forming, by ensuring that appliances achieve complete combustion (see Figure 3.1).

The following chemical equation, using methane, shows this reaction and the production of CO:

$1m^3\ CH_4 + 1.75m^3\ O_2 + 8m^3\ N_2 =$

$0.5m^3\ CO + 0.5m^3\ CO_2 + 2m^3\ H_2O + 8m^3\ N_2$

Methane + insufficient Oxygen + Nitrogen =

Carbon Monoxide + Carbon Dioxide + Water vapour + Nitrogen

Characteristics of gas

For combustion to be usable, it must be controlled. If uncontrolled, it will be dangerous and inefficient.

The next section and Table 3.1 are our guidelines to help you to:

- control combustion
- achieve fuel efficiency
- complete the combustion process

Relative Density (RD) – the weight or mass of the gas in relation to that of air

Air has a RD of 1.00. NG has an RD of 0.6 (a little over half the weight of air). Propane has an RD of 1.5 (one and a half times the weight of air). Butane has an RD of 2.0 (twice the weight of air), see **Explosive mixtures** in this Part.

Note: RD used to be referred to as Specific Gravity (SG).

Calorific Value (CV) – the heat given off by gases when they burn

This is measured in Megajoules per cubic metre (MJ/m^3) and is known as the Calorific Value (CV).

- NG has an average CV of 38.76 MJ/m^3
- Propane has an average CV of 93.1 MJ/m^3
- Butane has an average CV of 121.8 MJ/m^3.

The difference between 'Gross' and 'Net' Calorific Values (CVs)

The heat input of a gas appliance in the UK used to be given in terms of Gross CV. This changed to Net CV in 2000 – as part of the 'Europeanisation' of appliance safety standards.

So do make sure you know whether the manufacturer provides details of the heat input on the appliance data plate or appliance installation instructions, on the basis of:

- gross CV or net CV

The CV of a fuel is basically the amount of heat contained in a given quantity of that fuel. When fuels that contain hydrogen in their chemical make-up are burned, the hydrogen is converted to water vapour. The water vapour holds some of the heat produced as "latent heat" (in the case of NG, this is about 10% of the total heat).

It helps you to know that this latent heat:

- is included in the figure given for the gross CV of a fuel
- is excluded from the figure given for the net CV

See the explanation of Calorific Value (CV) in **Part 19 – Definitions.**

Table 3.1 Gas characteristics

Characteristic at standard temperature and pressure	Methane (Natural gas)	Commercial Propane	Commercial Butane	Observations
Relative Density (RD) (RD of air = 1)	0.6	1.5	2.0	Methane will rise, whereas Propane and Butane will gather at low level. Methane will diffuse into the air quicker than Propane or Butane.
Gross Calorific Value (CV)	38.76 MJ/m³	93.1 MJ/m³	121.8 MJ/m³	Appliances must only be used on the gas type for which they are designed.
Wobbe number (MJ/m³)	45.5 to 55.0	73.5 to 87.5	73.5 to 87.5	Approximate Wobbe number range.
Supply pressure (operating pressure)	21mbar	37mbar	28mbar	Natural gas appliances may have burner pressure adjustment. Generally LPG appliances do not.
Stoichiometric air requirements by volume/volume gas (m³)	9.76:1	23.76:1	30.00:1	LPG requires a greater amount of air to burn per volume than Natural gas.
Flame Speed	0.36 m/s	0.46 m/s	0.45 m/s	The speed at which a flame will burn along a gas/air mixture.
Flammability Limits	5 to 15% gas in air	2 to 10% gas in air	2 to 9% gas in air	Gas and air mixtures will only burn within these ranges.
Maximum flame temperature	1930°C	1980°C	1996°C	Approximate temperatures.
Ignition temperature	704°C	530°C	408°C	Approximate temperatures.
Note: These figures are approximate due to slight variations in gases.				

Wobbe number (index) – heat produced at a burner

This is used to give an indication of heat produced at a burner when fuelled by a particular gas. The amount of heat produced by a burner depends on the following factors:

- gross CV of the gas
- RD of the gas
- size of the injector orifice
- gas supply pressure to the injector

The wobbe number is calculated by dividing the CV by the square root of the RD:

$$\text{Wobbe number} = \frac{CV}{\sqrt{RD}}$$

Supply pressure (operating pressure)

Natural gas is supplied through a regulator, which normally maintains an outlet operating pressure of 21mbar measured at the meter. This is often reduced further at the appliance to provide the required pressure at the burner.

See **Part 7 – Checking and/or setting regulators** and **Part 11 – Checking/setting appliance burner pressures/gas rates** for further guidance.

LPG is supplied via storage vessels or cylinders and operates at a pressure of 37mbar for Propane and 28mbar for Butane. This can be metered if required, before passing onto the appliance(s). This pressure is not normally reduced at the appliance.

Gas rate – the amount (volume) of gas that an appliance burns

This depends mainly on:

- the supply pressure
- the orifice size of the burner injector(s)

Appliance 'data plates' may express the gas heat input as a gross or net figure. When you need an accurate rating, carry out a test at the gas meter. See **Part 11 – Checking/setting appliance burner pressures/gas rates**.

Flame speed

Different types of fuel gases burn at different rates:

The flame speed of:

- NG is 0.36m/second
- Propane is 0.46m/second
- Butane is 0.45m/second

To keep a flame stable on a burner, the pressure at which the gas is supplied and the size of the injector orifice needs to be carefully calculated, to ensure that the rate at which the gas is supplied matches the rate at which it is burned.

It's because:

- if gas were supplied faster than it could burn, the flame would lift off the burner
- if gas were supplied too slowly, it would ignite/light back into the burner mixing tube

Many burners have flame retention devices to keep the main flames stable. These are usually smaller flames burning with air/gas mixtures at lower speeds than the main flames. These retention flames continuously re-light the main flames as they try to lift off the burner.

Flammability limits – when fuel gases will burn in air

Fuel gases will only normally burn in air between certain limits, for:

- Methane the range is between 5% and 15% gas in air
- Propane the range is between 2% and 10% gas in air
- Butane the range is between 2% and 9% gas in air

Ignition temperature varies according to gas

Any form of ignition needs to be hot enough to ignite the gas being used:

- NG ignites at approximately 704°C
- Propane ignites at approximately 530°C
- Butane ignites at approximately 408°C

Burners

Types

Burners vary in type and design, but most domestic gas appliances use pre-aerated burners where some of the air required for combustion is mixed with the gas before it is burned. They include:

- atmospheric (natural draught) burners
- forced draught burners

Figure 3.2 Typical pre-aerated burner (some burners may incorporate lint arresters, gauze and baffles)

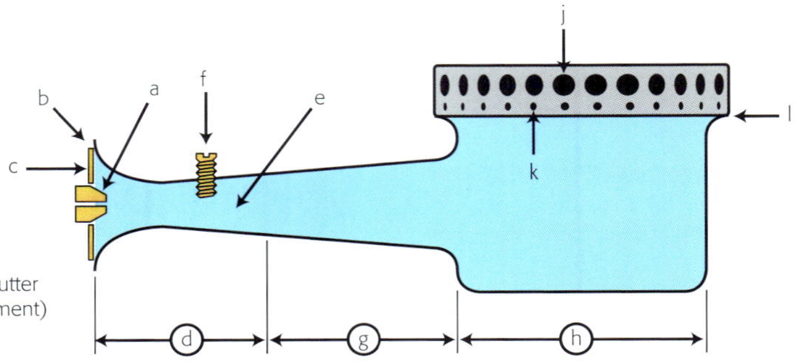

a) Injector
b) Primary airport
c) Aeration control shutter (primary air adjustment)
d) Venturi
e) Venturi throat
f) Throat restictor (alternative primary air adjustment)
g) Mixing tube (burner tube)
h) Burner head (burner body)
j) Burner ports (flame ports)
k) Retention ports (retention flame ports)
l) Alternative flame retention ring

Figure 3.3 Flame retention

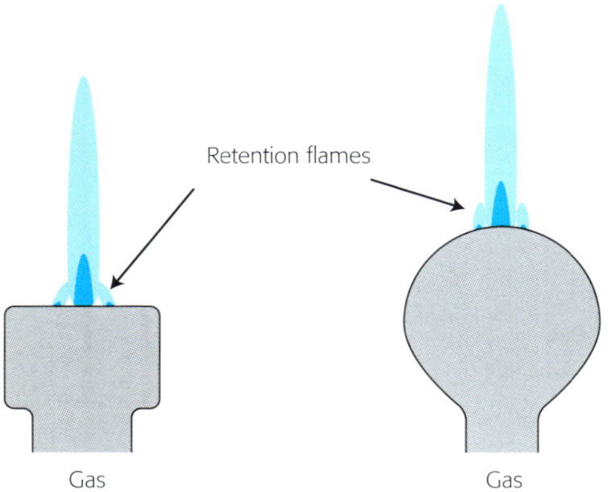

Forced draught burners are fed with supply air from a fan under pressure. They are commonly found in domestic high efficiency (condensing) boilers.

You need to be aware of:

1. The main components of a burner.
2. How each part works to create an efficient stable flame.

Atmospheric burners

Atmospheric burners (see Figure 3.2) generally comprise:

- injector
- primary air port
- primary aeration control (on some burners)
- venturi (to induce primary air flow)
- mixing tube (air and gas mixed together)
- some form of flame retention
- burner head including burner ports

Note: Some burners may incorporate lint arresters (gauze or similar) and baffles.

Injector

This is a fixed size orifice(s) allowing gas to pass to the burner for combustion. Burner design, use and heat considerations may include multiple injectors or multiple hole injectors.

Aeration

Atmospheric burners entrain air, known as primary air. This air is mixed with gas before combustion within the burner. To complete the combustion process, air is required around the flame. This is known as secondary air.

The makeup and amount of air affects the characteristics of the flame. The proportion of air mixed with gas before combustion has taken place (primary aeration) is generally between 40% and 50%.

Lint arresters

These are designed to stop atmospheric dust and lint particles getting into or around a burner via the primary air port(s). Any build up of these can adversely affect the flame picture (shape and colour).

Flame retention

On certain types of burners, the main flame at the burner head needs to be stabilised or prevented from lifting off. Retention flames continuously re-light or cross-light the main burner flame(s) – see Figure 3.3. Retention flames burn at a slower speed than the main burner flame. See **Flame speed** in this Part.

Flame picture – its appearance can help you fault-find

Some flames are designed to be softer and quieter to achieve a certain flame spread or picture e.g. the grill on a gas cooker, or the flame appearance on a decorative fuel effect (DFE) gas appliance. Some flames are designed to be fiercer e.g. some water heaters and gas cooker hobs which require a very high heat input.

Burner maintenance – what you need to do

- when you service or maintain a gas appliance, you need to clean the burner thoroughly
- you also need to refer to the appliance(s) manufacturer's service/maintenance instructions for cleaning details specific to the appliance(s)
- you need to dismantle (split) burners (where design allows) in order to clean inner filters or gauzes
- ignite burner and confirm flame picture is correct

Figure 3.4 Different flame pictures

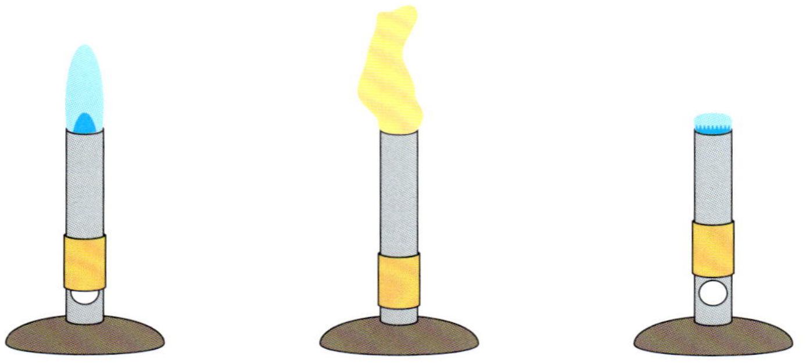

a) Complete combustion

b) Incomplete combustion (too little primary air)

c) Incomplete combustion (too much primary air)

Table 3.2 Possible causes of incomplete combustion

Possible fault	Location	Possible causes	Signs
Gas rate and pressure	Burner/heat exchanger	Gas rate or injector size incorrect	Incorrect flame picture. Soot staining/ deposits
Insufficient ventilation	Appliance	Insufficient air to replace that used during combustion	Incorrect flame picture or lift off, caused by blocked or undersized ventilation
Flame impingement	Burner/heat exchanger. Cooker grill fret	Incorrect burner position. Blocked or damaged burner ports. Blocked heat exchanger. Distorted cooker grill fret	Flames touching heat exchanger. Flame lift or incorrect flame picture. Soot deposits. Damage or distortion to burner or grill fret
Vitiation	Burner	Combustion air is deficient of oxygen due to possible spillage caused by a faulty flue	Insufficient ventilation. Incorrect flue design or damaged/blocked flue. Flame lift off. Soot staining/deposits
Aeration	Burner	Incorrect adjustment to primary air ports. Blocked lint arrester	Incorrect flame picture; flame noisy or floppy. Dust/lint deposits around burner tube or injector(s)

Note: Gas operatives have a legal duty to make safe or report appliances that show danger signs. It is essential that any warning signs of faulty combustion are not ignored when on site, or when advised by gas users (see **Part 2 – Gas emergency actions and procedures**).

Diagnosing burner or combustion faults – simple guidelines to help you

Good combustion: can mainly be identified by the presence of stable blue flames with distinct inner (bright green-blue colour) and outer (darker bluish-purple colour) cones.

Poor combustion: can affect the length, colour, stability and definition of the flame. See Figure 3.4.

Note: Some gas appliances are designed with flames representing solid fuel burning appliances. See 'Flame picture – its appearance can help you fault-find' in this Part.

Many modern burners are designed without the facility for alteration of the primary air port. Once the variable aspects of gas supply, (that is the gas pressure and gas rate), have been adjusted, you can assume that the only variables left are:

- the condition of the burner
- its position
- its supply of air
- the condition and size of the injector(s)

Table 3.2 gives you details of some of the causes of incomplete combustion.

Odd coloured gas flames

Most gas appliances have burner flames that are visible in use. In some cases, the burner flames are exposed to the air in the dwelling: this, generally, does not present a problem.

You need to be aware that nebulisers affect flame characteristics. Where a householder suffering with a respiratory condition uses a Nebuliser to relieve the symptoms, the gases given off by the device may affect the flame characteristics and hue. When concentrated, these gases may cause the burner flames to change colour.

The range may vary from a pale pink to a bright orange. The flames also appear to be much larger than normal, as salts in the gases expose the full outer mantle of the flame, which is normally not visible to the naked eye.

The flames will return to their normal characteristic size and colour once the room(s) affected by the gas from the Nebuliser have been purged with fresh air.

Explosive mixtures – a general explanation

You do need to know how a gas/air mixture can react when ignited under uncontrolled conditions:

Generally, gas will only burn if it is mixed in correct proportions with air. The percentage of gas to air for combustion to take place needs to be:

- 5-15% for NG
- 2-10% for Propane
- 2-9% for Butane

The lowest percentage content is called the Lower Flammability Limit (LFL). The highest percentage content is called the Upper Flammability Limit (UFL). Although gas can burn outside these limits, under normal conditions the flame will go out.

Gas/air mixtures with excess air for complete combustion are called gas lean or simply lean and will be combustible down to a gas content of:

- 5% (LFL) for NG
- 2% (LFL) for Propane and Butane

Gas/air mixtures with too little air for complete combustion are called gas rich or simply rich and will be combustible up to a gas content of:

- 15% (UFL) for NG
- 10% (UFL) for Propane
- 9% (UFL) for Butane

Ignition – factors that aid or hinder it

- well mixed volumes of gas and air will be easiest to ignite because each gas molecule will have the necessary oxygen molecules in close proximity. Ignition energy will not be wasted heating surrounding areas searching for gas or air molecules
- the further the mixtures are removed from stoichiometric the more difficult ignition becomes. Mixtures closest to the LFL or UFL are the most difficult to ignite
- a source of ignition needs to be above the ignition temperature of the air/gas mixture in order to initiate combustion
- a burning surface showing a colour less than bright red is not likely to cause ignition. A lit cigarette will not normally ignite an air/gas mixture because it is not hot enough, but the match/lighter used to light the cigarette certainly will
- a source of ignition also needs to provide energy. The electrical spark from a ringing telephone may not release enough energy to initiate ignition, but the spark from a mains voltage light switch will have sufficient energy

Gas explosions

A mixture slightly richer than stoichiometric will have the highest rate of propagation of flame (burn) through it. Consequently, the highest explosive pressure will be generated with this mixture.

Mixtures between LFL and stoichiometric will burn away quickly because all the necessary air is immediately available. There will be little flame lingering to ignite other substances.

On the other hand mixtures between UFL and stoichiometric mixture will burn away slowly because some of the necessary air needs to be drawn in from outside the mixture zone. The flame will therefore linger, enabling ignition of other substances.

Explosions can occur in three ways:

1. An explosion on its own with no significant fire.
2. An explosion following a fire.
3. An explosion followed by a fire.

To explain further:

1. An explosion on its own, in domestic premises with NG, will almost always be a 'lean mixture explosion'. It will probably result from a gas concentration between 6-8%.

   Mixtures between 5-6% gas in air are difficult to ignite, because of the high energy ignition source required and the variations in mixture concentrations that occur over short distances due to the imperfect mix.

   This means that a small volume of a combustible mixture can be surrounded by an incombustible mixture, forming as it were a 'fire barrier.'

Lean gas explosions will almost always be ignited by an ignition source which was already there before the explosion occurred, (e.g. a pilot light). An ignition source can be introduced at the right time (e.g. a match or an electric light being switched on), but this is an unusual coincidence.

2. An explosion which follows a fire usually occurs because the fire caused gas to escape, by affecting a gas fitting. As the fire creates the ignition source, the explosion is usually a 'lean mixture explosion'.

   However, a fire may cause a gas escape, but so contaminate (vitiate) the atmosphere of air that there is no sufficiently hot ignition source available to ignite the gas/air mixture when it reaches LFL. Later, when fresh air is introduced by opening a door, or by a window breaking due to heat, the fire may be fanned into new life and ignite the by-now rich mixture.

3. An explosion which is followed by a fire is almost always due to a gas concentration greater than 8% – and probably greater than stoichiometric mixture.

   Most combustible materials found in domestic premises require contact with hot gases for longer than one second, often considerably longer, before they ignite. This is only achieved by a rich mixture explosion.

   A gas-rich mixture is normally ignited by a source introduced to the mixture, rather than one that is 'ready and waiting' to ignite the mixture as soon as it becomes combustible.

   The ignition source may be introduced manually e.g. by operating an electric light switch, or automatically through a system thermostat calling for heat.

Air/gas mixing (explosive mixtures)

Giving you a general understanding of the effect of Relative Density – and the rate of increase in gas concentrations with time:

Natural gas

The RD of NG is lighter than air and therefore will rise due to its natural buoyancy.

- if the escaping gas is released into the open air it will readily disperse
- however, if the gas escapes into a property it will rise to the ceiling and form an ever increasing rich mixture layer, gradually growing thicker

As the gas is emitted from the point of escape, it is in a turbulent stream, which draws air into itself. The resulting air gas mixture has a higher RD than the neat gas and will rise more slowly. As it rises, it mixes with more air and becomes progressively less buoyant but of greater volume and so also spreads out horizontally.

As the mixture rises, it simultaneously displaces heavier mixtures above it, which fall and remix at low level with the newly escaping gas. To some extent, a convection current is formed.

Very quickly the concentration of gas becomes fairly uniform throughout the area above the level of the source of escape. Even if the escape propels a vertical stream of gas into a room, almost the same conditions will result.

The pressure of the escaping gas is added to the upward buoyancy motion. The resulting higher kinetic energy stream will draw in more air, reducing the gas concentrations and the upward motion more quickly.

LPG

As the RD of Propane and Butane is heavier than air, any escaping gas will, initially, fall to low level. However, over a period of time, the gas and air molecules will mix and the RD of the gas will reduce with the mixture. It will become increasingly buoyant, acting very similarly to a NG/air mixture contained in a room or building.

Gas escapes on LPG vessels and cylinders sited outside may release greater amounts of gas due to the higher pressure between the vessel and regulators. Under normal/windy conditions, the gas will mix and may disperse quickly. On a still day, concentrations may build-up around the location similar to that contained by a room.

Attention: If you are searching for gas escapes with a portable gas detector, you need to ensure the instrument being used is suitable to detect the type of fuel escaping (if in doubt check the maker's specifications).

Gas detectors

NG and LPG are extremely flexible sources of energy. Their uses are wide spread – cooking, heating in the home and in recreational vehicles, such as leisure accommodation vehicle (LAV) and boats.

Clearly the consequences of a serious gas escape can be devastating – and a number of people are injured or killed annually in such incidents.

There are devices on the market which give a loud audible alarm and warning light – to aid the gas user in detecting unburnt gases.

Other devices available have relay output and interconnect options. This enables connection to control panels, safety shut-off valves on gas supplies and remote alarms. Another facility enables a number of alarms to be linked together in series.

Advice for you to give

You must always advise gas users that any gas appliance needs to be installed correctly and serviced/checked for safety at regular intervals.

However, as problems can occur between service dates for various reasons and the safety of gas users is paramount, we do recommend the use of gas detection systems.

Gas alarms

NG and LPG detector alarms must be approved to meet the stringent requirements of the British Standard for domestic gas alarms – BS EN 50194.

BS EN 50194 supersedes BS 7348. Under this standard, units are calibrated to initiate an alarm at gas levels between 5% and 20% of LFL, which provides an early warning of a gas escape and minimises false alarms (the gas becomes explosively dangerous at 100% LFL (5% gas in air)). Both audible and visual warnings are given.

Models are also available with relay outputs, which link to gas shut-off valves and control systems. Always check with the manufacturer of the gas detector for full specifications.

Where you should site gas alarms

- units are designed for main electrical (110-230V a.c.) and 12-24V d.c. Gas detectors can be wall mounted and the location of the unit is important to enable it to operate effectively. For this reason, always site them in accordance with manufacturer's instructions
- for NG, this would normally be in the room with the most used gas appliance. With NG being lighter than air, it is important that you install the alarm at the top of the highest opening from the room: normally not more than 300mm from the ceiling and between 1m and 5m from the appliance (see Figure 3.5)

Figure 3.5 Gas detector position Natural gas

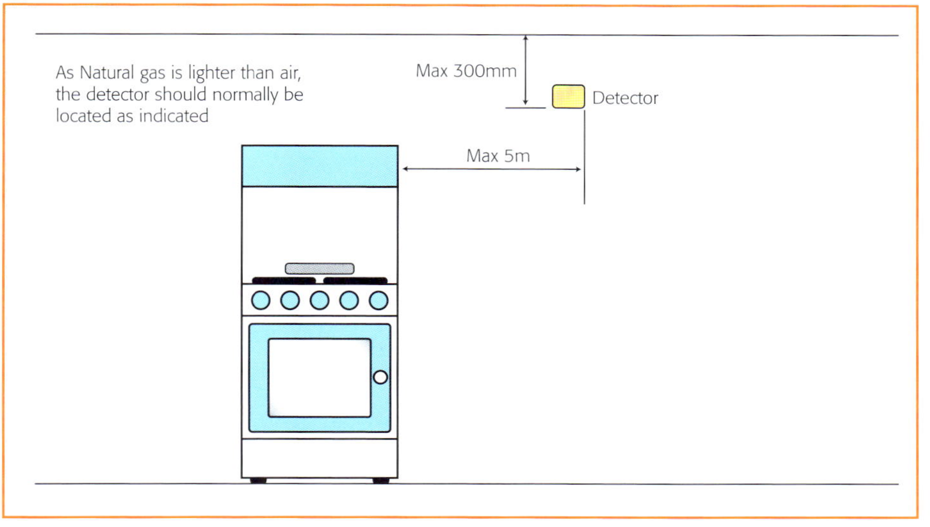

Figure 3.6 Gas detector position LPG

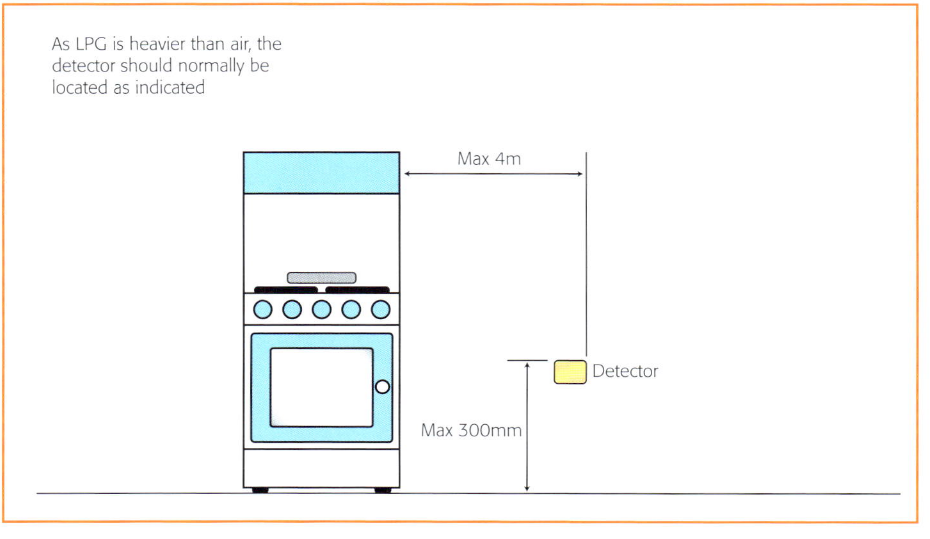

- with LPG being heavier than air, the normally site the alarm no more than 300mm from the floor and no more than 4m from the appliance (see Figure 3.6)

Carbon monoxide (CO)

Incomplete combustion forms toxic CO

Providing a gas appliance is installed, serviced and adjusted correctly and there is an adequate supply of fresh air, combustion of the gas will be satisfactory. The resulting products of combustion will consist mainly of Carbon Dioxide (CO_2) and water vapour (H_2O), both of which are non-toxic.

Note: The formation of CO_2 in large amounts can be a safety concern and typically serves as a precursor to the formation of CO - see 'Effects of Carbon Dioxide (CO_2)' in this Part.

When the process of combustion is interrupted, e.g. when there is insufficient air, or contaminated air (vitiation), incomplete combustion will occur. This will result in the formation of CO, which is a toxic, colourless, tasteless, odourless, and non-irritant gas, harmful to both humans and animals.

Why is CO toxic?

Oxygen (O_2) from the air we inhale passes through the lungs and "loosely" combines with haemoglobin (the oxygen-carrying molecule in the red blood cells) to form oxyhaemoglobin. This is then circulated through the arteries to the body tissues.

The "loose" combination allows the oxygen to be released upon arrival at the tissues. The deoxygenated haemoglobin then plays a part in transporting CO_2 (produced by the tissues) through the veins and back to the lungs. Here the CO_2 is exhaled and new oxygen is absorbed for the breathing process to be repeated.

The inhalation of CO disrupts this process dramatically. Haemoglobin has a greater affinity (200-300 times) for CO than for oxygen.

If CO is inhaled, the haemoglobin will preferentially combine with it to form carboxyhaemoglobin rather than oxyhaemoglobin. As the levels of CO increase, less oxygen is absorbed into the bloodstream.

Carboxyhaemoglobin is not only of no use to the tissues, but it restricts the remaining oxyhaemoglobin from giving up its oxygen. The result is that the tissues are deprived of the essential oxygen required to provide energy for the chemical reaction of the living cells.

What makes CO poisoning extremely dangerous is its insidious onset – and often fatal consequences. It affects mental ability and incapacitates people without them knowing anything is wrong. Any sudden exertion by an affected person may cause collapse, so that he or she cannot escape from the contaminated location. The result is often fatal.

Symptoms of CO poisoning

The effects of CO poisoning can be very much like flu:

- severe headaches
- nausea
- dizziness
- tiredness
- general body weakness
- victims' faces can also take on a cherry red coloured appearance

You need to be sure to note comments from gas users complaining of these symptoms, especially when you investigate reports of fumes (see Table 3.3).

Table 3.3 Effects of carboxyhaemoglobin on human beings

% CO	Parts per million (PPM)	Effects on adults	% Saturation of CO in blood stream
0.01	100	Slight headache in 2 to 3 hours	13%
0.02	200	Mild headache, dizziness, nausea and tiredness after 2 to 3 hours	20% – 30%
0.04	400	Frontal headache and nausea after 1 to 2 hours; risk to life if over 3 hours exposure	36%
0.08	800	Severe headaches, dizziness, convulsions within 45 minutes; unconsciousness and death possible after 2 to 3 hours	50%
0.16	1600	Headaches, dizziness and nausea within 20 minutes; collapse, unconsciousness and death possible within 1 to 2 hours	68%
0.32	3200	Headache, dizziness and nausea within 5 to 10 minutes; possible death after 15 minutes	70% – 75%
0.64	6400	Severe symptoms within 1 to 2 minutes; death within 15 minutes	80%
1.28	12800	Immediate symptoms; death within 1 to 3 minutes	85% – 90%

Health and safety advice

Despite haemoglobin's great affinity for CO (relative to oxygen), it can be displaced by adequate oxygenation, so the following treatment is key:

- the victim(s) should immediately leave/be removed from the contaminated area to the open air and inhale pure oxygen, if available
- the victim(s) should be kept at rest, avoiding any unnecessary exertion (this can be fatal)
- always seek medical advice

You must do everything to ensure your own and others' safety at all times.

Other precautions you need to take are:

- adequately ventilate all affected areas
- carry out local atmospheric checks with an electronic combustion gas analyser, as necessary, before you allow people access
- be aware that areas can be severely contaminated with high levels of CO, even at considerable distances from the source (see **Migration of CO** in this Part).

  CO is of approximately the same density as air at the same temperature, but as it will be produced during combustion, it will be lighter than the surrounding air and circulated by convection around the room
- seek permission from the Health & Safety Executive (HSE) when asked to work in a property where there has been a gas incident
- do not carry out any work until the HSE have confirmed that their investigation is complete

Migration of CO

When a gas appliance operates under fault conditions, the greatest and most rapid build-up of CO will occur within the room/space where the appliance is located.

However, CO movement through a building can occur even when any doors to the appliance room are closed, via gaps around the doors.

As a result, CO can build-up within adjacent rooms reflective of the amount of CO concentration in the appliance/source room.

The migration of CO between floors can proceed in a number of different ways:

- CO movement from an appliance room to an upper floor typically via a stairwell, will tend to cause CO to accumulate uniformly between floor and ceiling level in upstairs rooms.
- migration of CO to floors at a lower level than the source will tend to occur only if thermal conditions outside the appliance room promote mixing. This process does not take place so easily as the previous situation as dispersion of hot buoyant combustion products tends to be upwards.

CO migration can make detection of CO difficult, particularly in multi-occupancy buildings where CO produced at one location is able to move to another part of the building via roof spaces, other rooms, along ducting and shared flue systems.

Effects of Carbon Dioxide (CO_2)

CO_2 is formed during the combustion process and if it were allowed to build-up in a room – either by poor/faulty flueing and/or ventilation, or via appliance design in the case of flueless – would lead to a contaminated atmosphere possibly affecting the safety of people and animals in that atmosphere as well as adversely affecting the combustion process.

Additionally, CO_2 is produced as a by product of the breathing process, which could further exacerbate a heavily contaminated atmosphere.

It is therefore important to ensure correct installation and safe operation of gas appliances, as well as providing adequate ventilation to ensure both correct combustion and a healthy internal environment.

Table 3.4 CO_2 concentrations

Parts per million (PPM)	Effects on adults
250 – 375	Normal levels in external atmosphere
375 – 1000	Normal indoor levels (adequate ventilation)
1000 – 2000	Poor comfort level, feelings of drowsiness
2000 – 5000	Stagnant/stale air, causing headaches and sleepiness
>40,000	O_2 starvation; short term effect may induce coma or death; long term effect - brain damage

The level of CO_2 in the room will depend on:

- number of occupants
- type of appliances, i.e. flued or flueless
- external ambient conditions
- time of day (or night)
- amount of ventilation.

Normal CO_2 levels in the external atmosphere are likely to be approximately 250 – 375ppm, but can be higher than this if close to a busy road (see Table 3.4).

CO_2 concentrations of 1.5% - 2% (15,000 – 20,000ppm) are considered to be at a point where oxygen will be sufficiently reduced to interfere with the combustion process, with incomplete combustion and the associated risk of CO formation likely.

The limits of CO and CO_2 specified by the World Health Organisation (WHO) for occupied areas are as follows;

- CO – 10ppm
- CO_2 – 5000ppm above ambient

Note: Both are based on an 8 hour time weighted average (TWA).

If levels of CO_2 in habitable areas exceed 2800ppm (except for short term use of flueless appliances such as cookers), persons in those areas need to be evacuated, all fuel burning appliances should be switched off and windows and doors etc. opened until the level of CO_2 is reduced to acceptable levels.

Any apparatus used to carry out these measurements will need to be capable of reading within a range of 0 to 5000ppm with a resolution of 50ppm, or 5000 to 20,000 with a resolution of within 2%.

Portable electronic combustion gas analysers

The need to maintain a level of good clean combustion is essential for appliance efficiency and safety reasons. Appliances where combustion quality is poor may work inefficiently and can be potentially dangerous.

- for this reason, British and European standards set minimum standards for appliance combustion performance. There are an ever-increasing number of appliances on the market, including condensing boilers and range-cookers, that use forced draught premix burners. The manufacturers' instructions for this type of appliance frequently require the use of a combustion analyser to successfully undertake commissioning and servicing procedures
- the flame picture has historically been considered to give a reasonable indication of whether satisfactory combustion is taking place. A blue coloured flame is generally considered satisfactory while an orange or yellowing flame is normally considered unsatisfactory. However, a visual indication is not fully reliable and a more detailed visual examination of the appliance may be necessary. With some types of appliance it is not possible to see the flame picture

So when should you use them?

A portable electronic combustion gas analyser (ECGA) should be used when commissioning certain types of gas appliance (boilers for example) and as part of the servicing/maintenance procedures in accordance with the manufacturer's instructions.

Don't use a combustion analyser as an alternative to normal servicing procedures, which should be carried out in accordance with the appliance manufacturer's instructions (see also **Appliance servicing** in this Part). A number of gas appliances will require full servicing irrespective of the combustion performance reading.

You can use an ECGA (also referred to as a flue gas analyser), which should conform to BS EN 50379-3 (see Note) to determine the level/extent of servicing required for a given appliance, to confirm the safe performance of some gas appliances following servicing, or to help in the identification of unsafe gas appliances.

Because of particular appliance design restrictions, it may not always be possible to take a combustion performance reading as some appliances do not lend themselves easily to this.

These include some types of gas fires (that are installed using a closure plate) and some room-sealed appliances, where for lack of suitable sampling points or access to the flue terminal (i.e. it's located at heigh level/roof level) you may not be able to obtain a combustion performance reading.

Note: BS EN 50379-3 replaces the previous BS 7927 as the build standard for portable electronic combustion gas analysers.

You can use ECGAs to:

- verify the combustion characteristics of gas appliances
- determine the concentrations of CO within buildings
- identify the source of a CO emission

For many years the gas industry has used the ratio of the CO to CO_2 in the products of combustion to confirm safe appliance operation.

When tested under the special conditions (see Note) required by both European and British Standards during the appliance approval process, the permitted maximum ratio of CO to CO_2 is:

- for most modern appliances, 10 parts of CO to 1000 parts CO_2. By dividing 10 by 1000 we get a CO/CO_2 ratio of 0.01
- for some older appliances, 20 parts of CO to 1000 parts CO_2. By dividing 20 by 1000 we get a CO/CO_2 ratio of 0.02

The above figures apply to all appliances burning NG or LPG.

Note: It is not appropriate to test under these certification conditions in the field. See the section on 'Appliance testing' in this Part for further guidance.

The formula to calculate a CO/CO_2 ratio is:

$$CO/CO_2 \text{ ratio} = \frac{\text{\% vol. CO concentration}}{\text{\% vol. } CO_2 \text{ concentration}}$$

$$\text{or} \qquad = \frac{\text{ppm CO}}{\text{\% vol. } CO_2 \times 10000}$$

The analyser normally displays the CO level in parts per million (ppm) of the POC and the CO_2 as a percentage of the POC. The analyser also calculates the ratio.

The CO level in the POC would be expected to be very low. The higher the level of CO in the POC, the greater the degree of incomplete combustion.

Appliance testing – helping you assess safe operation/post-work check

You can carry out a combustion performance test by:

- sampling the POC from the flue of the appliance and measuring the CO (see Note) and CO_2 levels
- the CO/CO_2 ratio may be used to assist you to determine whether the appliance is operating safely

Note: The HHIC (the Heating and Hotwater Industry Council) who represent boiler manufacturers operating in the UK market have introduced a new requirement to measure and record CO levels, along with the combustion ratio when commissioning a new gas boiler installation.

From 1st April 2014, boiler manufacturers with the wide support of the industry require the CO and CO/CO_2 ratio to be measured and recorded on the Benchmark Log Book that accompanies the appliance and which forms part of the warranty conditions.

The test can help you identify those appliances that are in need of a service. You'll need to carry out all the necessary visual safety checks as well.

You can use a test as a post-work check after you have serviced or worked on a gas appliance.

Currently within the gas industry, the following trigger-value combustion ratios are used to assess an appliance's combustion performance:

- a reading greater than 0.004 is a good indication that an appliance is in need of a service

- a reading less than or equal to 0.008, following a service is a good indication that an appliance is operating safely and will continue to operate safely until the next planned service date, even though the reading is above the 0.004 trigger value
- for readings greater than 0.008, additional checks should be carried out to ensure that the appliance is assembled, cleaned and operating as the manufacturer would expect

Where it has not been possible to obtain a combustion performance result of 0.008 or less, then you need to do the following checks:

- operating pressure at the gas meter
- appliance operating pressure
- verify the gas rate at the meter test dial
- the size of pilot injector
- the pilot position (incorrect siting could cause flame impingement)
- internal gas escapes (escapes from splits in the burner, connections, etc.)
- burner clean, correctly assembled and positioned
- burner undamaged, check flame picture
- heat exchanger clean, especially low water content appliances with copper-finned heat exchangers (check if the fins could be restricted or even blocked)
- position of internal baffles in the heat exchanger
- open-flued appliance flue operation, ensure no spillage (can cause vitiation)
- room-sealed appliance operation, no POC re-circulating causing vitiation, flue exhaust duct sealed satisfactorily

What if all of the checks prove satisfactory, yet a ratio greater than 0.008 is still recorded?

You should seek further advice from the appliance's manufacturer.

Industry Standards for using flue gas analysers

The British Standard Code of Practice, BS 7967 entitled – 'Guide for the use of electronic portable combustion gas analysers for the measurement of carbon Monoxide in dwellings and the combustion performance of domestic gas-fired appliances' – provides guidance on how to use ECGAs, which meet the product standard BS 7927 (analysers that display CO, O_2, CO_2 and CO/CO_2 ratio):

Note: A 5th Part to the suite of standards deals with non-domestic installations and therefore, isn't discussed within this manual.

The aim of the Code of Practice (CoP) is to:

1. Give detailed guidance on the use of combustion gas analysers for the safety testing of gas appliances.
2. Provide typical combustion performance characteristics for a range of appliance types and where appropriate remedial action to be taken.
3. Assess the build-up of POC in buildings.
4. Provide guidance on the use of combustion gas analysers when undertaking the servicing and maintenance of gas appliances.

Note: CORGI*direct* has produced two publications and one form specifically designed to cover the use of ECGAs, both for servicing/maintenance visits and for when investigating reports of fumes. See Part 22 – CORGI*direct* Publications.

In addition to the use of analysers and where applicable, you need to always carry out the relevant flue-flow and spillage safety checks in accordance with the appliance manufacturer's instructions and/or BS 5440-1. For further guidance, see **Part 14 – Flue testing**.

Carbon monoxide detectors

Providing a gas appliance is installed, serviced and adjusted correctly and there is an adequate supply of fresh air, combustion of the gas will normally be satisfactory.

However, under adverse weather conditions or where an alteration to the building has taken place, this could have an adverse effect on the safe operation of an open-flued gas appliance.

Also the appliance may simply break down or the flue could become blocked in between normal service dates.

The use of CO detector alarms provides gas users with added protection and the early detection of such gases can prevent a gas related incident from taking place (see Figure 3.7).

Note: CO detector alarms should not be used as a first line of defence against the effects of CO. Regular gas safety checks/maintenance and correct installation of appliances is of greatest importance.

There are two types of CO detectors generally available:

1. Colour change "spot" detectors.
2. Mains/battery powered detector alarms.

Colour change "spot" detectors

This type of detector operates without the aid of an electrical source. It is not an alarm, but provides a visual warning only. An orange coloured spot turns grey or black as CO is detected (at levels as low as 100 ppm of CO).

Note: Reaction times can vary with humidity.

The shelf life for an unopened pack can be up to 3 years. When exposed to the atmosphere, the spot can remain active for up to 3 months.

Figure 3.7 CO detector

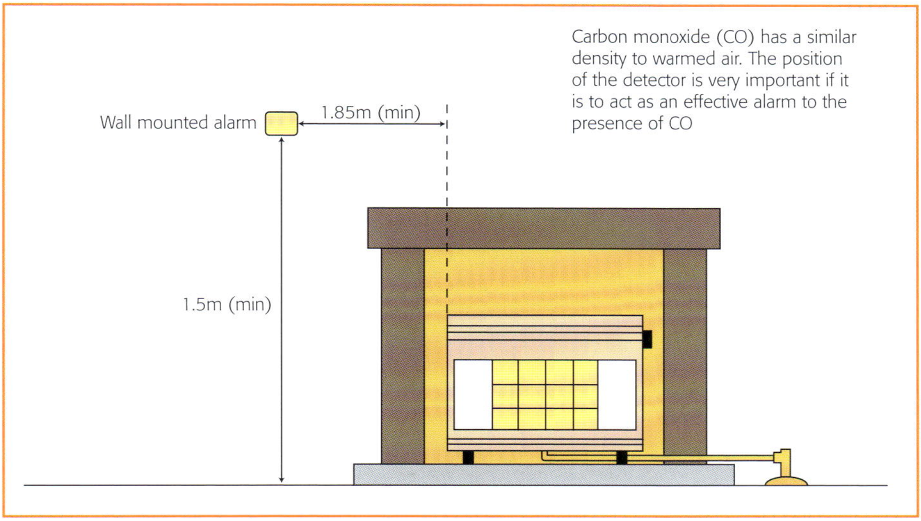

Attention: This type of detector can be damaged or indeed neutralised by halogens, nitrous gases and ammonia (e.g. babies' nappies and cat litter). For this reason, the detector should be sited away from cleaning materials, solvents and other contaminants. If the spot detector steadily becomes darker/bleaches over time it should be replaced and the gas appliance checked.

Spot detectors can be attached to the wall by a self-adhesive patch and are normally positioned at eye level near to the appliance. They should be regularly checked to see if the spot has darkened.

If the spot has darkened general advice is to:

- turn off the appliance and ventilate the room(s)
- get the appliance checked for safety by a registered gas operative before further use

CORGI*direct* does NOT recommend the use of colour change "spot" detectors for the following reasons:

- they provide a visual indication only that CO has been detected
- their life span is limited
- their effectiveness can be affected by household chemicals and other substances
- they do not provide an audible alarm and because of this their existence can be forgotten by the user

Mains/battery powered detector alarms

Where CO detector alarms are required or specified, they should be manufactured to BS EN 50291 and display a BSI Kitemark.

Note: The previous BS 7860 has been superseded by BS EN 50291.

Mains/battery powered detector alarms can be:

- mains powered hard wired (fitted by a competent electrical contractor)
- mains powered to a standard 3 pin mains socket
- 9 volt battery operated

There are currently three different types of CO sensors available, each having different characteristics:

1. Semiconductor sensor (mains powered).
2. Gel cell sensor (battery powered).
3. Electrochemical cell sensor (battery or mains powered).

Semiconductor sensors

These consist of heated dioxide pellets with precious metals.

The better quality sensors use a unique surface cleaning process. This helps the gas being absorbed at a relatively low temperature on the surface, before being re-absorbed and measured at the same time at a higher temperature.

The surface is then cleaned at the higher temperature before being cooled to repeat the cycle of absorption and measurement. A carbon filter also helps to improve selectivity.

Note: Some low cost CO detectors use a more basic semiconductor sensor which can be much more prone to false alarms.

Gel cell sensor

The gel cell sensor is normally combined with a battery in a single pack. This needs to be replaced every 1 to 2 years in order to achieve the required level of effectiveness. The gel cell consists of a porous material impregnated with an active material.

Carbon monoxide reacts with the active material and darkens it, which is detected electronically by sending a beam of light through the cell.

These sensors darken slowly and are slow to recover when CO is present. It is possible to have low levels of CO (e.g. 15ppm), which are not normally regarded as being damaging to health, but can ultimately cause the cell to become ultra sensitive and eventually alarm.

Electrochemical cell sensor

This is reputed to be the best type of CO detector available. It works by electrochemically oxidising the CO to form CO_2. As this takes place, hydrogen ions and electrons are generated at the working electrode cell.

The hydrogen ions travel through the electrolyte to the counter electrode where they are consumed by reacting with oxygen from the surrounding air. The electrons are taken via the contact pins into the external circuit and fuel the reaction.

The current generated is directly proportional to the concentration of CO at the working electrode.

The advantages of the best quality cells:

- neither the electrodes nor the electrolyte are consumed in the process, providing maintenance free devices which can remain operational for many years
- the cell requires no external power (e.g. for heating) to operate – and generates its own power in the form of an electrical current from the CO

- a special membrane ensures stability of output by controlling the diffusion of gas to the electrodes. It also minimises environmental effects
- fast, repeatable response, with minimal humidity effects
- a non-consumable filter maintains the high selectivity of the cell

Note: Many cells now have a life span of 5+ years under normal operating conditions.

Wide range of models available

They can be installed free-standing or wall mounted; be battery operated or be hard wired with a battery backup. Operating voltages are generally 230V a.c., 24V d.c. or 12V d.c.

Detector alarms are available with relay output and interconnect options. These allow connection to control panels, gas safety shut-off valves and remote alarms.

Be aware that positioning is very important

BS EN 50291 requires that detector alarms are positioned in the room where the appliance is located and, if wall mounted, at a height which is greater than the height of any door or window, but at least 150mm from the ceiling.

Individual units can also be linked together in series.

You always need to refer to the manufacturer's installation instructions

Alarm levels as required by BS EN 50291 are:

- at 50ppm the alarm must sound between 60-90 minutes
- at 100ppm the alarm must sound between 10-40 minutes
- at 300ppm the alarm must sound within 3 minutes

See Table 3.3 for a list of the effects of carboxyhaemoglobin on human beings.

Ventilation – 4

4 – Ventilation

Figures

Tables

4 – Ventilation

Installing air vents for combustion air and compartment ventilation for new gas appliances – guidance to help you

Can you refer to this section for previously used or second hand gas appliances, not carrying the 'CE' mark or BS approval?

Yes, if they have a data plate bearing information ensuring that the appliance is suitable for the gas type and pressure (i.e. NG at an appliance inlet pressure of 20mbar, Propane and Butane where the supply regulator has been set to provide an operating pressure of 37mbar and 28mbar respectively).

The Standards and Regulations that cover this subject – and the competence you need

BS 5440-2 specifies the ventilation requirements for the majority of NG/LPG, domestic, new or replacement appliances carrying the CE mark and with a rated heat input not exceeding 70kW based on the net Calorific Value (CV).

This British Standard can also be used to calculate the ventilation requirements for existing installations of gas appliances. You will find areas not covered by BS 5440-2 and additional guidance in the following:

- gas installations in caravan holiday homes and residential park homes (RPH) constructed to conform to BS EN 721 or BS 3632, as appropriate
- LPG installations in boats, which are covered under PD 5482: Part 3 and/or ISO BS EN 10239
- decorative fuel effect (DFE) gas appliances covered under BS 5871-3
- domestic direct gas-fired tumble dryers BS 7624
- domestic gas-fired warm air heaters BS 5864
- domestic flueless gas fires and greenhouse heaters BS 5871-4

When you carry out any work relating to gas appliances and other gas fittings covered by this part:

- you must be competent and hold a valid certificate of competence for each work activity that you wish to undertake (see **Part 1 – Gas Safety legislation – Competence** for further guidance)
- you must install all gas appliances and other gas fittings in accordance with the Gas Safety (Installation and Use) Regulations, British Standards, Building Regulations and Regulations for Electrical Installations, or in accordance with those Regulations appropriate to the geographical region in which they are to be installed
- you must also refer to the manufacturer's installation instructions for the appliance

Clean, fresh, adequate supply of air required for complete combustion

Every gas burning appliance needs clean, fresh air to achieve complete combustion. An adequate air supply also assists the performance of an open-flued chimney system and where necessary, provides cooling air for compartments.

Air is generally supplied by purpose provided ventilation opening(s) in the outer wall of the building.

The size of the air vent(s) is determined by taking into account:

- the type of gas appliance
- the heat input to the appliance(s); and
- the adventitious ventilation that occurs in every building

The GSIUR stipulate that no person shall install a gas appliance unless it can be used without constituting a danger to any person. So, when required, you must ensure that there is a sufficient permanent supply of air available to the appliance for proper combustion and correct operation of the chimney system.

Check if gross or net Calorific Values (CV) are used to quote heat input

It is important you know whether the heat input of an appliance is quoted in gross or net CV terms when determining the ventilation openings required for a gas appliance using BS 5440-2.

The CV information you need for

1. New installations

Where the manufacturer's installation instructions quote BS 5440-2 and the appliance data badge is quoted as a net CV, then follow the information provided in the manufacturer's installation instructions.

Imported gas appliances will generally be badged on the basis of net CV, so apply the requirements of BS 5440-2 – unless otherwise specified in the manufacturer's installation instructions.

Note 1: If there is any doubt regarding the basis of the CV, consult the manufacturer or UK agent.

Note 2: Manufacturers of certain fixed flueless gas fires may state ventilation requirements that are not in accordance with BS 5440-2. Where this is the case, those manufacturers will not state compliance with BS 5440-2 but their ventilation requirements remain valid.

2. Existing installations

If the manufacturer does not express the heat input as a net CV, but rather as a gross CV then you'll need to convert a gross CV heat input into the equivalent net CV and apply the ventilation requirements in accordance with BS 5440-2.

The ratio of gross to net CV heat input is dependent on the fuel type and is approximately as follows:

- NG 1.11:1
- Propane 1.09:1
- Butane 1.08:1

The difference between gross and net CV between Natural, Propane and Butane gas is negligible so to avoid confusion, use a general figure of 1.1 for all of the above fuel types when converting from a gross CV to a net CV, or vice versa, when calculating ventilation requirements.

Note: How you determine net CV for ventilation purposes is different to the way you determine gas rating requirements for gas appliances. See Part 11 – Checking/setting appliance burner pressures and gas rates.

Example: An open-flued appliance with a gross CV heat input of 27.5kW using NG would have a net CV heat input of:

$27.5 \div 1.1 = 25$kW(net).

Therefore the combustion ventilation requirement would be:

$25 - 7^* = 18$

$18 \times 5\text{cm}^2 = 90$

Therefore 90cm^2 free area is required.

*See **Adventitious ventilation – the 'natural ventilation' present in rooms/buildings** in this Part.

Note: All heat inputs referred to in this part are given in net terms unless stated otherwise.

Adventitious ventilation – the 'natural ventilation' present in rooms/buildings

The term 'adventitious' ventilation refers to natural ventilation and occurs as a result of air entrainment from cracks and gaps around doors, windows, suspended wooden floors, etc.

An air vent is not normally required for the first 7kW (gross or net) of the rated heat input of an open-flued appliance.

This allowance is due to the amount of ventilation that is present in all buildings and rooms even after considerable effort has been made to reduce air changes, i.e. draught proofing and double glazing, etc.

Research shows it is virtually impossible to reduce the amount of this ventilation in habitable rooms to less than 35cm^2: the figure used by the British Standards Institution to determine the allowance in the calculation for appliances with a rated heat input up to 70kW (net).

Attention: Do not take adventitious ventilation into account when you calculate the air requirements for:

- flueless appliances
- appliances in compartments
- decorative fuel effect (DFE) gas appliances

Ventilation openings and grilles – guidance to help you

As currently there is no British Standard for the design or production of propriety air vents used with gas appliances, Advantica Technology have developed a procedure to encourage air vent manufacturers to be consistent about the characteristics of their air vents. This is published by BG Technology as: 'A dynamic test procedure for characterising ventilator free-area'.

Apply the guidance below to new or replacement installations – and if you find existing installations which do not meet these requirements:

- advise the person responsible for the premises
- follow the current Gas Industry Unsafe Situations Procedure (GIUSP) see **Part 8 – Gas Industry Unsafe Situations Procedure**

Purpose-provided ventilation should take the form of a proprietary air vent

Air vents provide combustion and cooling air – and are vital for the performance and safe operation of open-flued appliances. Ensure they comply with the following requirements:

- air vents should be non-closable and should not incorporate any additional gauze or screens
- the size of the ventilation openings should prevent the entry of a 10mm diameter ball but should allow the entry of a 5mm diameter ball
- air vents should not be located externally where they can become easily blocked or flooded, or where contaminated air would be entrained, e.g. in a car port
- any air vent in an internal wall, other than from a compartment ventilator, should be located no more than 450mm above floor level to prevent the spread of smoke in the event of a fire occurring
- the minimum separation distances of an air vent from a chimney terminal should be in accordance with the guidelines in Table 4.1 and Figure 4.1

Note: These separation distances from appliance chimney terminals apply equally to the sitting of room extract fan outlets – except cooker hood outlets and tumble dryers with a reverse flow non-return flap. You need to connect the extraction terminal via a duct to the extract hood or tumble dryer.

Table 4.1 Minimum separation distances between air vents and appliance terminals

Air vent position	Appliance heat input (kW) (net)	Room-sealed & open-flued	
		Natural draught	Fanned draught
		Separation (mm)	
A Above a terminal	0 to 7	300	300
	>7 to 14	600	300
	>14 to 32	1500	300
	>32 to 70	2000	300
B Below a terminal	0 to 7	300	300
	>7 to 14	300	300
	>14 to 32	300	300
	>32 to 70	600	300
C Horizontal to a terminal	0 to 7	300	300
	>7 to 14	400	300
	>14 to 32	600	300
	>32 to 70	600	300
Note: > = greater than.		See Figure 4.1	

- where air vents are installed in a series from room to room, the air vents on internal walls should be at least 50% greater in free area than the air vent to outside air (see **Where air vents are in series – what you need to do** in this Part)
- no air vent should penetrate a protected shaft or stairway
- air vents should not communicate with a ventilated roof or underfloor space if that space communicates with other premises
- consideration should be given to the presence of Radon gas (see advice on **Air supply from under-floor (Radon gas) – particular care needed regarding ventilation** in this Part)
- an air vent which is to supply air to an open-flued appliance should not communicate with any room/space containing a bath or shower

Figure 4.1 Minimum separation distances between air vents and appliance terminals

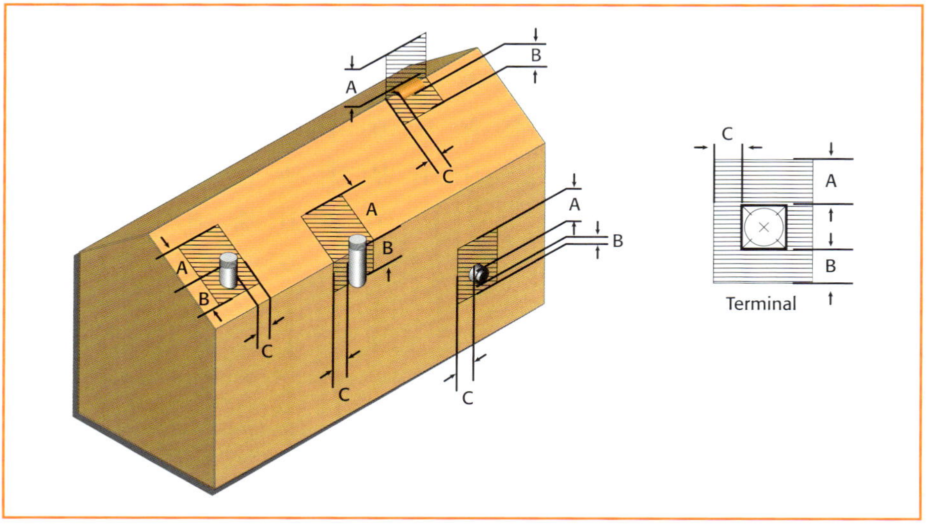

- air vents installed in a cavity wall should include a continuous duct/sleeve across the cavity. The cross-sectional area of the duct/sleeve should not be less than the area of the opening required
- air vents installed through a cavity wall should not be staggered
- air vent ducts should normally be limited to 3m in length and incorporate no more than two 90° bends
- an intumescent air vent (special air vents, designed to prevent the spread of smoke in the event of a fire) may be used provided that it complies with the relevant Building Regulations and as long as it does not incorporate a manual closing device (see **Intumescent air vents – expand and close to prevent the spread of smoke in a fire** in this Part)
- do NOT install an air vent which is to supply air to a combined gas fire/back boiler/back circulator unit within the builder's opening. Air passing through vents installed in this location could affect the flame pattern of the fire and its safe operation
- to reduce the risk of illegal entry, do not install air vents in openable windows on ground floors or where there is access to the window

When you have determined the ventilation requirement, the physical size of the vent is dependent on the free air passing through the unobstructed cross-sectional areas of the grille, i.e. the collective individual openings in the grille.

Due to their aesthetic appearance, terracotta air bricks are often specified by architects for the ventilation of gas appliances.

However, problems may arise when air vents are specified by their physical dimensions rather than their unobstructed cross-sectional openings, e.g. '220mm x 145mm air vent required'. This gives no indication of the actual 'free area' of the air brick.

Note: Even if you are not involved in the installation of the air brick or ventilation opening, you still need to ensure:

- **that it has the correct free area**
- **that it is sleeved across the cavity**
- **that it has suitable grilles fitted to the internal and external wall, before commissioning a gas appliance**

See also **Maintenance of air vents** in this Part.

How you calculate the vent free area to satisfy ventilation requirements

Manufacturers of air bricks will normally specify the free area of the bricks in their literature. If you do not have this, you should:

- measure the dimensions (cross-sectional area) of one hole in the air brick
- multiply this by the total number of holes to give the combined free area of the air brick
- take care to ensure you measure the smallest dimension to give the true cross-sectional area of the hole

Figure 4.2 shows a typical air brick as viewed from the front. Figure 4.2 also shows that same air brick in side profile, showing a considerable reduction in the hole size due to tapering. It is this smaller, internal dimension that you measure to calculate the vent free area.

Example

A room containing an open-flued appliance has a combustion air requirement of 80cm^2.

A builder has provided one single air brick measuring 220mm x 220mm, containing 49 square holes. On the external face of the air brick, the holes measure 10mm x 10mm, but they taper to 8mm x 8mm on the inside. Therefore the free area can be calculated as follows:

49 x 8mm x 8mm = 3136mm^2 or 31cm^2.

Figure 4.2 Typical terracotta air brick

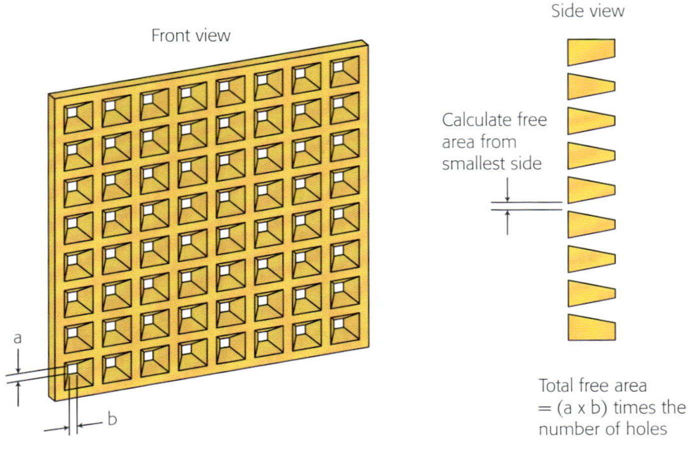

To satisfy the ventilation requirements of the appliance, three air bricks of this type and size would be required.

Manufacturers have made advancements in air brick design. Air bricks are now available with rectangular holes which allow a greater free flow of air.

In all cases, you need to be satisfied that the installation complies with the GSIUR before commissioning the appliance.

Intumescent air vents – expand and close to prevent the spread of smoke in a fire

These air vents are specially designed to do this. Generally they incorporate a latticework of holes to allow for continuous ventilation – but in the event of extreme heat and in a fire, they will expand and close.

You must therefore check (especially in existing installations) that air vents of this type are fully open (not closeable) and that they are able to supply the amount of ventilation air required. See also **Maintenance of air vents** in this Part.

Air requirements for open-flued appliances

Guidelines to help you provide the air requirements in each case are as follows.

Single appliance

Calculate the ventilation requirements for open-flued appliances using the maximum rated heat input of the appliance, i.e.

- the maximum rated heat input of a range-rated central heating boiler
- the maximum rated domestic hot water heat input of a combination boiler

Table 4.2 Ventilation for open-flued appliances

Relates to permanent openings for appliances up to a maximum rated heat input of 70kW (net) in accordance with BS 5440-2.

Maximum total rated heat input	Appliance(s) location	Direct to outside air
One appliance under 7kW (net or gross) (including gas fires)	In a single room, space, through room, lounge diner etc.	Ventilation not normally required, as adventitious air usually provides sufficient air for combustion
One or more appliances totalling in excess of 7kW (net)	In a single room, internal space, through room, lounge diner etc.	5cm^2 per kW (net) of total rated heat input above 7kW
Two or more gas fires, up to a total rated heat input of 7kW each (net or gross) (14kW total)	In a through room, lounge diner, etc.	Ventilation not normally required as adventitious ventilation will usually provide sufficient air for combustion. For a higher kW rating, allow an additional 5cm^2/kW (net) above 14kW
Two or more appliances. Add the total rated heat input of all appliances	Single room or internal space	Calculate the total ventilation requirement of all the appliances based on their maximum heat input rating. Provide 5cm^2/kW (net) in excess of 7kW*

* For multi-appliance installations, only one appliance should be considered for the adventitious air allowance.

Note: Should the appliance fail the spillage test, it may be necessary to install an air vent to rectify the fault, e.g. a gas fire fitted in a property with solid floors, double glazing and cavity wall insulation, may not have sufficient adventitious air to allow for the correct operation of an open-flued chimney system.

- the total maximum heat input of a combined appliance i.e. fire/back boiler, fire/back circulator or warm air heater/circulator unit

Table 4.2 provides you with further guidance on the ventilation requirements of open-flued appliances in rooms or internal spaces.

Open-flued appliance installed in a room or internal space

Where an open-flued appliance with a heat input exceeding 7kW (gross or net) is installed in a room or internal space, you need:

- to provide that room or internal space with a purpose provided permanent ventilation opening with a minimum free area of 5cm^2 for every kW of appliance maximum rated heat input in excess of 7kW (gross or net)
- to ensure that the air vent should be direct to outside air, or to an adjacent room that is itself ventilated direct to outside air to at least the same minimum free area

Example

An open-flued appliance with a maximum rated heat input of 16kW (net) is installed in a room. The heat input exceeds 7kW by 9kW (16 – 7 (for adventitious air) = 9); 5cm^2 of free area is required for every kW above 7kW.

The calculation is therefore 5 x 9 = 45cm^2.

You can see that 45cm^2 of ventilation will be required for this appliance.

Multi-appliance ventilation – more factors to take into account

Where there are two or more gas appliances installed in the same room, it is described as a 'multi-appliance installation'.

Working out the ventilation requirement for such installations can be confusing. Your aim is to assess such installations and provide ventilation for the appliance, which is likely to be operating and generating the maximum demand.

You calculate the air vent free area requirement using the greatest of the following:

- the total rated heat input of flueless space heating appliances (see note 1)
- the total rated heat input of open-flued space heating appliances (see note 2)
- the maximum rated heat input of any other type of appliance

Note 1: A space heating appliance is taken to be a central heating boiler, warm air heater, fire or space heater.

Note 2: Where the interconnecting wall between two rooms has been removed and there are two similar chimneys with similar gas appliances fitted then, providing the heat input rating of each appliance does not exceed 7kW (gross or net), no additional ventilation is normally required.

Example

An open-flued gas boiler and open-flued gas fire are installed in the same room. The gas boiler has a heat input of 20kW (net) and the gas fire has a heat input of 4kW (net).

You calculate the ventilation requirement as follows:

20kW + 4kW = 24kW

24kW – 7kW (adventitious air) = 17kW

17kW x 5cm^2 = 85cm^2

Therefore the ventilation requirement for the multi-appliance installation is 85cm^2 free area.

Note: Where applicable, only one appliance should be considered for the adventitious air allowance.

If the room also contains an oil or solid fuel burning appliance, treat them separately i.e. one ventilation provision for the gas appliance and another for the solid fuel or oil appliance. If the heat input of such an appliance is unknown, estimate it using the following calculation:

$$\text{Input} = \frac{\text{Output} \times 10}{6}$$

Note: For a solid fuel open fire or small closed stove of unknown heat input, the ventilation requirement should be taken to be 100cm^2. Where a DFE is fitted, unless otherwise stated by the manufacturer's instructions, 100cm^2 of ventilation should be fitted.

For compartment installations, see 'Ventilation requirements for appliances in compartments' in this Part.

Where air vents are in series – what you need to do

When three or more air vents are installed in series to vent from an internal space to outside air via a number of rooms:

- increase all internal air vents in size by 50%.

See Figure 4.3 for further guidance.

For aesthetic reasons, ventilation grilles are generally fitted in pairs (i.e. one on each side of a wall). A pair of grilles is therefore classed as:

- one air vent; and
- the first air vent is the pair fitted in the external wall

The additional requirements for air vents in series do not apply to the calculation of air vent sizes for a compartment.

Figure 4.3 Air vents in series

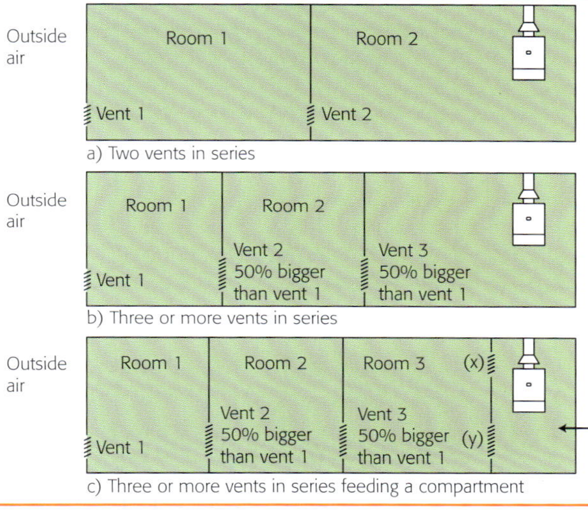

Note: For aesthetic reasons ventilation grilles are generally fitted in pairs i.e. one on each side of a wall. A pair of grilles will therefore be classed as one vent and numbered with the first grille being the pair fitted in the external wall, in contact with outside air. However, when three or more air vents are installed in series to vent from an internal space to outside air via a number of rooms, all internal air vents should be increased by 50%. The "air vent in series" requirements do not apply to the calculation of air vent sizes for a compartment (see x and y).

See also **Ventilation requirements for appliances in compartments** in this Part.

Where there is a decorative fuel effect (DFE) gas appliance – what you need to do

Unless the manufacturer's installation instructions say otherwise, install 100cm^2 of permanent ventilation for decorative fuel effect (DFE) gas appliances up to 20kW (net).

For a DFE fitted in a fireplace without a throat e.g. a DFE under a canopy, the Building Regulations Approved Document 'J' for England and Wales (Combustion Appliances and Fuel Storage Systems) states that the ventilation is sized at 50% of the cross-sectional area of the flue.

For detailed ventilation information on DFE requirements, including air entering a builder's opening from below floor area, see the current manual entitled 'Gas Fires and Space Heaters' from the Gas Installer Manual Series – Domestic.

See also Table 4.3

Therefore as an example the ventilation requirements for two DFEs plus an open-flued boiler of 30kW in a room would be as follows.

2 x DFE = 200cm^2

1 x Adventitious air 35cm^2

1 x 30kW boiler = 30 x 5 = 150cm^2

Total ventilation = 200 + 35 + 150 = 385cm^2

Is ventilation required for Inset Live Fuel Effect (ILFE) gas fires?

Ventilation is not normally required for ILFE gas fires up to 7kW (gross or net) unless the manufacturer's installation instructions state otherwise.

Table 4.3 Ventilation for DFE gas appliances*

(sizing of permanent air vents for approved DFEs up to a maximum rated heat input of 20kW (net) supplied with Natural gas or LPG in accordance with BS 5871-3)

No	Installation rated input	Appliance(s) location	Ventilation requirement to outside air
1	One DFE ≤7kW	room	normally 100cm² minimum, but check manufacturer's instructions, as some models may not require any additional ventilation
2	One DFE >7kW - ≤20kW	room	normally 100cm² minimum (or as per manufacturer's instructions)
3	Two DFEs ≤20kW	room or through room	normally 200cm² (or as per manufacturer's instructions) plus 35cm² (adventitious air)
4	One or two DFEs plus other appliance(s)†	through room or space with two open-flues	DFE requirement based on options 1-3 plus whichever is the greatest of: a) 5cm² per kW heat input (net) of all open-flued space heating appliances, such as central heating boilers; or b) the total flueless space heating requirement; or c) 5cm² per kW heat input (net) of a single, open-flued non-space heating appliance, such as a water heater; or d) the greatest individual requirement of any other type of appliance

Note: If the appliance fails a spillage test, it may be necessary to provide additional ventilation. A DFE appliance installed in a property with a solid floor, double glazing and cavity wall insulation may not have sufficient adventitious air to allow for the correct operation of the flue.

* For a DFE fitted in a fireplace without a throat e.g. a DFE under a canopy, the Building Regulations Approved Document 'J' for England and Wales requires that the ventilation is sized at 50% of the cross-sectional area of the flue.

† Treat oil and solid fuel appliances as if they were gas appliances and calculate input as being 166% of the heat output. For a small solid fuel fire or small closed fire of unknown input, provide a permanent opening of 100cm².

Warm air heaters – how to introduce air for combustion

In today's modern, air tight constructed dwellings (which often include draught and sound proofing material) an alternative method of providing air for combustion to an open-flued warm air heater is by drawing air from the roof space into the warm air ducting system.

This eliminates the need for openings in walls, doors or windows and reduces the risk of draughts.

Depending on the appliance manufacturer's particular requirements, the construction of the dwelling and the position of the warm air heater, it may be advantageous to use the following fan assisted method of introducing air for combustion:

1. Connect a duct (fitted with a bird guard) from a ventilated roof space, or from a waterproof grille on an outside wall, to the return air duct or return air plenum on the warm air heater.
2. Install a lockable damper in the fresh air duct and adjust it to provide the necessary volume of air. A minimum airflow rate of $2.2m^3$/hr should be drawn into the air chamber for every 1kW of the appliance maximum input rating (net). You need to make this adjustment. When you have completed the commissioning procedure, lock the damper in position (to prevent unqualified persons making adjustments).

When the warm air heater operates, its fan will draw in external air, mix it with the return air and circulate it throughout the warm air distribution ducts. Warm air should be circulated into the room/space where the heater is installed using a non-closing register fitted in that area.

Note: Where this method of air supply is used, refer to the appliance manufacturer's installation instructions for further guidance.

Domestic tumble dryers – new installations must be flued to outside air

Modern domestic tumble dryers are now available with heat input ratings up to and including 6kW (net).

Tumble dryers have been reclassified by PD CEN/TR 1749 as a type 'B' appliance. This means that all new installations have to be flued directly to outside air.

The methods available are:

- via a wall or window using a proprietary termination device
- where appliances do not exceed 3kW (net) rated heat input, as above; or
- via a hose hung directly out of a window

In addition to an openable window, for a gas tumble dryer installation where the room volumes (m^3)/appliance heat input rating (kW net) is less than $3.7m^3$/kW, purpose provided ventilation of $100cm^2$ will be required.

Example 1.

If a gas-fired tumble dryer of 6kW is installed in a room measuring 3.5m long by 2.5m wide and 2.4m high. In this location, as well as the requirement for an openable window, an air vent of $100cm^2$ needs to be provided.

$3.5 \times 2.5 \times 2.4 = 21(m^3) \div 6(kW) = 3.5m^3/kW$

Example 2.

If a gas-fired tumble dryer of 4.5kW is installed in a room measuring 3.5m long by 2.5m wide and 2.4m high. In this location, only an openable window needs to be provided.

$3.5 \times 2.5 \times 2.4 = 21(m^3) \div 4.5(kW) = 4.6m^3/kW$

Effects of extract fans on open-flued appliances – why additional ventilation may be required

Where an extract fan is fitted in the same room as an open-flued appliance, additional ventilation may be required to prevent the depressurising effect of the fan affecting the performance of the flue. As a general guide, an additional 50cm^2 will normally be sufficient.

Extraction systems include:

- room-extract fans
- fan assisted open-flued appliances
- cooker hoods (at all speeds)
- warm air heaters (gas and electric)
- Radon gas extraction systems*
- landfill gas extraction systems

* See also **Air supply from under-floor (Radon gas) – particular care needed regarding ventilation** in this Part.

As a result of modern building and draught-proofing methods, extract fans installed outside a room containing an open-flued appliance may also cause a depressurising effect on the flue serving that appliance.

If this is the case, check the appliance flue performance (spillage test) with the door(s) separating the appliance from the extract fan open and all other doors and windows closed.

Is the flue affected by the extract fan?

If so, you need to provide sufficient ventilation to overcome the effect in the room containing the appliance.

Risk of spillage of flue gases could create an 'Immediately Dangerous' situation

Building Regulations warn that where there is an open-flued appliance installed in a kitchen with mechanical ventilation or extraction capabilities, there may be a risk of spillage of flue gases, which could create an 'Immediately Dangerous' situation.

See **Part 8 – Gas Industry Unsafe Situations Procedure** and also see **Part 14 – Flue testing** for further guidance.

It may be necessary to provide less than the set requirement of 60 litres per second mechanical ventilation in kitchens containing open-flued appliances. The Building Regulations do not give specific guidance as to the amount of reduction which may be necessary, but it must be sufficient:

- to allow successful completion of spillage tests
- to ensure clearance of combustion products at all times

This indicates that there is a possibility of spillage problems in some conditions if a kitchen contains a production model extract fan and an open-flued appliance.

To overcome these spillage problems, additional ventilation may be required. Achieve this by:

- partially opening a window and measuring the free area of the opening at a point when spillage is no longer evident during the test
- calculating the additional ventilation requirement
- adding it to any previously calculated requirement
- installing a permanent air vent

Figure 4.4 Passive stack ventilation

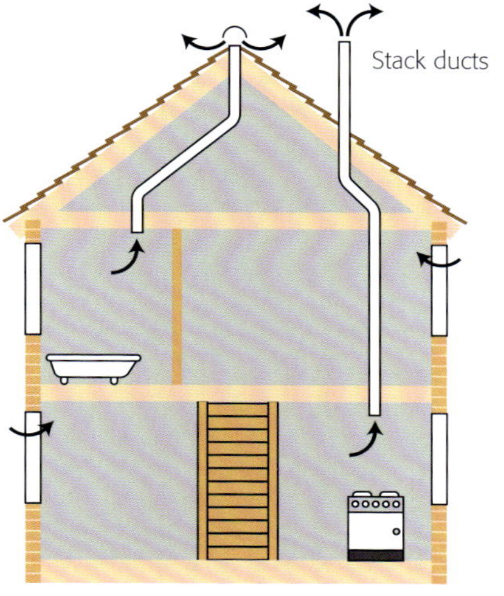

Note: Where you are newly installing a gas appliance and the requirements are not met:

- you must not leave the appliance connected to the gas supply
- you must seal the gas supply with an appropriate fitting and label it
- you must issue a Warning Notice to the person responsible for the property (see also Part 8 – 'Gas Industry Unsafe Situations Procedure' and Part 10 – 'Emergency notices, warning labels and reporting forms')

Passive stack ventilation – a method of extracting warm, stale air from a building

The Passive Stack Ventilation (PSV) system consists of a series of ducts typically installed in wet rooms i.e. kitchen/bathroom etc. (see Figure 4.4), located in each room to be ventilated. These ducts pass vertically through the building and terminate above roof level.

The system operates by a combination of the natural stack effect (i.e. the movement of air due to the difference in densities between warm buoyant air within the building and cooler, heavier surrounding air outside, see **Part 14 – Flue testing – Thermal inversion**) and the effect of the wind passing over the roof of the dwelling.

In modern homes, PSV systems may be fitted. In this case, where an open-flued appliance is installed that requires permanent combustion ventilation to be provided, then you need to consider the PSV ducts independently from the provision of ventilation for gas appliances i.e. do NOT use them for this purpose.

Locations where the use of air vents are restricted

Protected shafts prevent the spread of fire and smoke – make sure an air vent does NOT penetrate them

Protected shafts (see **Part 19 – Definitions**) are stairs or other shafts passing directly from one compartment floor to another and are constructed and enclosed in such a way as to prevent the spread of fire or smoke.

An example of a protected shaft can be found in flats over two storeys high with individual accommodation on each floor, where the means of access and exit is via the protected stairway.

Do not install ventilation from a compartment or living area into a protected shaft.

Air supply from under-floor (Radon gas) – particular care needed regarding ventilation

An air vent should not communicate with a ventilated under-floor space, if that space communicates with other premises.

In the past, gas operatives have utilised the ventilated under-floor space between a suspended floor and the ground below as a source of combustion air. This method has become popular because there is much less of a draught effect than there would be from a ventilator fitted in an outside wall.

However, research has highlighted the risks to health from naturally occurring Radon gas, which percolates upwards through the ground and into the atmosphere.

Radon gas is a colourless, odourless gas, which is radioactive and formed where uranium and radium are present. It can move through cracks and fissures in the subsoil and collect in spaces both underneath and within dwellings. Where it occurs in high concentration, it can pose a risk to health.

Whilst it is recognised that the air inside every house contains Radon gas, some dwellings built in certain defined areas of the country may have unacceptably high concentrations unless suitable precautions are taken.

In the UK, the granite areas of South-West England are of principal concern, but high concentrations of Radon gas are also found in some other parts of the country.

So you must take particular care when installing under-floor ventilation and ensure that the installation of any ventilation to a gas appliance:

- does not put occupants of the dwelling at risk; or
- does not compromise any measures which are already in place to prevent Radon gas entering habitable areas of the dwelling

For further guidance refer to the BRE publication 'Radon: guidance on protective measures for new dwellings'. Visit www.brebookshop.com to obtain a copy of this publication.

Air supply from roof spaces – varying requirements

Do not install an air vent in this location, if the roof space communicates with other premises.

1. When utilising roof space ventilation, care should be taken to ensure that the roof space is also ventilated in accordance with the appropriate Building Regulations.

A roof-space ventilated in accordance with the Regulations, i.e. by means of permanent ventilation grilles giving a free area of not less than 0.3% of the roof plan area, may be deemed as outside air (see Note).

2. Where a roof-space is not constructed in accordance with the appropriate Building Regulations, it should only be deemed as another internal space and you have to treat it as such when calculating the ventilation requirement. In addition, be aware of problems relating to condensation forming in inadequately ventilated roof-spaces with this method of ventilation.

In both cases, any air vent(s) into roof-spaces should be designed or located with openings not less than 300mm above ceiling joists or 150mm above the level of insulation (which ever is greater) to avoid blockage, e.g. by insulation material.

Note: New dwellings built post 1985 should meet the appropriate Regulations. Dwellings that have been re-roofed post 1985 should also have been installed to comply with the appropriate Regulations, but double check.

Ventilation requirements for appliances in compartments

How to remove heat from compartments – so that an appliance operates safely

The following information deals solely with the ventilation requirements of compartments. It does not include information on the fire resistance requirements. You can find this in the specific appliance manuals in the current Gas Installer Manual Series – Domestic.

When operational, all gas appliances emit radiant and convected heat to the surrounding area. The amount of heat emitted varies, dependent on the model and manufacturer. Boilers that incorporate a large cast iron heat exchanger can often emit as much as 0.75kW/hour.

Where a gas appliance is installed in a room or internal space, the heat is dissipated to the surrounding area and will contribute to the heat input for that room or space.

However, when an appliance is installed in a compartment, this heat needs to be removed. Otherwise the compartment temperature will increase to an extent that may ultimately affect the safe operation of the appliance and its controls.

This heat is removed by providing the compartment with ventilation grilles at both high and low level. Be aware that:

- if ventilation grilles are direct to outside air, the heat is lost and in extreme weather conditions, could cause freezing of the appliance and auxiliary equipment
- if the ventilation grilles are installed in the compartment walls or door to an internal room, the heat is not lost and will contribute to the general heating load for the dwelling

As outside air is much cooler than that found inside the dwelling, ventilation free areas to outside air are sized at 50% less than those to internal spaces, but generally ventilation to outside air is more difficult and expensive to install.

What is a compartment?

An appliance compartment is defined (by Approved Document 'J' to the Building Regulations) as an **"enclosure specifically constructed or adapted to house one or more combustion appliances"**.

BS 6798 for gas boilers not exceeding 70kW net provides a similar definition, but also goes on to state **"and which is not a habitable space"**.

Therefore and depending on the appliance(s) installed, a compartment may require air for combustion and ventilation at both high and low level for cooling purposes.

Other small spaces, e.g. lobbies or enclosed porches, which house gas appliances but have not been specifically adapted, may also require compartment style ventilation.

In fact, any space in which you are unlikely to spend any length of time and is fitted with a gas appliance is likely to overheat unless adequate ventilation is provided (see also the appliance manufacturer's installation instructions for guidance).

Good practice for compartment ventilation can be covered in the following general guidelines:

1. Where applicable, ventilation needs to be provided for both combustion air and cooling air. It may be via ventilation grilles only or through ventilation grilles connected by ductwork to another suitably ventilated space.

Note: Louvred doors may be used – providing that the total free area measured across the smallest dimension of the slots is not less than the total for the high and low level ventilation requirements of the appliance(s).

2. Ventilation should be taken from the same source, i.e. internally through a room or from an external wall or roof space. It should never be a mix of both, which could cause cross-ventilation and affect flue performance.
3. Air vents should be positioned at both high and low level and as far apart as possible to assist air circulation and heat dissipation.
4. Combustion air ventilation should be sited at low level.
5. Ducting at high and low level should consist of separate ducts and should not be joined.
6 The lower duct supplying combustion or cooling air should be insulated to promote air circulation.
7. A duct should not be installed from low to high level.
8. Air vents should pass through the same side wall; ducts should come from the same outside air space.
9. When more than one appliance is installed in a compartment, the combined maximum rated heat input of the appliances should be used to calculate the ventilation requirement.
10. When calculating the ventilation requirements for appliances installed in compartments, no allowance should be made for adventitious air. The total maximum rated heat input is used for this purpose.

Open-flued appliances in compartments need both high and low level ventilation

They need this for:

- complete combustion
- correct operation of the flue; and
- compartment cooling

Both high and low level air vents need to communicate with the same air source, e.g. both to outside air on the same wall, or both to a ventilated internal room/space. It should never be a mix of both, which could cause cross-ventilation and affect the safe operation of the flue.

Attention: As the use of ducts from a higher level is a non-preferred method for new installations (BS 5440-2), you need to find an alternative method of providing combustion/compartment ventilation.

Where an open-flued appliance is installed in a compartment and the ventilation will be taken direct from outside air:

- calculate the free area of the air vents at 5cm^2 per kW of the maximum rated heat input at high level and 10cm^2 per kW of the maximum rated heat input at low level

Table 4.4 Compartment ventilation for open-flued appliances

Vent position	Into a room or internal space	Direct to outside air
High level	10cm^2 per kW of heat input	5cm^2 per kW of heat input
Low level	20cm^2 per kW of heat input Then as below.	10cm^2 per kW of heat input
Ventilation for a room or internal space (permanent openings, based on a total maximum rated heat input (net)) See **Air requirements for open-flued appliances** in this Part.		

Figure 4.5 Ventilation summary for open-flued appliances

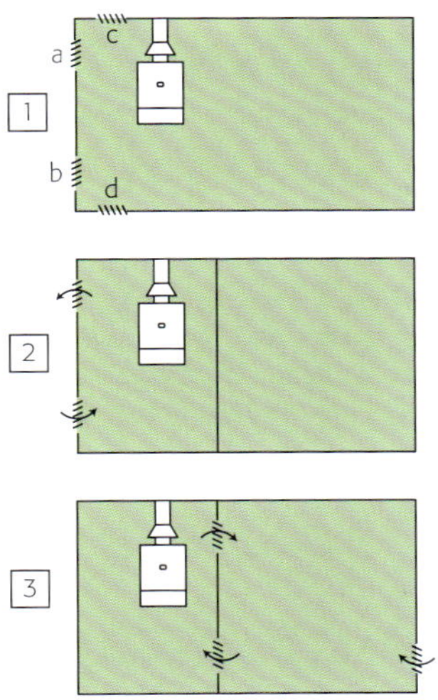

IN A ROOM
Either a, b, c, or d direct to outside air at 5cm^2/kW of rated heat input above 7kW

IN A COMPARTMENT
Ventilated direct to outside at 5cm^2/kW of rated heat input at high level and 10cm^2/kW of rated heat input at low level

IN A COMPARTMENT
Ventilated via a room or internal space 10cm^2/kW of rated heat input at high level and 20cm^2/kW of rated heat input at low level. The room is ventilated as in 1 above

Where an open-flued appliance is installed in a compartment and the ventilation will be taken from a room or internal space:

- calculate the free area of the air vents at 10cm^2 per kW of the maximum rated heat input at high level and 20cm^2 per kW of the maximum rated heat input at low level

Also refer to Table 4.4 and Figure 4.5.

Where the ventilation communicates with a room, the room itself needs to be ventilated sufficiently to satisfy the combustion air requirement for the appliance (see also **Air requirements for open-flued appliances** in this Part).

If the ventilation has to be taken through a series of rooms, do refer to **Where air vents are in series – what you need to do** in this Part.

Table 4.5 Compartment ventilation for room-sealed appliances
(permanent openings, based on a total maximum rated heat input (net))

Vent position	Into a room or internal space	Direct to outside air
High level	10cm² per kW of heat input	5cm² per kW of heat input
Low level	10cm² per kW of heat input	5cm² per kW of heat input
In a room or internal space	No requirement for additional ventilation	

Note: Any air vent, which is to supply air to an open-flued appliance, should not communicate with any room/space containing a bath or shower.

Room-sealed appliances in compartments – which need additional ventilation?

Due to advancements in appliance design, some room-sealed appliances can now be installed in compartments without the need for additional ventilation. Where this is the case, the appliance manufacturer's installation instructions will have the information you need.

When you install a room-sealed appliance that uses ventilation from outside air:

- carefully consider the position of the ventilation openings in relation to the flue terminal to prevent the products of combustion from re-entering the property
- apply the separation distances outlined in Table 4.1 and Figure 4.1

Where you install a room-sealed appliance in a compartment that does require high and low level ventilation:

- calculate the free area of the ventilation grilles at 5cm2 per kW of the maximum rated heat input of the appliance(s)

See Table 4.5 'When direct to outside air.'

Calculate ventilation to an internal space at 10cm2 per kW of the maximum rated heat input of the appliance(s) See Table 4.5.

For ventilation paths see Figure 4.6.

Airing cupboard ventilation – ensure vents are not blocked

Ensure the air vent sizes comply with the requirements for compartments and install so that both high and low vents are in the appliance compartment.

It means partitions and clothing above, below or to the side of the appliance should not block or restrict the circulation of air between the air vents.

Note: Where the appliance is open-flued, the ventilation sizes and restrictions should comply with those for 'Open-flued appliances in compartments need for both high and low level ventilation' in this Part.

Under-stairs cupboard ventilation – number of storeys in premises has an impact

The air vent sizes should comply with the requirements for compartments.

Where the premises in which the cupboard is located has more than two storeys, the air vents need to be direct to outside air. In premises up to and including two storeys, air vents can communicate with a room or internal space.

See advice on high level ventilation openings in **What is a compartment?** in this Part.

Figure 4.6 Ventilation summary for room-sealed appliances

1

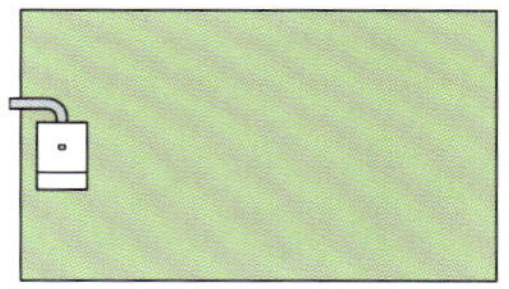

IN A ROOM
No vent required

2

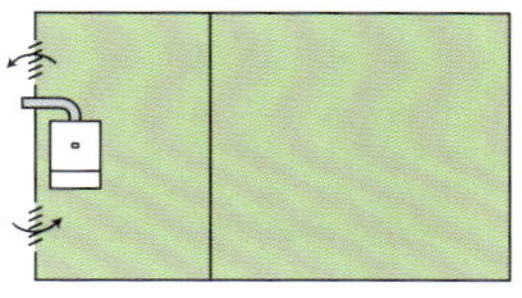

IN A COMPARTMENT
Ventilated direct to outside at 5cm^2/kW of rated heat input at both high and low level*

3

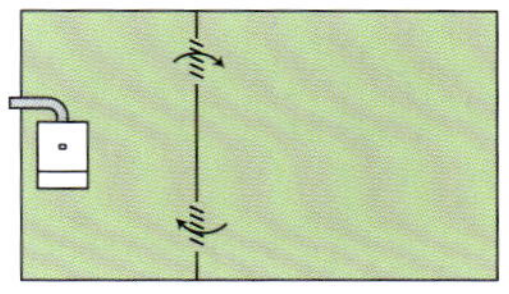

IN A COMPARTMENT
Ventilated via a room or internal space 10cm^2/kW of rated heat input at both high and low level*

* Where manufacturers specify

External installations – advice for you to follow

Where you install a boiler in an outside enclosure you must fit this with air vents direct to outside air, at both high and low level.

The air vents should be sized in accordance with **Ventilation requirements for appliances in compartments** in this Part.

Ventilation for flueless appliances – advice for you to follow

The following information is based on the Building Regulations (England and Wales) Approved Document 'F'. The regulations in Scotland, Northern Ireland and the Isle of Man may differ from these.

Where you undertake installation work in Scotland, Northern Ireland and the Isle of Man, refer to the appropriate Building Regulations.

Where you install a flueless appliance in a room or internal space

- you must provide that room or internal space with an air vent that communicates directly with outside air and is sized in accordance with Table 4.6 – a gas cooker would be a typical flueless appliance
- you must ensure there is an openable window, or similar

Table 4.6 Minimum permanent ventilation opening free areas for flueless appliances

Type of appliance	Maximum rated heat input limit (net)	Room volume (m^3)				Openable window required†
		Below 5	>5-10	>10-20	Above 20	
Domestic oven, hotplate, grill or a combination of the above**	None	$100cm^2$	$50cm^2$*	Nil	Nil	Yes
Instantaneous water heater	11kW	Not allowed	$100cm^2$	$50cm^2$	Nil	Yes
Fixed space heater in a room	$45W/m^3$ of heated space	$100cm^2$ plus $55cm^2$/kW (net) by which the rated input exceeds 2.7kW (net)				Yes
Fixed space heater in an internal space	$90W/m^3$ of heated space	$100cm^2$ plus $27.5cm^2$/kW (net) by which the rated input exceeds 5.4kW (net)				Yes
Space heater conforming to BS EN 449 in a room	$50W/m^3$ of heated space	>$15m^3$ – $25cm^2$/kW with a minimum of $50cm^2$ at both high and low level				Yes
Space heater conforming to BS EN 449 or a greenhouse heater in an internal space	$100W/m^3$ of heated space	>$15m^3$ – $25cm^2$/kW with a minimum of $50cm^2$ at both high and low level				Yes
Independent greenhouse heater (see BS 5871-4)	4.2kW	High and low level ventilation comprising of $39cm^2$/kW for every kW above 2.7kW				Yes
Refrigerator	None	Nil				No
Single boiling ring (2 or more should be treated as a hotplate)	None	Nil				No

* If the room or internal space containing these appliances has a door which opens directly to the outside, no permanent ventilation opening is required.

** The appliance, unless a single burner hotplate/boiling ring, shall not be installed in a bed/sitting room of less than $20m^3$ volume.

† Alternative acceptable forms of ventilation opening include any adjustable louvre, hinged panel or other means of ventilation that opens directly to outside air. This is in addition to the permanent air vent requirement.

What to do when installing a flueless appliance in habitable rooms – through other rooms and spaces with no direct ventilation to outside air

You need to refer to the following special requirements:

1. Two habitable rooms may be regarded as a single room for ventilation purposes, if there is a permanent opening between them, equivalent to at least 1⁄20th of their combined floor areas and there is also purge ventilation equivalent to at least 1⁄20th of the combined floor areas plus a minimum of 8000mm² of background ventilation (see Figure 4.7).
2. A habitable room ventilated through a conservatory (see Figure 4.8) would satisfy the requirement if:
    a) there is provision from the conservatory to outside for both:
        i) purge ventilation – one or more ventilation openings with a total area of at least 1⁄20th of the combined floor area of the habitable room and space, and
        ii) background ventilation – a ventilation opening (or openings), of at least 8000mm² (see Figures 4.9 and 4.10)
    b) there are openings (which need to be closable) from the habitable room and the conservatory for:
        i) purge ventilation – one or more ventilation openings with a total area of at least 1⁄20th of the combined floor area of the habitable room and space, and
        ii) background ventilation – a ventilation opening (or openings), of at least 8000mm², which should be at least 1.7m above floor level (see Figures 4.9 and 4.10)

Ventilation requirement for non-habitable rooms not containing openable windows

In kitchens and utility rooms not containing openable windows (i.e. internal rooms) the requirement will be satisfied if there is one of the following:

a) Mechanical extract ventilation (see Note), which can be either of the following:
    - 30l/s where the extract is adjacent to hob, or
    - 60l/s when the extract is not adjacent to hob

Note: Mechanical extract ventilation can be controlled either manually or automatically; manual would be where in rooms with no natural light it would be appropriate for the extract fan to be controlled by the operation of the light switch and automatic is where the extract fan is controlled by a humidity sensor/switch.

b) Passive stack ventilation (PSV).

c) An open-flued appliance with the appropriate ventilation requirement.

For a) to c), an air inlet of 7600mm² needs to be provided e.g. a 10mm gap (under cut) under a standard door.

For further guidance on gas cookers in internal kitchens, refer to Gas Safe Register's TB 'Gas cookers in internal kitchens':

- TB 005(A) Ventilation requirements for England, Wales, Isle of Man and Guernsey
- TB 005(B) Ventilation requirements Scotland; and
- TB 005(C) Ventilation requirements Northern Ireland

Figure 4.7 Two rooms treated as a single room for ventilation purposes

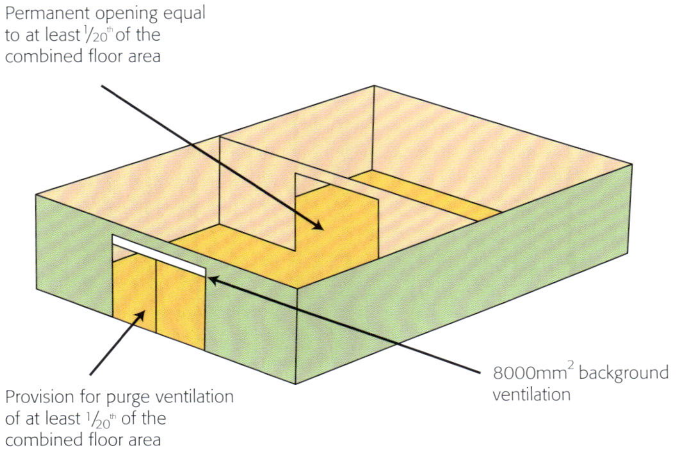

Figure 4.8 Habitable room ventilated through an adjoining space

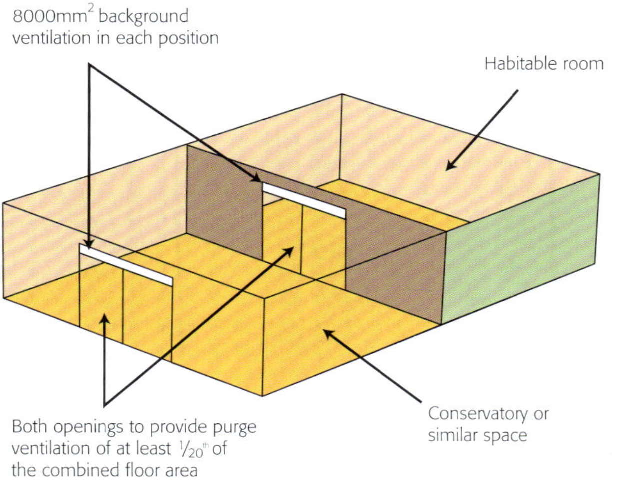

Figure 4.9 Provision for background ventilation

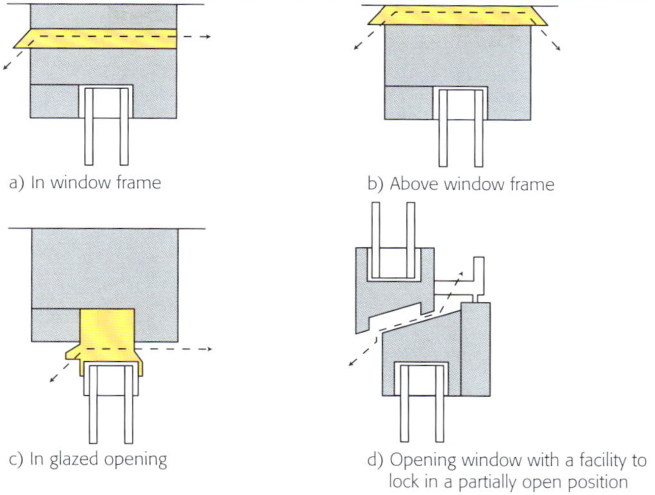

Figure 4.10 Provision for background ventilation

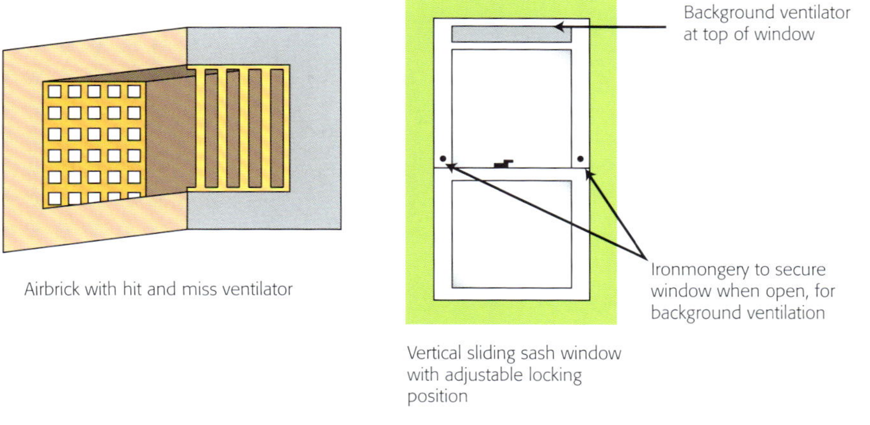

Maintenance of air vents

When you are called to a property to carry out maintenance work on the installation, you need to check the ventilation requirement.

Carry out both a visual and physical inspection of the air vents to ensure that they are correctly located and unobstructed, both internally and externally.

Do this by:

- removing fly screens
- if the air vent is of the closable type, either replacing it with an appropriate air vent, or fixing it permanently in the open position
- checking (as part of the physical inspection) to ensure that the air vent is not blocked with mortar extrusions, cavity insulation, fly screens or even a clear polythene/plastic sheet – a common occurrence where users experience drafts, which is generally in response to poor placement of the air vent (use of a screwdriver, graduated probe etc inserted at various points along the louvres or similar will assist in this regard)
- checking the cross-sectional free area of the air vent to ensure that it is of adequate size for the appliance(s) installed, see Note

Note: Purpose provided ventilation for non-room-sealed appliances (open-flued and flueless) where the ventilation is 0 to 89% of the requirement, should be classed as 'At Risk' (AR) in accordance with the current Gas Industry Unsafe Situations Procedure. Where the ventilation is between 90 to 100% of requirement, this is acceptable under current standards.

Installation of pipework and fittings – Section 5

5 – Installation of pipework and fittings

Installation of pipework and fittings – Section 5

5 – Installation of pipework and fittings

Installation of pipework and fittings – 5

5 – Installation of pipework and fittings

Figures

5 – Installation of pipework and fittings

Tables

Essential requirements and differences between installation of Natural gas (NG) and Liquefied Petroleum Gas (LPG) that you need to know

Although many of the requirements for the installation of NG and LPG systems are the same, there are some important differences.

Information in this Part covers the essential requirements for both NG and LPG installations and:

- is relevant to design parameters, materials and methods of installing copper, steel, pliable corrugated (stainless steel) tubing and polyethylene (PE) gas pipework, in sizes up to 35mm (DN35 / R1¼ steel), for the supply of 2nd and 3rd family gases (NG and LPG, respectively) on premises
- applies to low pressure gas pipework where the nominal operating pressure is 21mbar for NG, 37mbar for Propane gas and 28mbar for Butane gas

Note: The term pliable corrugated (stainless-steel) tubing is sometimes abbreviated to corrugated stainless-steel tubing (CSST) or pliable tubing (PLT) and will be refered to as CSST / PLT throughout this Part.

You will also find further information for domestic gas meter installations in **Part 16 – Domestic gas meters** and for LPG installations, **Part 18 – LPG – General requirements**.

The Regulations that apply and the competence you need to carry out this work:

When you carry out any work (see **Part 19 – Definitions**) in relation to gas pipework, appliances and other gas fittings covered by this Part, you must be competent and hold valid competency certification for each work activity that you wish to undertake (see **Part 1 – Gas safety legislation – Competence** and with regards to gas meters, **Part 16 – Domestic gas meters** – for further guidance).

The GSIUR require you to ensure that gas installation pipework and fittings are installed safely with due regard to:

- the position of other pipes, pipe supports, drains, sewers, cables, conduits and electrical apparatus
- any parts of the structure of any premises in which it is installed which might affect its safe use

Additionally, where you connect installation pipework to a primary gas meter without a main protective bonding conductor, you need to inform the responsible person that bonding may be required and should be carried out by a competent person
(see **Equipotential bonding – minimises hazards from electrical shock** in this Part and **Part 16 – Domestic gas meters**).

Note: CORGI*direct* recommends that when you identify any defect:

- you first speak about it to the gas user
- you then also document it (see Part 10 – Emergency notices, warning labels and forms)

What is installation pipework?

It is defined as any pipework or fitting on the outlet of the primary meter installation to points at which appliances are to be connected.

For installations where a gas meter is not fitted (i.e. bulk storage vessel/cylinder(s) for LPG) it is defined as any pipework or fitting on the outlet of the ECV to points at which appliances are to be connected.
(see **Part 18 – LPG – General requirements**).

5 – Installation of pipework and fittings

The factors to consider when you install pipework

1. At the planning stage, verify that the pipework will be adequate:

- for current requirements; and
- any future planned extensions

Also important to note and where necessary, provide drawings to include:

- the position of any ducts and channels when installation pipes are to be concealed
- special requirements, e.g. limitations with regard to the proximity to other services

2. Then consider:

- safety precautions
- materials to be used
- size of pipe required
- route and appearance

Safety precautions you must take

You must ensure that:

- all exposed metallic materials in the proposed work area are tested to detect the presence of stray electrical current which could be with the use of a non-contact voltage indicator which is capable of showing voltages of 50V or more. (see Figure 5.0) If voltage is detected or there is any doubt as to the integrity of the electrical system then the assistance of a suitably competent electrician should be sought.
- the installation pipework is physically protected, or
- it is located where it is not liable to be subjected to mechanical damage
- the bore of the pipe is not restricted by kinks or burrs
- you prevent the ingress of foreign matter, (water, dirt etc.) into the pipes
- if you are likely to leave the pipework unattended at any time, you plug or cap all open ends of pipework with an appropriate fitting
- if you leave the general location for any period during the installation work, you consider the risk of persons restoring the gas supply at the primary meter or storage vessel/cylinder(s)
- when you solder capillary joints with a blow lamp on pipework that contains gas, you always isolate/disconnect the gas supply to that part of the pipework and ensure it is adequately purged of gas
- you keep naked flames away from all open ends of pipework or fittings
- you do not use oxy-gas flame cutting equipment on any meter, pipe or fitting containing gas

Figure 5.0 Non-contact voltage detector (single pole)

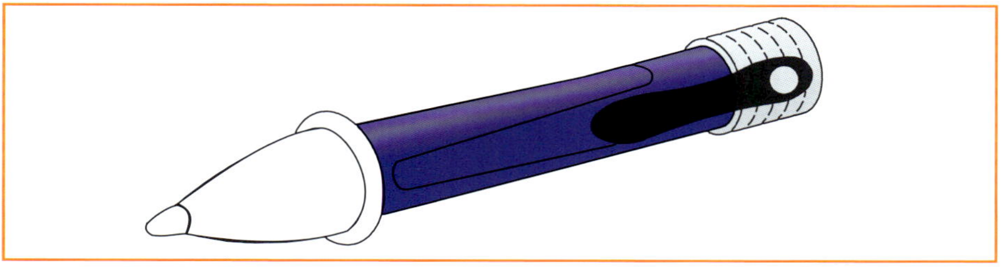

Figure 5.1 Temporary continuity bond

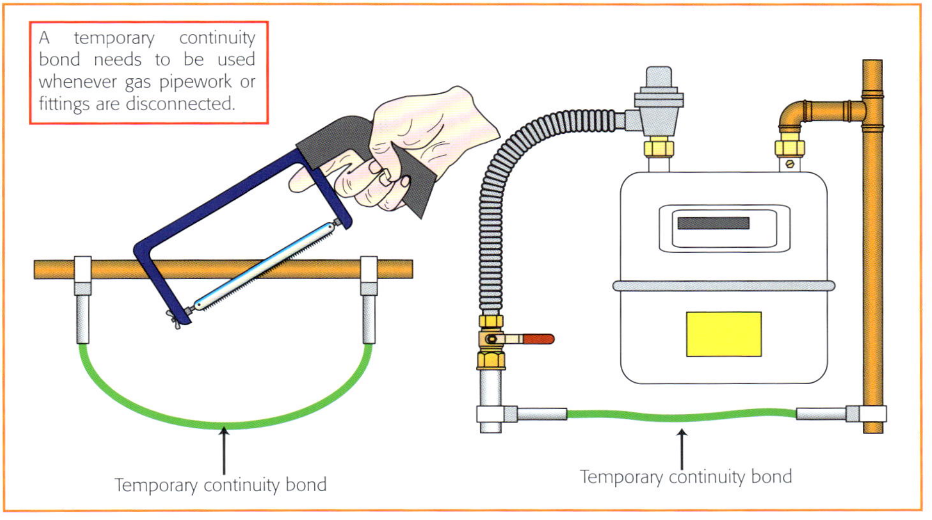

Extending installation pipework? Additional safety information you need

If installation pipework is already connected to a gas meter and you are extending it, you must:

- disconnect the meter and seal both the open ends with dust caps to prevent any 'flash back' to the air/gas mixture present in the meter
- place the meter in a safe position, to prevent any accidents

Is any installation pipework no longer required?

Then you must:

- disconnect this as close to the point of supply as is practicable
- make sure you seal all open ends, using an appropriate fitting (including redundant pipework)
- ensure that any part of the installation that may contain gas is labelled

Temporary continuity bond – use for your own safety before you start work

Before you start work connecting or disconnecting any installation pipework, pipe fitting or gas meter, attach a temporary continuity bond to the pipework (as shown in Figure 5.1). This is to maintain electrical continuity and to prevent production of a spark or shock that could prove hazardous to you.

Use a temporary continuity bond whether or not a main protective bonding conductor is present (see **Equipotential bonding – minimises hazards from electrical shock** in this Part).

The recommended procedure for using a temporary continuity bond is:

1. Where appropriate, it is advisable that you isolate any associated gas appliances from the main electrical supply.

2. Fit the first clamp to the upstream side (service pipe side) of the gas meter, pipe fitting or pipework, ensuring a good metallic contact is achieved. This may involve cleaning the pipework or removing rust, paint etc.
3. Fit the second clamp to the downstream side (appliance side) of the gas meter, pipe fitting or pipework, again ensuring a good metallic contact is achieved.
4. Carry out any work that is required without disturbing the temporary continuity bond.
5. When you have completed the work, disconnect the temporary continuity bond in reverse order, i.e. downstream side first.

Undertake periodical inspection of your temporary continuity bond and replace the lead if you find it is faulty or damaged.

Note: It is recommended that:

- the temporary continuity bond is at least 1.2m long and comprises a single-core insulation flexible cable or equivalent of at least 250V rating.
- the cable should have a minimum cross-sectional area of 10mm multi-strand construction generally in accordance with BS 6004, BS 6007 or BS 6231 - with a robust clip or clamp firmly attached at each end.

Materials used for NG and LPG – some differences

Generally, materials used for NG are also suitable for use with LPG, but there are some exceptions.

If you are installing pipework and fittings for LPG installations, consider this section in conjunction with **Part 18 – LPG – General requirements**.

Selecting materials for use as installation pipework:

You need to consider:

- mechanical strength
- the type and grade of pipe for a particular installation
- its appearance
- the need for protection against corrosion

New gas pipework and fittings installed inside dwellings must not be made of lead, lead alloy or any non-metallic substance unless it is a flexible pipe connected to a readily moveable flueless gas appliance.

Gas pipework on domestic premises should be either copper, steel, polyethylene (PE) or CSST / PLT and should conform to a recognised British Standard (see Table 5.1).

Copper pipe should be jointed using either capillary solder, compression joints or press-end fittings, all of which comply with the relevant British Standards (see Table 5.1).

Screwed steel pipes and fittings should comply with the relevant British Standards (see Table 5.1).

Where the copper, steel or CSST / PLT has a factory applied plastic covering this is not deemed to be a sleeve and you will need to use additional sleeving when the pipework passes through a wall/floor of solid construction (see **Sleeves** in this Part).

PE gas installation pipework and fittings shall only be used for external pipework. They are generally used underground and may be beneficial where it is required to run externally from one location to another.

Note: Do not under any circumstances use other plastic pipes that may be suitable for water supplies and use push-fit or crimp-fit fittings used for water for gas installation pipework.

Table 5.1 Suitable materials for gas installation work

Pipework	Compliance Standards
Steel pipe	BS EN 10255 BS EN 10217-2 BS EN 10216-1 or BS EN 10217-1 and/or BS EN 10216-2
Rigid stainless steel	BS EN 10216-5, BS EN 10217-7, BS EN 10312 or BS 6362
Malleable iron fittings	BS 143 or BS 1256 and BS EN 13349
CSST / PLT	BS 7838 or BS EN 15266
Copper pipe	BS EN 1057 or BS EN 13349
Copper capillary and compression fittings	BS EN 1254-1, BS EN 1254-2, BS EN 1254-4 or BS EN 1254-5
Copper press-end fittings	BS 8537
Thread sealing	BS EN 751-1, BS EN 751-2 and BS EN 751-3
PE pipe	BS EN 1555-1 and BS EN 1555-2
PE fittings	BS 5114 and BS EN 1553-3
Flexible hoses	BS 669-1 BS 669-2
Valves	BS 1552, BS EN331, PRS1/E(24) or GIS/V7-3(25)

Jointing – facts to help you

Copper pipe/fittings

Copper pipe should conform to BS EN 1057 or BS EN 13349 and should be jointed using either capillary solder, compression joints or press-end fittings.

Note: Press-end fittings should be assembled in line with the manufacturer's instruction's and should not be painted with oil or solvent based paints unless allowed by the manufacturer. The yellow ochre on the body of the fitting should remain visible.

Capillary action soldered joints

These can be solder ring or end feed type fittings and rely on solder being drawn into the gap between the fluxed pipe and fitting (see Note).

Note: Copper pipework may be jointed without conventional fittings, by forming the pipe ends with the use of a purpose-made tool in accordance with BS EN 1254-1 or BS EN 1254-5.

Flux – provides protection from oxidation

The function of a flux is to protect cleaned metallic surfaces from attack by oxygen in the atmosphere (oxidation). If the surfaces to be soldered are not perfectly clean and free from oxidation, you may find it difficult to make a satisfactory joint.

Two types of flux:

1. **An 'active' (or self-cleaning) flux** is one where the flux reacts with surface oxides and heat to provide a clean surface. In addition, the metal surfaces are now 'wetted' allowing a free flow of solder.

In order to achieve this, flux is generally corrosive. It is important that it is only corrosive during the heating process, so only apply it to the external pipework and **never** to the inside of fittings, where it may cause corrosion at a later date.

2. **An inactive flux** will require the use of an abrasive cleaning pad such as wire wool to ensure all surfaces are clean. Follow this by an application of flux to prevent further oxidation occurring until the joint is made.

Preparing and making a capillary soldered joint using soft solder wire

There are two types of fitting:

End feed – where solder is applied to the mouth of the fitting.

Solder ring – where solder is contained in an annular ring inside the fitting.

The procedure for making a soldered joint is basically the same for both types of fitting (apart from the use of solder being applied to make the joint with end feed fittings) and is as follows:

1. Ensure that the pipe and fitting sizes are compatible (there are still some existing imperial copper pipes in use which require an adaptor fitting for metric conversion).
2. Ensure that the pipework is not under strain, that the pipe ends are cut square and that any burrs are removed from the internal bore of the pipe.
3. Depending on the flux type being used, either, thoroughly clean the outside of the pipe ends and the internal surface of the fitting, or apply an active flux.
4. Assemble the joint ensuring that the pipework is fully inserted into the fitting.
5. Apply heat evenly to the fitting. When the joint has reached the correct temperature apply the solder. If the joint is hot enough, the solder will run into the fitting. Additional solder is not normally required on solder ring type fittings.
6. Allow the joint to cool, ensuring that no movement occurs.
7. Remove any remaining flux on the pipe or fitting with a damp cloth.

Do not allow flux and its residues to come into contact with stainless steel pipe. Remove flux and clean the pipework to prevent corrosion.

Warning: If you apply solder to a joint at the same time as heat from a blow torch and before the pipe and fitting have had time to reach the correct temperature this may result in localised heat. This melts the solder but only forms a solder ring around the mouth of the fitting with no capillary action. This is likely to result in a weak joint prone to failure.

Important: You must apply the correct amount of solder to the fitting.

Excess solder can run into the fitting, which may cause an obstruction within the pipework. In most cases, the diameter of the pipe is an indication of the amount of solder required, e.g. if the pipe size is 15mm, form a 90° bend 15mm along the solder wire followed by, in the case of a socket, another 90° bend. There should be sufficient solder in the two 90° bends to make a satisfactory joint. Use a 22mm length of solder for each joint in a 22mm fitting.

When finished, visually examine all joints to confirm that the joints are sound and that the solder has run.

Note: When using flux (especially active/self-cleaning flux) it is important you apply it:

- **sparingly to the pipework only (never inside the fitting); and**
- **in accordance with the flux manufacturer's instructions**

Compression fittings – where and how to assemble them:

Only use compression fittings in accessible positions where they can be readily tightened.

Do not use them:

- under floors, or
- in ducts or voids.

Compression joints, if assembled correctly, ensure a perfect seal, which can be dismantled and re-made. You do not normally need to add jointing compound to ensure a good seal.

When using compression fittings, the pipework should not be under strain, pipe ends should be cut square and remove any burrs.

Assemble the fittings as follows:

1. Ensure that both the pipe and fitting are compatible in size.
2. Slide the tubing nut and olive (a compression ring usually made of brass or copper) onto the copper pipe.
3. Insert the copper pipe into the body of the compression fitting until the pipe end rests on the abutment in the fitting (take care that the pipe does not enter at an angle).
4. Hand-tighten the nut and use a suitable tool to tighten approximately 1 to 1½ turns to form a seal. Experience will tell when a perfect joint has been made: the number of turns depends on the manufacturer's tolerances.

A correctly tightened olive will be bowed, shaped and crimped to the pipe with no sign of undue distortion and it will make a good seal. If the pipe is not inserted fully into the fitting and the olive is incorrectly positioned, it will not make a seal.

Jointing procedure with press fittings on copper pipes

Elements of the system

The correct integrity of the completed joint relies upon:

1. The materials selected for the pipe.
2. The fitting and its sealing 'O' ring.
3. The correct selection of the tool and its related parts (such as jaws, collars and ability to complete the joint in a single process, without release of the jaws until the joint is completed).

In more detail, ensure that:

- you select copper pipes to comply with BS EN 1057 or BS EN 13349
- the press fitting complies with BS 8537
- the press tool, jaws and collars need to be suitable for the fitting and pipe to be assembled
- the press tool needs to be able to develop sufficient force to complete the pressing of the fitting
- the press tool needs to be such that once a pressing cycle has commenced, that cycle cannot be abandoned until the completion of the pressing cycle (except in an emergency where your safety is involved). If you have to abandon the cycle before completing the pressing action, discard the joint and fitting and repeat the complete process
- ideally the tool should be able to leave an imprint on the completed joint
- the tool is maintained as specified by the manufacturer and kept in a clean condition

When installing, take into account the manufacturer's recommendations, including the following:

- select the tool including jaws and collars to match the selected fittings

- check the tool for correct operation and cleanliness
- check the fitting for suitability and the correct insertion of a gas 'O' ring (normally yellow or grey in colour, with a yellow mark on the fitting)
- the pipe should be marked as required with the correct insertion depth
- check the pipe and fitting for cleanliness
- insert the pipe with a slight twisting action into the joint
- locate the tool and its jaws over the bead at the mouth of the fitting and complete the pressing cycle without dismantling or abandoning the process (see Note)
- on completion, clean the tool and check its tolerance, as specified by the manufacturer

Note: Completion normally occurs when the jaws fully enclose the mouth of the fitting

Attention: For gas installation work do not use push-fit and crimp-fit fittings designed for use with pipework intended to carry water.

Steel pipe/fittings – points to remember:

Joints should be screwed, ground face or compression type unions.

Use a suitable jointing compound as shown in Table 5.1.

Ensure all threads are clean and remove any cutting oil that has been used from the thread and the inside of the pipe before you apply any jointing compound.

Only apply jointing compound to external threads of the pipe/fitting and remove any excess compound once the joint has been made.

For NG, it is recommended (BS 6891) that hemp should only be used on the back nut of an existing long screw fitting in conjunction with a suitable jointing compound.

For LPG installations do not use long screw fittings, fibre washers or hemp.

Polytetrafluoroethylene (PTFE) tape/string for use on domestic gas pipework installations

You can use PTFE tape/string providing:

- it is the correct type (normally supplied on a reel to BS EN 751-3); and
- it is suitable for NG and LPG

Wind PTFE tape with a 50% overlap to give double cover starting from the thread run out in a direction counter to the thread form (see Figure 5.2). Do not use PTFE tape with other jointing compounds.

Apply PTFE string in accordance with the manufacturer's instructions and do not use it with other jointing compounds.

Figure 5.2 Thread wrapping method

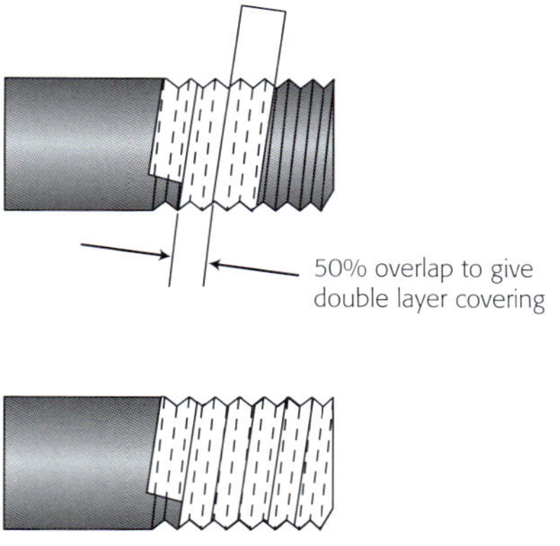

CSST / PLT – improves resistance of pipework to external factors

CSST / PLT tube used for gas should conform to:

- BS 7838: Corrugated stainless steel semi-rigid pipe ans associated fittings for low-pressure gas pipework of up to DN 50 or
- BS EN 15266: Stainless steel pliable corrugated tubing kits in buildings for gas with an operating pressure up to 0.5 bar.

It should have a tightly fitting coating or cover and be colour coded yellow.

The coating or cover is designed to improve the resistance of the pipework to external corrosion and mechanical damage.

You normally install it in one length from a gas meter (see Note) to the appliance – the yellow cover is marked at one metre intervals indicating how much pipe is left on the reel. There are dedicated mechanical fittings and adaptors you need to use in conjunction with the appropriate tube, but these fittings are simple to assemble (see **CSST / PLT – mechanical fittings** in this Part).

Important: Do not use CSST / PLT as a flexible connecter for removable appliances.

Note: Unless the meter is restrained securely (e.g. by a meter bracket), make the gas pipework connection to the gas meter installation using securely fixed rigid pipework for at least the first 600mm.

Install and size CSST / PLT in accordance with the manufacturer's installation instructions.

CSST / PLT – mechanical fittings

The fittings used for CSST / PLT are specific to the product but are simple to assemble, making for fast installation.

The advantage of this system is similar to that of compression and press-fit fittings in that no 'hot' work is required to make a gas tight connection; a regular requirement for contract work.

The following describes how to make a gas-tight connection between CSST / PLT and a mechanical fitting. The method shown is for products which are produced by OmegaFlex Ltd, the manufacturer of TracPipe®.

Although TracPipe® is discussed here, there may be other manufacturers who produce similar products in which case follow their specific guidance.

Making a mechanical connection

The connection of TracPipe® CSST / PLT to a mechanical fitting requires no jointing compounds – a gas tight seal is made by a metal-to-metal seal using the properties of both the CSST / PLT and the special design of the 'AutoFlare®' fitting.

No special tools are required – you will need a wheel cutter with a stainless steel blade, craft knife (or similar) and a pair of spanners (adjustable spanners offers greater flexibility).

Using your wheel cutter, position the tool so that the wheel sits between two corrugations. Applying light pressure, cut through the outer protective cover and CSST / PLT one whole turn at a time – ensure you go in one direction only.

On each full turn, increase the depth of cut by a quarter turn until you have a clean cut through the material (see Figure 5.3).

Next you need to strip back the protective yellow cover to allow the assembly of the fitting. Using a craft knife or similar cut the coating approximately 25mm back from the end of the tube (see Figure 5.4).

Slide the back nut on to the tube and insert the two split rings in to the first corrugation. Pull the back nut forward over the rings, which will keep them locked in position (see Figure 5.5).

Position the fitting in front of the nut and engage the threads – you will note that the fitting has a beveled face and protruding insert (see Figure 5.6). Tighten the back nut using your adjustable spanners – resistance will increase as you tighten, flaring the CSST / PLT within the fitting.

To make a gas-tight joint and once physically tight – with the hex flats in-line, turn a further two hex flats (⅓ turn) to correctly torque the fitting (see Figure 5.7).

Once all joints are completed, pressure test the installation – see **Part 6 – Testing for gas tightness (Natural gas)** and/or **Part 18 – LPG – General requirements** in this manual.

Once you have proved the installation gas tight, all that's left to do is to protect the now exposed portion of CSST / PLT at each fitting.

OmegaFlex Ltd produce a yellow silicone tape, which adheres to itself allowing easy installation. Starting at the tube, wind the tape toward the fitting and on to half the back nut (see Figure 5.8). Repeat for the other side of the fitting – don't cover the fitting completely.

Figure 5.3 Cutting to length

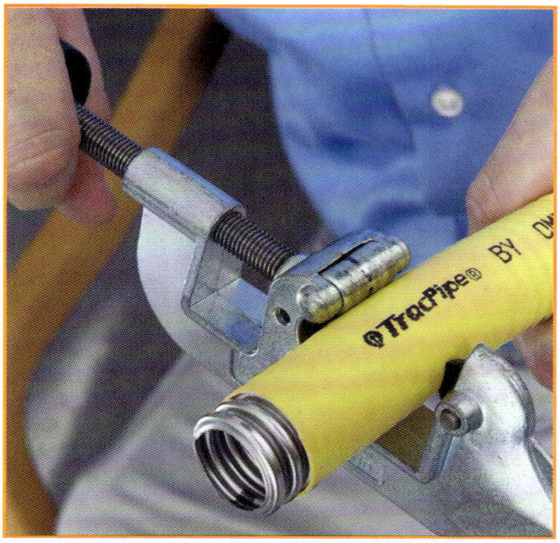

Figure 5.4 Cutting back cover

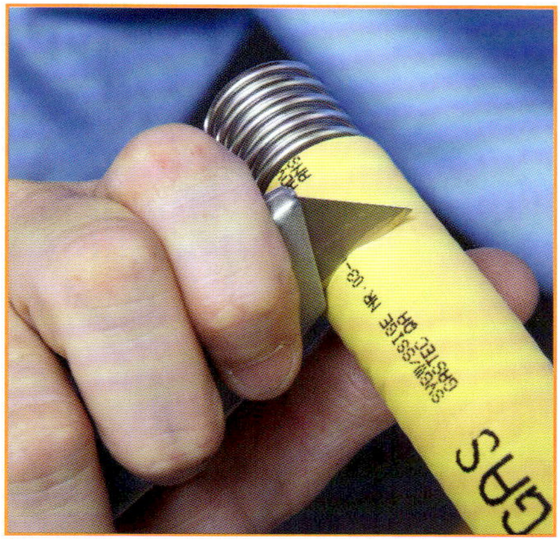

Figure 5.5 Installing split rings

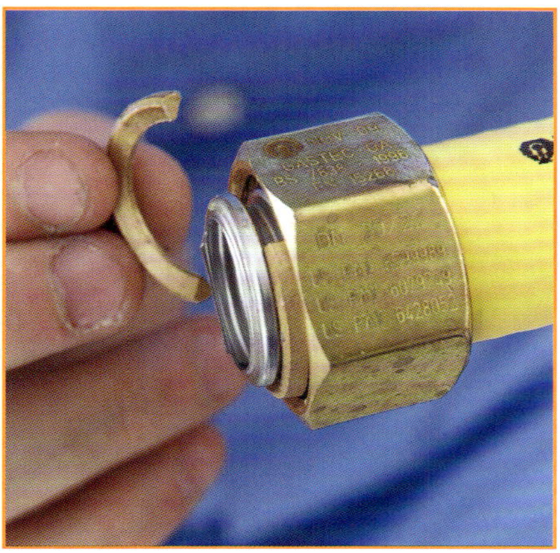

Figure 5.6 Installing Autoflare® fitting

Figure 5.7 Torquing the fitting

Figure 5.8 Applying protective silicone wrap

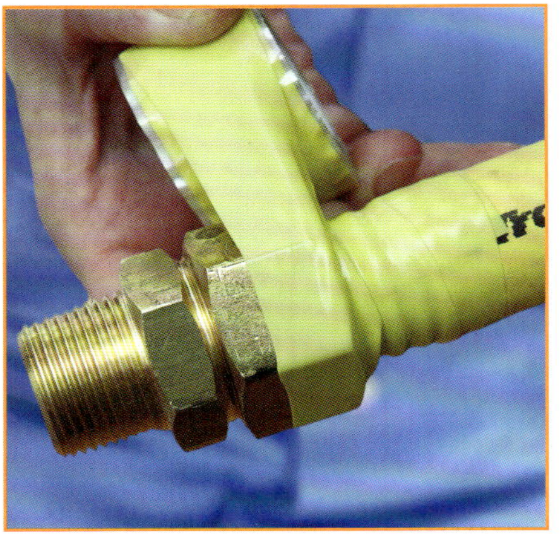

Polyethylene pipe (PE) – may be used where an installation pipe runs externally from one location to another

PE pipe and fittings are suitable for use with NG and LPG.

There are a number of important points to remember:

1. PE installation pipework and fittings used for gas should conform to BS EN 1555-1 and BS EN 1555-2 for pipe, BS 5114 and BS EN 1553-3 for fittings and be coloured yellow for identification as a gas supply.
2. Primarily PE pipework is for use below ground. Above ground, keep PE pipework as short as possible and protect against sunlight and mechanical damage. Using a glass reinforced plastic (GRP) sleeve usually achieves this.
3. Buried PE pipework should have a depth of cover when buried in accordance with Table 5.2. Consider using a suitable back-fill material and a tracer mesh laid above the route of the supply, colour-coded yellow and carrying a printed warning (e.g. 'GAS PIPE BELOW') .

   If the conditions of the ground is such that the specified minimum depth of cover cannot be met then a mechanical protection method could be used, such as concrete slabs installed below ground level & approximately 100mm above the pipe.

   The GSIUR prevents the installation of gas pipework beneath buildings – unless adequate steps are taken to prevent damage to the installation pipework should there be any movement of the structure or the ground.

   This is not a preferred method of installation and where possible, route pipework around buildings (see also **Exterior pipework – fittings, protection and other requirements** in this Part).

Table 5.2 Buried external pipework depth of cover

Location of pipework	Minimum depth of cover
Private gardens and pathways with no vehicular traffic	375 mm
Private drives with light vehicular traffic	450 mm
Private drives where there may be heavy vehicular traffic (e.g. for LPG deliveries)	600 mm
Agricultural land and fields	1100 mm
Other private ground	600 mm

4. Cover PE pipes and fittings stored externally and likely to be exposed to periods of prolonged sunlight.

Jointing – two methods generally used for PE pipework:

1. The heat fusion process – which needs specialist training

Heat fusion requires sophisticated tools and equipment. Joints may be of the butt fusion, electro fusion or hot iron types.

All fittings should comply with BS EN 1555-3.

The joint surfaces of the pipe and fittings are raised to a specific temperature by applying a heating tool over a specified time, to form molten surfaces. They are then brought together with firm, constant pressure and held rigid for a short time, whilst cooling takes place. The molten material on each surface fuses together, forming a completely homogeneous joint.

The use of fusion tools and equipment provides the means for quick, easy and accurate jointing. Correct use ensures that the pipe and fittings are well prepared and that the fusion surfaces are brought to the correct temperature for sound fusion joints.

The guidance above gives only a brief description, as specialist training is required to successfully carry out fusion jointing on PE pipework and fittings. You should only carry out this type of work if you are competent to do so.

The ACS assessments VESLP2 or EFJLP1, as appropriate would cover the requirements for electro fusion.

2. Mechanical fittings – may be used when pipe fusion is impractical

Mechanical fittings should comply with BS EN 1555-3 or BS 5114

You may use them where pipe fusion is impractical or inconvenient due to lack of knowledge and/or the correct equipment.

Metallic compression fittings have been specially designed for use with PE pipework and provide a sound and acceptable alternative method of jointing the pipework.

Mechanical fittings will be required:

- at any transition point between metallic and PE installation pipes; and
- at riser connections from underground to above ground pipework where entering a building

If the fittings are not made of corrosion protected material, you need to apply additional protection on site.

Note: For gas installation pipework you must not use other plastic pipes that may be suitable for water systems including their push-fit fittings.

Lead (lead composition) pipework – not to be used for new installations

The GSIUR state that lead or lead alloy should not be used for new gas installations.

However in the past, lead pipework was used for gas meter installations and lead alloy pipes (known as lead composition pipes) were installed as installation pipework. These may still be found in old housing stock.

What do you do if you find existing lead pipe in a customer's premises?

You may leave it in use as long as the material:

- is in a good sound condition;
- is well supported; and
- is of an adequate size for the appliance or appliances that it serves

Attention: Lead (lead composition) pipework is not suitable for use with LPG.

Note: It may be appropriate that you recommend to the person responsible for the premises that they should replace lead pipework with a suitable alternative.

Pipe sizing (NG) – the right diameter essential to supply the right amount of gas

General requirements – what you must do

This pipe sizing section is specific for the requirements of NG. For information relating to LPG see **Part 18 – LPG – General requirements**.

When designing gas installations, you must take great care to ensure that the pipe size selected is of adequate diameter to supply sufficient gas to all appliances at the same time (see also **Part 16 – Domestic gas meters – Meter sizing – and 'diversity of use factor'**).

Failure to do this may result in pitfalls:

- at best, the appliance performance may fail to achieve customer expectations, e.g. inadequate hot water temperatures and flow rates for combination boilers
- at worst, it may lead to insufficient gas pressure at appliance burners
- in extreme cases, inadequate gas pressure could ultimately cause pilot or burner flames to extinguish

How do you avoid these pitfalls?

1. When designing an installation and calculating pipe sizes, it is important that you take into account the required gas rates of all appliances connected to it. (see Note)

   Also make an allowance for:

- any pressure loss attributable to frictional resistance caused by pipe-walls and fittings
- possible future known extensions, especially if the pipework is to be concealed

Note: For range rated appliances use the maximum heat input. For variable rated appliances use the heat input required to meet the requirements of the heating installation.

2. Gas appliances are designed to operate at a nominal operating pressure of 20mbar at the inlet connection to the appliance. It is important that this pressure is maintained to ensure stable flames and that sufficient gas is available for each appliance.

   You achieve this by:

- fitting a regulator to the gas meter inlet that is designed and adjusted to provide a nominal operating pressure of 21mbar at the meter outlet connection (see **Part 7 – Checking and/or setting regulators** and **Part 16 – Domestic gas meters**)

Figure 5.9 Pipe sizing example (NG) for a typical copper pipe gas installation

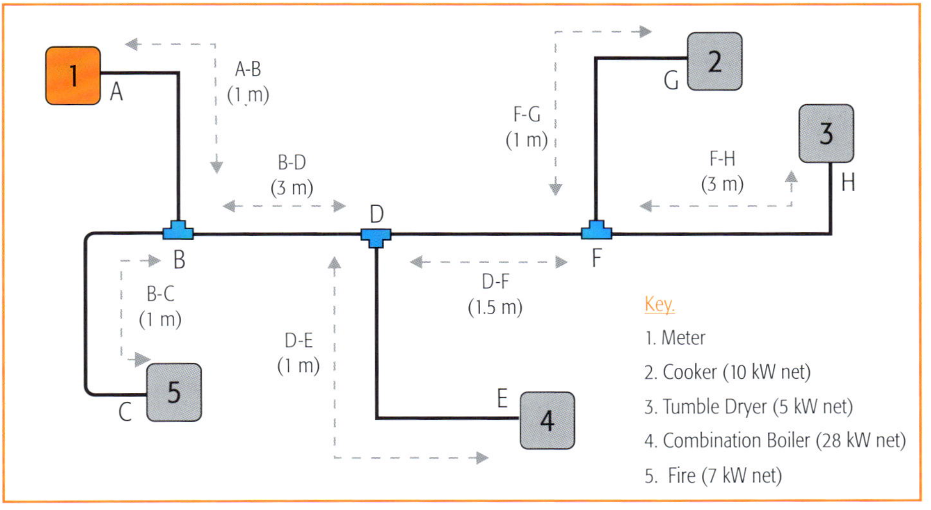

- designing the gas installation pipework to ensure that the maximum pressure loss between the meter outlet and the point the appliance(s) is connected does not exceed 1mbar.

Note: Where you use a wheel cutter to cut pipework, take care to ensure that you remove all burrs from the internal bore of the pipe. Failure to do so may result in increased friction at each joint or fitting, resulting in a subsequent pressure loss across the installation in excess of the design losses.

Guide for sizing gas supplies (NG)

Tables 5.3 and 5.4 show the pressure loss data for copper pipe and pressure loss in fittings.

An allowance is made for fittings to take into account their frictional resistance. With elbows, tees 45° or 90° bends fitted, you also need to add the equivalent lengths to the actual length of the installation pipe in line with Table 5.4.

Figure 5.9 gives an example of a typical copper pipe installation using NG. Figure 5.9 also shows the sections of pipework and the input ratings of the appliances. The pipework has been sized using Table 5.3 and 5.4 with the results shown in Table 5.5

When sizing gas pipes, always take into account the permissible pressure loss in each section of the installation pipework.

Table 5.3 Approximate natural gas pressure loss data (mbar/m) - copper pipe

Flow Rate	Heat Input		Nominal Pipe Size (mm) A)						
(m3/h)	Gross (kW)	Net (kW)	8 (6)	10 (8)	12 (10)	15 (13)	22 (19)	28 (25)	35 (32)
0.25	2.70	2.46	0.2675	0.0710	0.0255	0.0077	0.0014	0.0004	0.0001
0.50	5.40	4.91	0.8348	0.2188	0.0777	0.0231	0.0040	0.0011	0.0004
0.75	8.10	7.37		0.4285	0.1514	0.0447	0.0077	0.0022	0.0007
1.00	10.81	9.82		0.6940	0.2444	0.0719	0.0123	0.0035	0.0011
1.25	13.51	12.28			0.3553	0.1042	0.0178	0.0050	0.0016
1.50	16.21	14.73			0.4833	0.1414	0.0240	0.0067	0.0021
1.75	18.91	17.19			0.6276	0.1832	0.0311	0.0086	0.0027
2.00	21.61	19.65			0.7877	0.2296	0.0388	0.0108	0.0034
2.25	24.31	22.10			0.9630	0.2804	0.0473	0.0131	0.0042
2.50	27.01	24.56				0.3353	0.0565	0.0156	0.0049
2.75	29.72	27.01				0.3945	0.0663	0.0183	0.0058
3.00	32.42	29.47				0.4577	0.0769	0.0212	0.0067
3.25	35.12	31.93				0.5249	0.0880	0.0243	0.0077
3.50	37.82	34.38				0.5960	0.0998	0.0275	0.0087
3.75	40.52	36.84				0.6709	0.1123	0.0309	0.0097
4.00	43.22	39.29				0.7496	0.1253	0.0345	0.0108
4.25	45.92	41.75				0.8321	0.1390	0.0382	0.0120
4.50	48.63	44.20				0.9182	0.1533	0.0421	0.0132
4.75	51.33	46.66					0.1681	0.0462	0.0145
5.00	54.03	49.12					0.1836	0.0504	0.0158
5.25	56.73	51.57					0.1996	0.0548	0.0172
5.50	59.43	54.03					0.2162	0.0593	0.0186
5.75	62.13	56.48					0.2334	0.0640	0.0200

Table 5.4 Typical allowances for pressure loss in fittings

Nominal Pipe Size			Equivalent Length (m)				
			Bends		Fittings		
Steel		Copper	45^0	90^0	90^0 elbow	Tee A) (Entering Branch)	Tee A) (Exiting Branch)
≤ 15 mm	R 1/2	≤ 15 mm	0.15	0.20	0.40	0.75	1.20
20 mm	R 3/4	22 mm	0.20	0.30	0.60	1.20	1.80
25 mm	R 1	28 mm	0.25	0.40	0.80	1.50	2.30
32 mm	R 1 1/4	35 mm	0.30	0.50	1.00	2.00	3.00

A) Use the largest connection size on the fitting that is not necessarily the branch size

Note: For pressure losses associated with press end fittings, CSST / PLT tube & fittings, flexible connections, secondary meters, check valves, regulators, etc. reference needs to be made to the manufacturer's data.

Table 5.5 Pipe sizing results

1	2	3	4	5		6	7	8
Pipework Section	Maximum Heat Input A) kW (net)	Estimated Pipe Diameter mm / DN	Pipe Length m	Fittings Allowance B)		Adjusted Length m	Pressure Loss/Metre mbar / m	Pressure loss C) mbar
				Type	Equivalent Length m			
A-B	50	28	1	Elbow x 1 Tee(Enter) x 1	0.80 1.50	3.3	0.0548	0.19
B-C	7	15	1	90° Bend x 2	0.40	1.4	0.0447	0.07
B-D	43	28	3	N/A	0	3.0	0.0421	0.13
D-E	28	22	1	Tee (Exit) x 1 Elbow x 1	2.30 0.60	3.9	0.0769	0.30
D-F	15	22	1.5	N/A	0	1.5	0.0311	0.05
F-G	10	15	1	Tee (Exit) x 1 Elbow x 1	1.80 0.40	3.2	0.1042	0.34
F-H	5	15	3	Elbow x1	0.40	3.4	0.0447	0.16

Pressure loss to points to be connected for each appliance

Appliance	Maximum Heat Input kW (Net)	Pipework Sections and Associated Pressure Loss					Total Pressure Loss (mbar)	Satisfactory
Cooker	10	A-B	B-D	D-F	F-G		0.71	Yes
		0.19	0.13	0.05	0.34			
Tumble Dryer	5	A-B	B-D	D-F	F-H		0.53	Yes
		0.19	0.13	0.05	0.16			
Boiler	28	A-B	B-D	D-E			0.62	Yes
		0.19	0.13	0.30				
Fire	7	A-B	B-C				0.26	Yes
		0.19	0.07					

If we take the pressure loss to the boiler in Table 5.5 the pressure loss between A and E should not exceed 1 mbar, this section is made up of three different sections of pipework, A-B, B-D and D-E. Each pipework run from meter to appliance should be treated in a similar way to determine the pressure loss.

Step 1

The pipework layout should be divided into sections and inputted into the Pipework Section column in Table 5.5 column1.

Step 2

Enter the net heat input/flow rate requirement for each pipework section into column 2. This must take into account any appliance which is served by the section identified e.g. section A-B would have a maximum net heat input of 50 kW as it serves all appliances

Step 3

Enter the estimated pipe size that may be required at this stage in column 3, you may need to revisit this section at a later date should the overall pressure loss be excessive.

Step 4

Enter the actual pipe length into column 4.

Step 5

Enter the allowance for fittings or bends into column 5, Equivalent Length, using the figures gained from Table 5.4 e.g. section A-B has one elbow with an equivalent length of 0.8m and a tee (entering branch) rated at 1.5m giving a total equivalent length of 2.3m. Due regard should be given to the position of tees and they should only be counted once in the sizing exercise.

Step 6

Add the equivalent length to the actual length and enter this into column 6, Adjusted Length, e.g. section A-B would be the total of 2.3m plus the actual length of 1m giving an Adjusted length of 3.3m.

Step7

Utilizing Table 5.3 input the relevant pressure loss (mbar/metre) based on the flow rate or net heat requirement and estimated pipe diameter you have chosen. If the exact net heat input or flow rate is not available go to the next figure upwards e.g. section A-B has a net input requirement of 50 kW and an estimated pipe diameter of 28mm but 50 kW is not on Table 5.3 so the next figure upwards would be 51.57 kW with a respective pressure loss per metre of 0.0548 mbar/m this figure should then be entered into column 7.

Step 8

Enter into column 8 the actual pressure loss which is determined by multiplying the values entered in column 6 and 7 together, e.g. section A-B has an adjusted length of 3.3m and a 0.0548 pressure loss per metre which gives a pressure loss for the section of 0.19 mbar.

Step 9

Determine the pressure loss in each section from the meter outlet to each individual appliance ensuring it does not exceed 1 mbar e.g. the pressure loss between A (meter) and E (combination boiler) in Figure 5.9 is made up of three sections of pipe, which are separated by tees, A-B, B-D and D-E each having a separate pressure loss which when added together give the overall pressure loss for this section of 0.62mbar, This needs to be repeated for each run from meter to appliance. The results of the example are shown in Table 5.5.

Note: At this stage you may need to revisit your initial estimations of pipe size due to a high pressure loss and increase the pipe size to reduce the overall pressure loss.

Route and appearance – always take construction of the building into account

When planning the route for gas installation pipework, consider the construction of the building, including the location of any damp proof membranes, lintels and reinforced structures.

This section deals with the many obstacles likely to be encountered during the installation of gas pipework and fittings in domestic premises. This includes the supporting of pipework, see Table 5.6.

Pipework Support

All installation pipes shall be adequately supported in line with manufacturer's instructions or in the absence of these the guidance in Table 5.6 can be followed.

The support shall be:

- designed to prevent the pipework from coming into contact with surfaces which are likely to cause corrosion (e.g. masonry, concrete, plaster).
- suitable for the environment and if used externally, corrosion and UV resistant.
- metallic in construction if used in protected areas where they need to be fire resistant.

Table 5.6 Maximum interval between pipe supports

Material	Nominal size	Interval for vertical run (m)	Interval for horizontal run (m)
Copper	Up to 15 mm 22mm 28 mm 35 mm	2.0 2.5 2.5 2.5	1.5 2.0 2.0 2.5
Steel	Up to DN 15 (R 1/2) DN 20 (R 3/4) DN 25 (R1) DN 32 (R1 1/4)	2.5 3.0 3.0 3.3	2.0 2.5 2.5 2.7
CSST / PLT	DN 15 DN 20/22 DN 25/28 DN 32	1.2 1.8 1.8 2.5	1.2 1.8 1.8 2.5

Sleeves

A sleeve (see Figure 5.10) is required to protect the pipework against any possible movement in the wall and corrosion from cement/mortar.

The sleeve shall be of a material:

- resistant to corrosion;
- impermeable to gas, e.g. copper, steel, PE or polyvinyl chloride (PVC); and
- capable of protecting the pipework against failure caused by movement of the structure.

Note: Take care to ensure that PVC does not come into contact with stainless steel because of the risk of corrosion occurring.

Where copper pipework is sleeved with steel or vice-versa, take great care to prevent the two materials from coming into contact with each other, as the combination could cause corrosion to occur by electrolysis.

The internal diameter of the sleeve should allow for an annular space around the pipe to ensure you can satisfactorily insert and withdraw the pipe and also apply adequate sealing material.

Note: Factory-sheathed/wrapped pipes are not considered to be a suitable alternative to a pipe sleeve.

Sleeves should:

- span the full width of the wall
- be continuous (not split and placed in position after the gas pipe has been fitted)
- be sealed at one end only between the sleeve and the pipe, with a flexible (non-setting) fire resistant material
- wherever possible, be vented to outside air. However, in the case of a meter box, this will not be possible as the additional purpose of the seal within the box is to prevent accumulations of gas from entering the building
- be sealed at each end to the structure (wall) with a suitable building material (cement, mortar, plaster, etc.)
- shall not impair the fire resistance or compromise the integrity of the structure.

Do not fit screwed, quick release, compression or press end connections within the sleeve.

Figure 5.10 Pipe sleeve

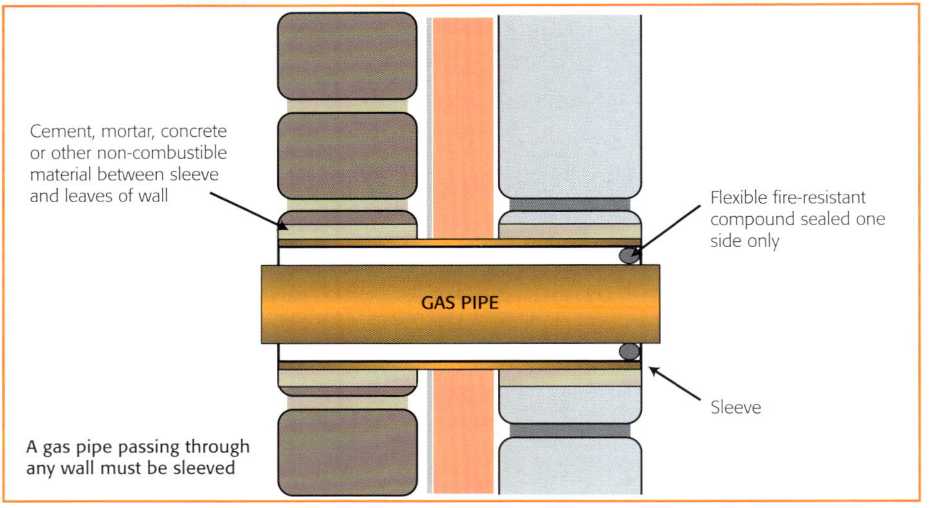

Location of pipes – and the care you must take to prevent damage or injury

Installing pipework in joisted floors & roof spaces

Prior to running pipework below suspended floors, carry out a visual inspection to note the position of any electrical cables, junction boxes and ancillary equipment in order to prevent any damage or injury when you insert the pipework.

- Pipes need to be correctly supported in line with Table 5.6 and follow the line of the joist or be at 90 degrees to the joist direction with no diagonal runs.
- The upper and lower flanges of metal web and timber engineered joists shall not be breached or damaged by drilling or notching. (see Figure 5.11)
- If you install pipes through metal web joists the pipework shall not come into contact with any metallic part of the joist.
- Where you lay pipework across solid timber joists fitted with flooring, locate the pipe in purpose-made notches or circular holes.
- For pipes that are installed in roof spaces the notching of joists, purlins, trusses or braces shall only be done with the prior approval of a structural engineer.
- Depth & span between end supports and notches can be seen in Figures 5.12, 5.13, 5,14, 5.15, 5.16 and 5.17.
- Do not cut notches in solid timber joists less than 100mm deep or greater than 250mm unless it is deemed acceptable by a structural engineer.
- Notches should have rounded corners and it is essential that they do not extend across the joist between floorboards. Notches and drillings in the same joist should be at least 100mm apart horizontally.

- Take care when re-fitting flooring with nails or screws, to prevent damaging the pipework. Where possible, the flooring should be marked to warn others that a gas pipe is directly below.

Note: Where manufactured joists are encountered that are of metallic, timber engineered or laminated design, the advice of the manufacturer shall be sought prior to the installation of any pipework that is to pass through the joist.

Figure 5.11 Notching and drilling of metal web joists and timber - engineered joists

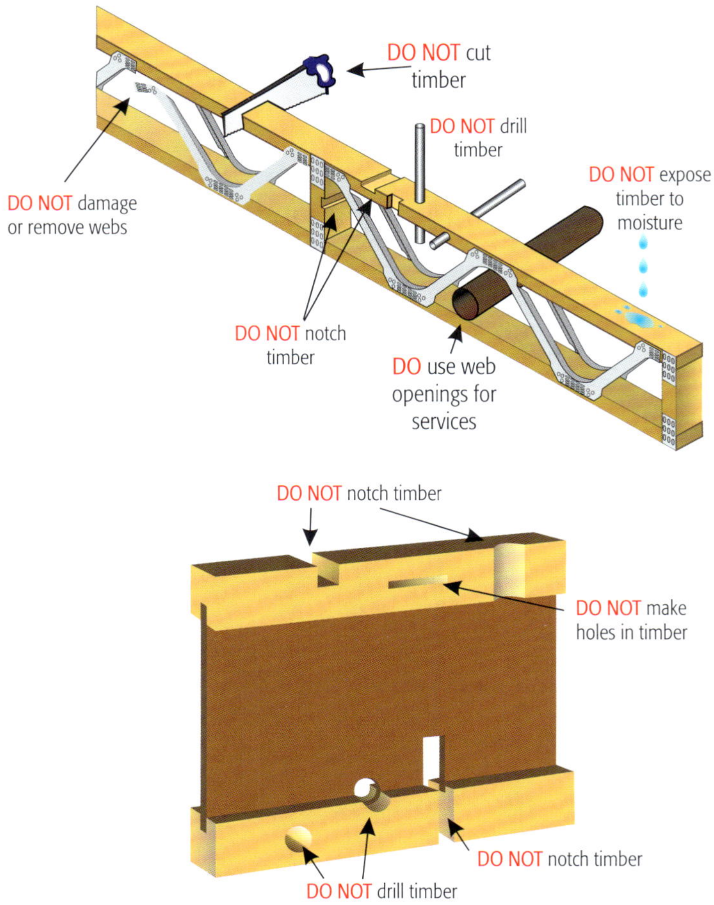

Figure 5.12 Limits for notching top of joist where drilling and notching occur

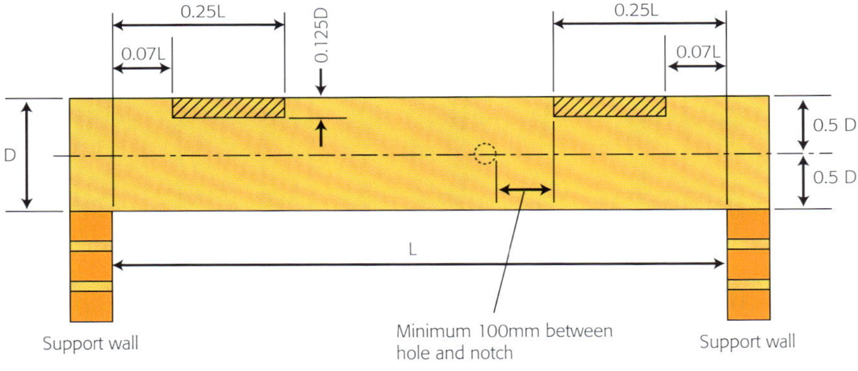

Figure 5.13 Alternative limits for notching top of joist where drilling & notching occur

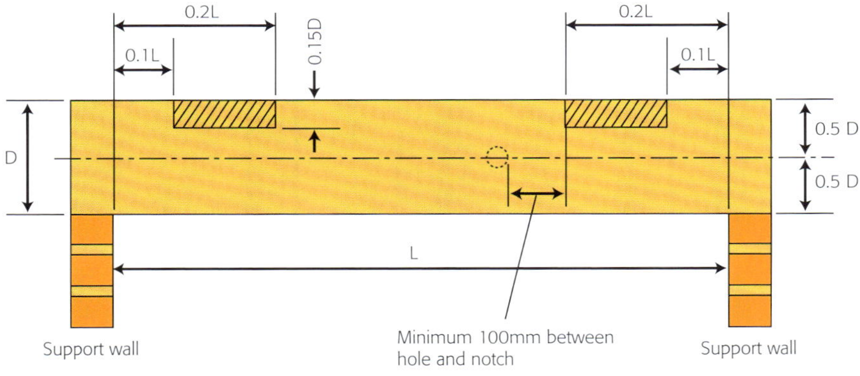

Figure 5.14 Limits for drilling of joists where drilling and notching occur

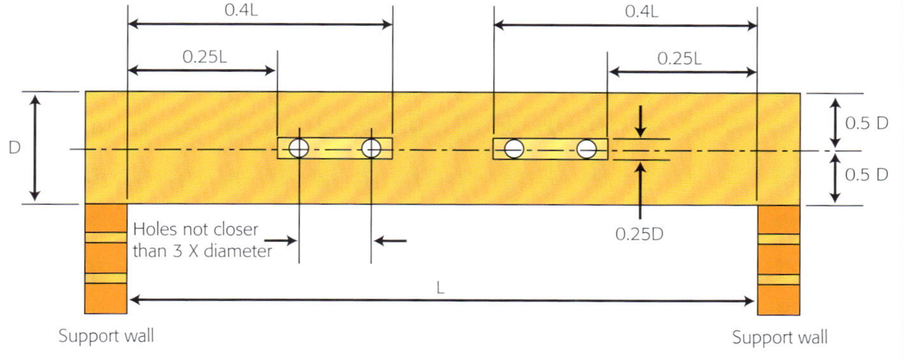

Figure 5.15 Limits for notching bottom of joist where drilling and notching occur

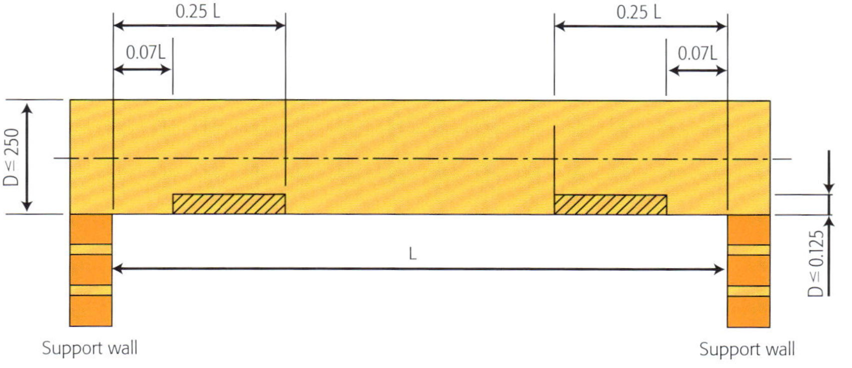

Figure 5.16 Alternative limits for notching bottom of joist where drilling and notching occur

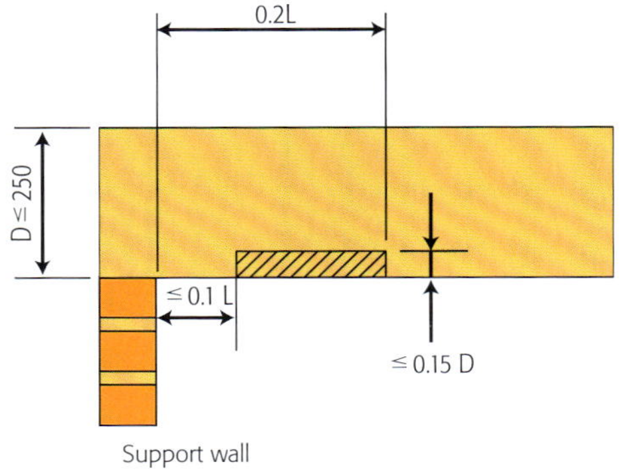

Figure 5.17 Alternative drilling zone where no notching occurs

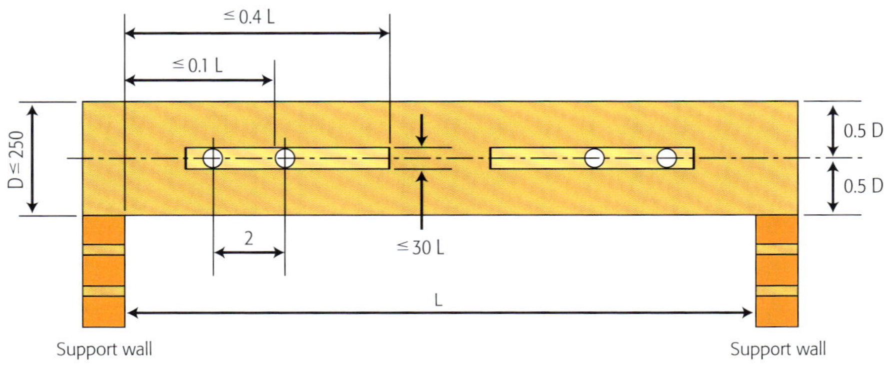

Figure 5.18 Typical examples of pipework laid in solid floors

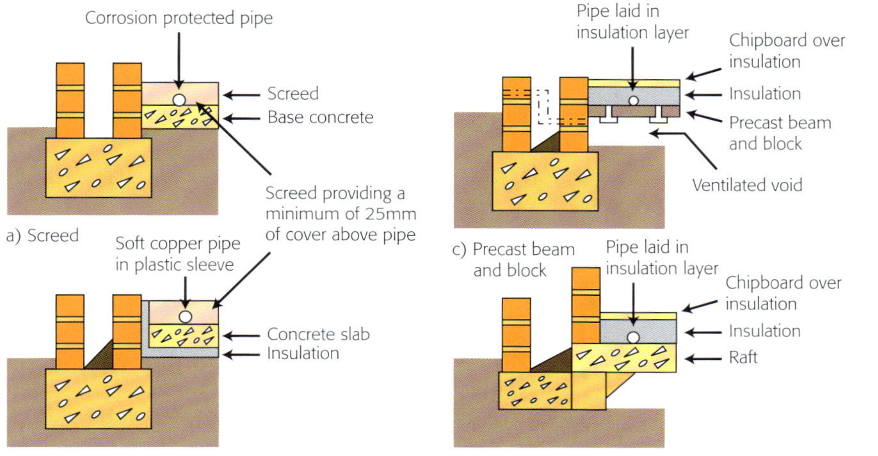

Laying pipework in solid floors – the types suitable and methods of protection needed

- steel and copper pipework can be laid in solid floors providing it is suitably protected against corrosion and mechanical damage. Keep joints to a minimum and do not use compression fittings.
- CSST / PLT to BS 7838 or BS EN 15266 can be laid in solid floors in accordance with manufacturers instructions or contained with a sleeve (see **Sleeves** in this Part).
- rigid stainless steel pipework should not be laid in solid floors.

See Figure 5.18 for examples of pipework laid in solid floors.

Note: Pipework can also be installed within preformed ducts within the solid floor with suitable covers - not shown in Figure 5.18.

Suitable methods include:

1. Fully annealed factory-sheathed soft copper tube or CSST / PLT passed through a larger sized plastic tube sleeving previously set into the floor slab and base hard core. Do not locate joints within the larger plastic tube sleeving.
2. Pipes laid on top of base concrete and subsequently covered by a screed (minimum covering of 25mm above the pipe) should be factory-sheathed or protected on site, using an appropriate wrapping material.
3. Steel or copper pipe laid in preformed ducts with protective covers.

4. Steel or copper pipe fitted with additional soft covering material. The coverings should be soft and thick enough to provide movement, yet resilient enough to support the concrete cover while it is setting. The covering should be at least 5mm thick and resilient to concrete ingress which would negate its ability to allow movement.

Note: Refer also to 'Corrosion protection – a number of forms, including applying on site' in this Part.

Passing pipework through solid floors

Where pipework passes vertically through solid floors, it should take the shortest practicable route and should be fully enclosed in a sleeve. For sleeve material and installation requirements see **Sleeves** in this Part.

Compartment floors – gas installation pipework

Natural gas installation pipework may be installed in intermediate floors without the need for additional ventilation as the adventitious ventilation (i.e. natural ventilation via cracks/openings in the structure) is sufficient to ensure any minor escapes of gas that may occur do not present a danger.

An intermediate floor is defined as a floor that separates one living space from another in the same dwelling, however, there are some instances where a property may have been split into separate flats and the regulations current at the time did not require the installation of compartment floors or ceilings for fire safety purposes. In these instances the gas pipes would not require additional ventilation.

Where there are compartment floors and ceilings, which separates individual dwellings in multi-occupancy properties (flats) these spaces should be treated as ducts and additional ventilation sized as per Table 5.7 will be required (see **Pipes in ducts – a guide to the differing types** in this Part).

Any ventilation provision must not reduce the fire provisions already provided for the building and consideration should be given to ventilating directly to outside the property.

A compartment floor is defined as a floor that separates one dwelling from the rest of the building as per the Building Regulations Part B covering fire safety.

Pipework in walls

General information to help you:

- installation pipework should, where possible, be vertical and placed in ducts with convenient access points; or
- provided adequate wall thickness is available, place pipework in chases. Avoid horizontal pipe runs
- prior to the chasing of a solid wall, inspect and note the proximity of any electrical cables/socket outlets, which may be already buried
- in the interest of safety, consider using a hand held electrical cable/pipe locater to trace hidden services concealed in the wall. You can also use these instruments to locate services beneath solid floors
- pipes need to be secured using suitable clips and should be protected against corrosion (see **Corrosion protection – a number of forms, including applying on site** in this Part)
- keep joints to a minimum
- do not use compression fittings
- test the installation for tightness before you apply any covering material (see **Part 6 – Testing for gas tightness (Natural gas)**). Figure 5.19 shows a typical example of pipework concealed in a masonry wall

Figure 5.19 Typical example of gas pipework in masonry walls

Pipe chase in masonry wall with a plaster finish

Maximum chase depth:

– Horizontal chase = $\frac{T}{6}$

– Vertical chase = $\frac{T}{3}$

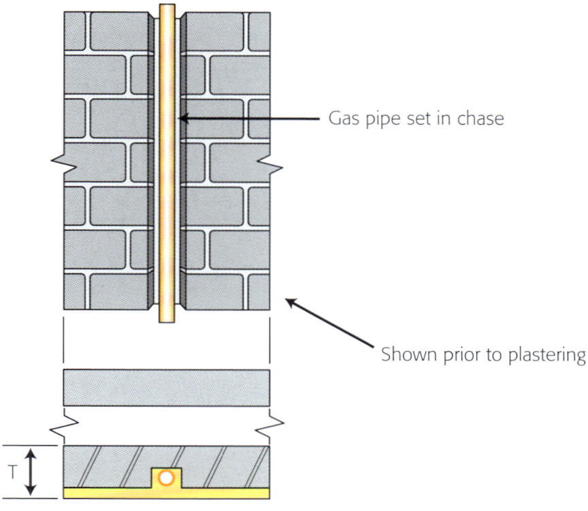

Cavity walls – restrictions affecting installation

The GSIUR restrict any person from installing any part of any gas installation pipework in a cavity wall unless the pipe is to pass through from one side to the other.

In this case, the pipe needs to take the shortest practicable route and must be fully enclosed in a gas tight sleeve. For sleeve material and installation refer to **Sleeves** in this Part.

However, because of the operational controls on certain 'living flame effect gas fires', it can be difficult to install pipework neatly. So for the sake of neatness, the GSIUR enable the final section of an installation pipe to run within a wall cavity recess.

The GSIUR define a 'living flame effect gas fire' as one that is:

- designed to simulate the effect of a solid fuel fire
- designed to operate with a fanned flue system
- installed within the inner leaf cavity wall

In this case, safety will not be compromised as the section of pipework within the wall cavity needs to be enclosed in a gas tight sleeve and sealed at the point where the sleeved installation pipe enters the fire, preventing gas accumulating in the cavity if the pipe becomes perforated (see Figure 5.20).

Figure 5.20 Pipework in cavity walls to connect flame effect gas fire (fan flued)

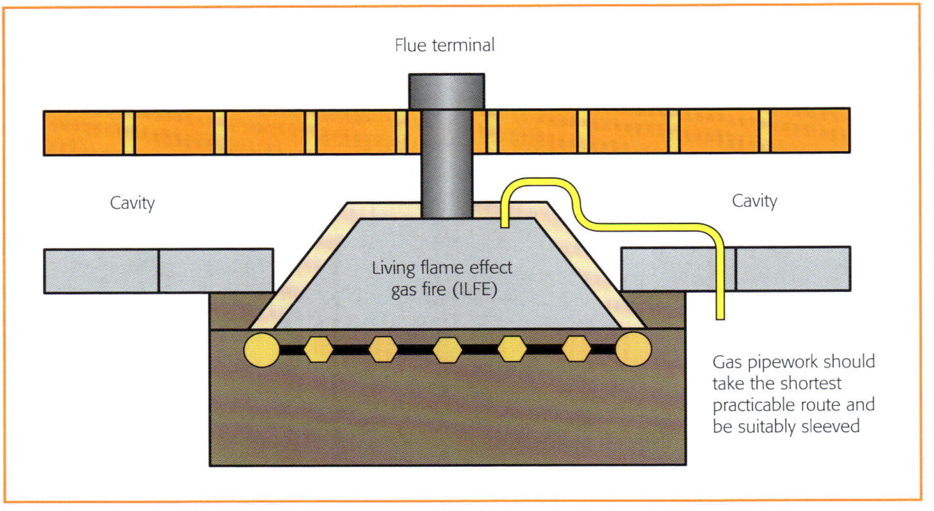

Figure 5.21 Typical examples of pipework in dry lined, stud and timber frame walls

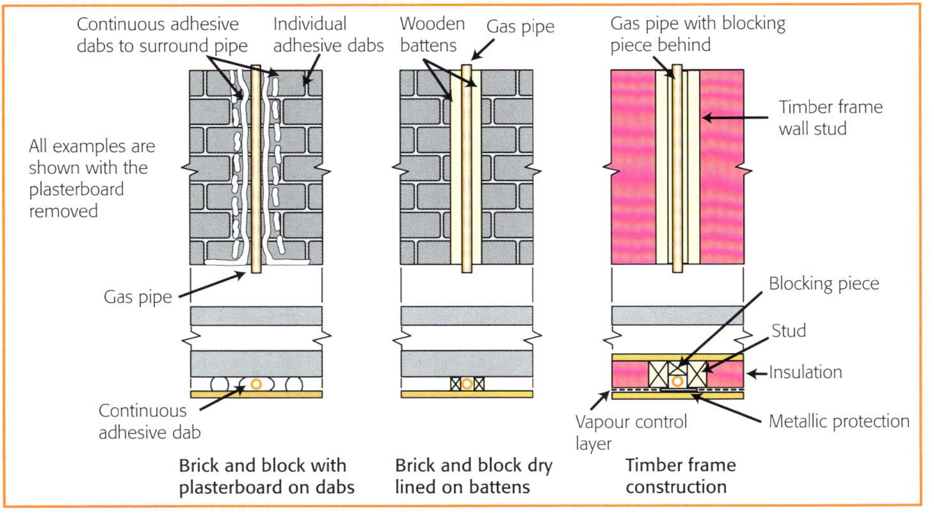

Dry lined walls – how you install

Installation pipework rising vertically should be surface mounted, corrosion protected and then encased completely with continuous adhesive dabs (mortar/plaster) to surround the pipe.

This is to ensure that when the plasterboard is applied, no gas can escape into the void between the inner wall surface and the dry lining (see Figure 5.21).

If plasterboards are to be fitted onto battens, vertically rising installation pipes should be surface mounted and corrosion-protected with a batten closely fitted on either side of the pipe (see Figure 5.21) with the addition of a 1mm thick steel plate to protect the pipe if it is within 50mm of the face of the plasterboard.

The joints between the batten and wall or wall studs should be sealed with mastic to avoid gas tracking around the batten.

Timber frame and light steel-framed construction walls

When installing installation pipes in timber framed and light steel-framed construction walls, it is important that you do not damage the vapour control barrier or interfere with any structural members of the house frame. Any installed pipes shall be installed giving due regard to the fire resistance of the structure through which they pass

Typical methods of installing pipes are shown in Figure 5.21.

Pipes should:

1. be run vertically within purpose designed channels, ducts or purpose provided holes.
2. be adequately secured to studding.
3. be run through the timber as close to the centre line as possible.
4. be protected against damage if it is run within 50mm of the face of the wall.
5. have a minimum number of joints with no compression fittings.
6. be protected where necessary from mechanical and corrosion damage and if the bottom plate or sole plate are cut notched or drilled these openings shall be treated with a wood preservative.
7. be separated from electrical services observing the minimum separation distances required.

Solid walls

Every pipe passing through a solid wall needs to be sleeved (see **Sleeves** in this Part).

Pipes in ducts – a guide to the differing types

Ventilation – requirements vary according to duct

A duct is a purpose-designed enclosure containing gas installation pipework.

Ducts may be vertical or horizontal and should be ventilated to ensure that any minor gas escape does not cause the atmosphere within the duct to become unsafe.

The level of ventilation is not intended to clear a major gas escape arising from damage or failure of a gas pipe, but to allow the gas to be detected.

The fire resistance of any duct shall have a fire rating equal to or greater than any void through which it passes.

The duct may run through a number of storeys or may be contained at each storey level.

- where ducts are continuous, running through a number of storeys (e.g. in flats or maisonettes) ventilation can be achieved by providing openings at both high and low level communicating with the outside atmosphere (see Figure 5.22) and size in accordance with Table 5.7
- where the duct is contained at each storey level, ventilation openings are required at high and low level into the property (see Figure 5.23)
- ducts with a small cross-sectional area (i.e. 0.01m^2 or less) and volume (i.e. 0.1m^3 or less) do not require any additional openings, as they are considered to be adequately ventilated by adventitious means.

If you need further guidance on the installation of gas pipework in timber frame properties, please refer to CORGI*directs* manual entitled Gas Installations In Timber/Light Steel Frame Buildings from the current Gas Installer Manual Series – Domestic.

Figure 5.22 Installation pipework ventilated duct

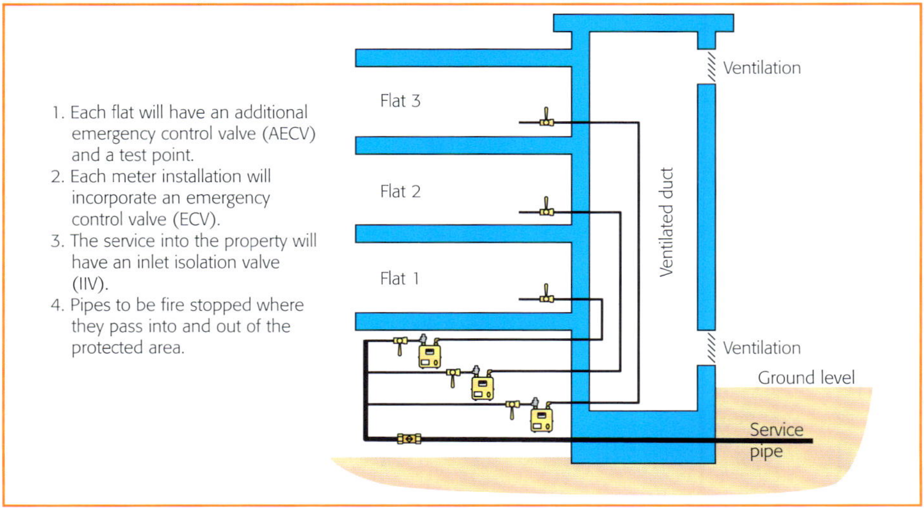

Table 5.7 Free area of ventilation openings for ducts

Cross sectional area of duct (m^2)	Minimum free area of each opening (m^2)
Not exceeding 0.01	no purpose provided openings required*
0.01 and not exceeding 0.05	Cross-sectional area of duct
0.05 and not exceeding 7.5	0.05
Exceeding 7.5	$^1/_{150}$th of the cross-sectional area of duct

* Regulation 19(6) of the GSIUR states "Where any installation pipework is not itself contained in a ventilated duct, no person shall install any installation pipework in any shaft, duct or void which is not adequately ventilated". Therefore where the cross sectional area of the duct is 0.01m² or less, to comply with the GSIUR at least one end of the duct should be left open.

Figure 5.23 Installation pipework fire stopped and ventilated in an enclosed area

1. Each flat will have an additional emergency control valve (AECV) and a test point.
2. Pipework fire stopped as it passes through each floor level.
3. Each enclosure to be ventilated at high and low level.
4. Each meter installation will incorporate an emergency control valve (ECV).
5. The service into the property will have an inlet isolation valve (IIV).

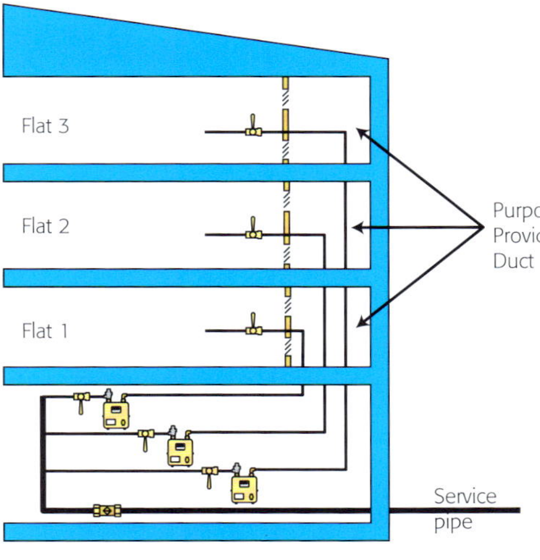

Fire stopping – how to achieve and maintain this

Gas installation pipes fitted in buildings containing flats or maisonettes need to be fire-stopped as they pass from one floor to another (see Note), unless they are fitted in their own continuous duct which is ventilated at high and low level to the outside atmosphere.

Where installation pipes from a continuous duct enter a flat or maisonette, they need to be fire-stopped at the point of entry (see Figure 5.22).

Note: You need to route installation pipework for multi-occupancy type dwellings through the dwelling for which the gas appliance(s) concerned is located. Installation pipework for other dwellings needs to be in purpose provided ducts which need to be built in accordance with the Building Regulations.

Where gas installation pipework passes through the structure (i.e. compartment walls or floors), the openings should be as small as possible and fire-stopped in such a manner as to allow thermal movement of the pipe and to ensure that fire resistance is not impaired.

To prevent the fire stopping material from being displaced, it should be supported by, or reinforced with, materials of limited combustibility.

Any fire stopping should, when tested in accordance with the appropriate part of BS 476, achieve the relevant periods of fire resistance for the structure in respect of load bearing capacity, integrity and insulation.

Protected areas – prevent the spread of fire and smoke

Protected areas include protected corridors, lobbies, halls, landings and stairways. The way they are enclosed prevents the spread of fire or smoke to or from another part of the same building.

Their design allows persons, air or objects to pass from one compartment to another, enclosed within a fire-resistant construction.

The protected area forms a complete barrier to fire between compartments in the building and they are used as a means of escape in the event of a fire.

A 'compartment' is a building or part of a building, comprising one or more rooms, spaces or storeys. A roof space above the top storey of a compartment is included in that compartment.

Fire resistance is the ability of a component or construction of a building to satisfy, for a stated period of time, some or all of the appropriate criteria specified in the relevant part of BS 476.

The requirements given do not normally apply to one or two storey domestic dwellings.

In addition to the requirements for fire resistance and fire stopping (see **Fire stopping – how to achieve and maintain this** in this Part) gas installation pipework installed to flats in protected areas (e.g. stairs, lifts, escalators, chutes and ducts) should adhere to the following guidelines:

- they should be made of steel pipework with screwed or welded joints; a continuous length of copper (no joints) or CSST / PLT manufactured to withstand fire test A of BS EN 1775:2007.
- they should be adequately ventilated direct to outside air by ventilation openings at high and low level in the protected area in accordance with Table 5.7
- they should not impair the fire resistance of the walls or floors of the structure

Pipes may be exposed or run in specially designed fire rated ducts (see Figure 5.24 and 5.25).

Note: A pipe is not considered to be within a protected area if it is contained within a fire-resistant duct that is itself ventilated direct to outside air.

Note: (see Figure 5.24). CSST / PLT manufactured to BS 7838 meets fire test A. For CSST / PLT manufactured to BS EN 15266 the manufacturer should be consulted

Figure 5.24 Pipework supplying gas to flats from remote meters via a protected area

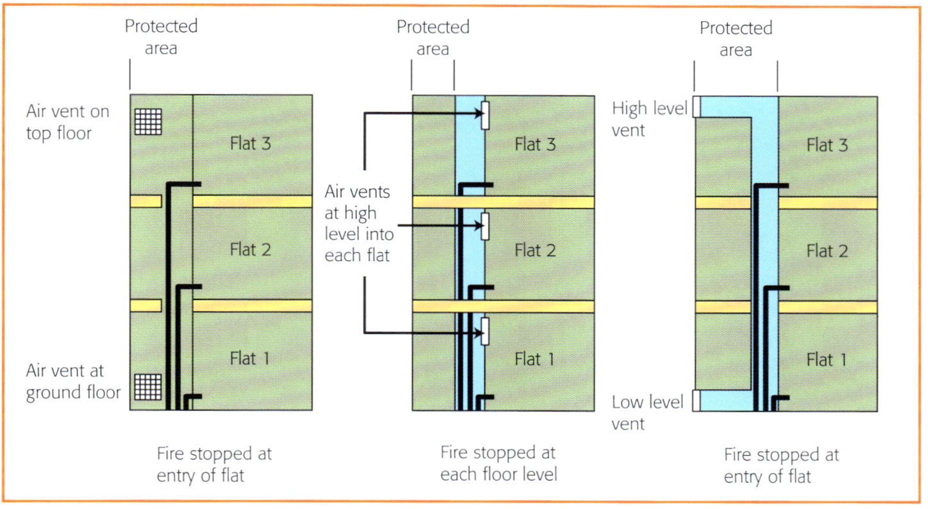

Gas installation pipework in voids

Where gas installation pipework is to be located in voids, refer to the following as appropriate:

- the space below a traditional floor e.g. floorboards, see **Installation pipework in wooden joisted floors** in this Part
- in a compartment floor e.g. 'cassette' constructed using 'I' joist see **Compartment floors – gas installation pipework** in this Part
- in a wall of hollow construction, see **Pipework in walls** in this Part
- in a purpose provided duct, see **Pipes in ducts – a guide to the different types** in this Part
- for any other void that is not purpose provided but could be treated as a duct, see **Pipes in ducts – a guide to the different types** in this Part

Figure 5.25 Copper and steel pipe in a protected area

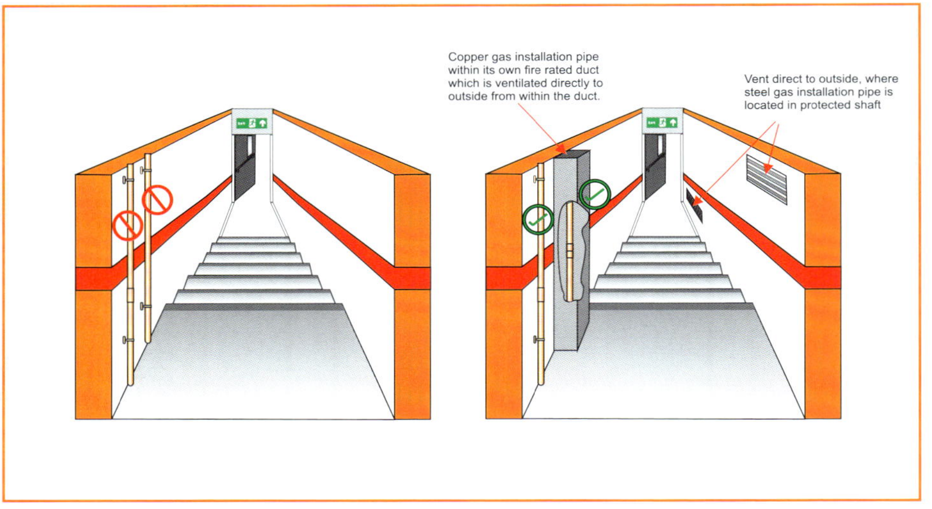

Figure 5.26 Typical exterior pipework

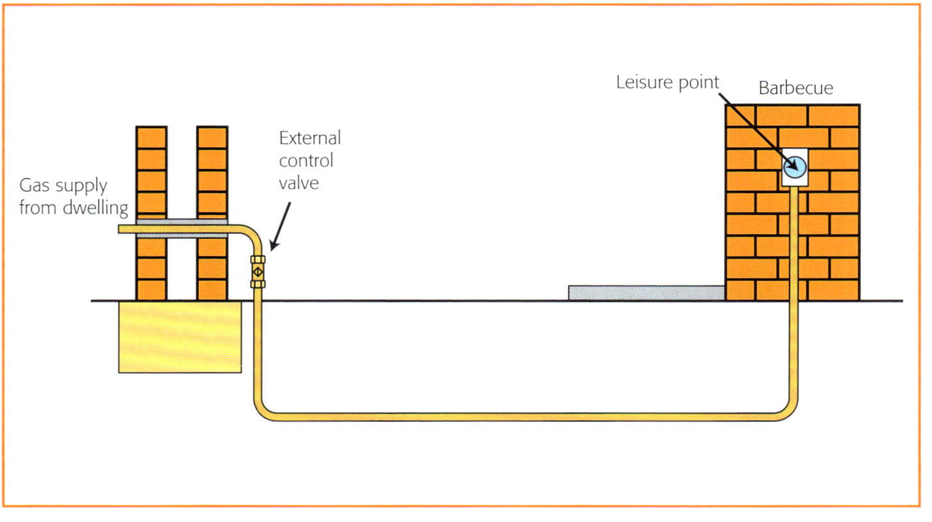

Figure 5.27 Exterior buried pipework (see Table 5.2 for minimum depth of cover)

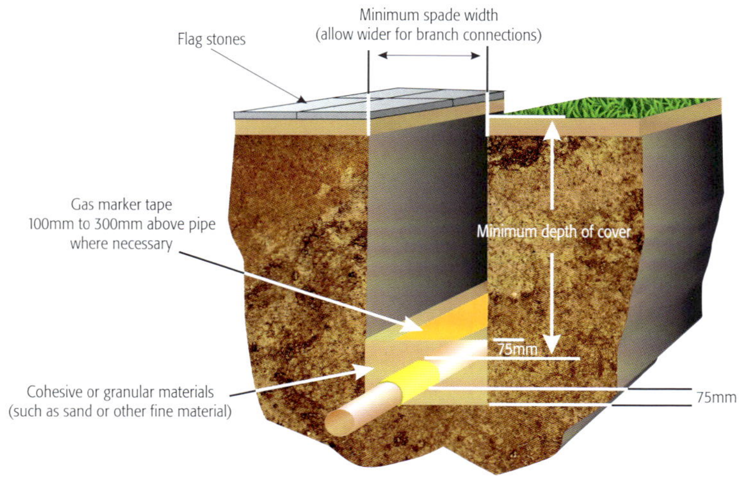

Exterior pipework – fittings, protection and other requirements

Buried external pipework shall be inherently resistant to, or otherwise protected against, corrosion. Steel and rigid stainless steel tubes shall not be buried.

PE pipe can be used if the pipework is buried. Metallic pipework can be buried providing a risk assessment has been undertaken and an inspection strategy developed and provided to the gas consumer.

Where metallic pipework is to be buried it shall be pre-insulated copper pipe, CSST / PLT installed to manufacturers instructions or contained within a sleeve (see **Sleeves** in this Part).

For more details of the requirements for buried pipework see Figure 5.26 and 5.27.

You need to fit an external control valve where the gas supply leaves the dwelling and fit sleeves as required.

Keep use of fittings to a minimum and use bends rather than elbows. Do not bury compression fittings – except where an appropriate fitting is used on PE pipework.

Where gas pipework is run externally above ground level and clipped directly onto the buildings surface, it should be factory-sheathed, wrapped, or painted with bituminous paint on site.

However, if copper pipe is installed using stand-off clips, additional corrosion protection is not normally required.

Weather conditions in the UK do not normally require copper pipework to be protected unless it may be subjected to additional corrosive sources.

Assess the route of the gas supply to ensure that it will not be damaged, either accidentally or deliberately. Provide additional protection where the pipework is located in an area where it could be subjected to mechanical damage in, for example, a carport.

Figure 5.28 Micropoint fittings

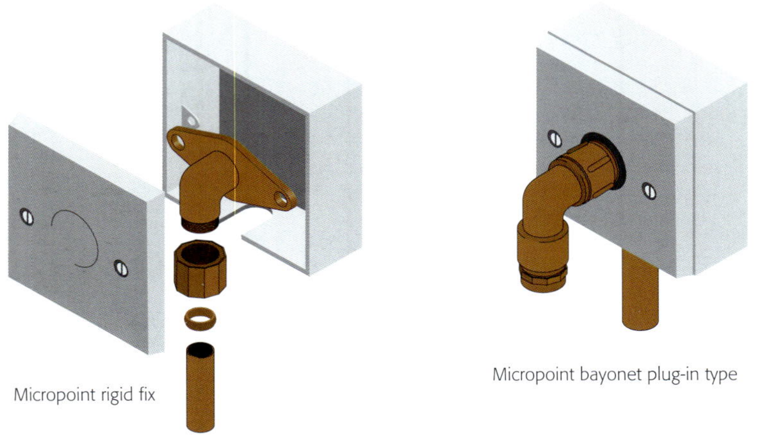

Micropoints

Micropoint systems utilise 10mm or 12mm copper pipe. Fittings are also available with 8mm and 6mm connections.

Purpose-designed termination points in recessed or surface-mounted plastic boxes, which contain a rigid fix connection are ideal as a carcass point are more attractive than a plugged or capped copper pipe for connecting gas fires etc. (see Figure 5.28).

Alternatively, they can incorporate a safety shut-off valve for connection to a bayonet type hose for connecting gas cookers, gas tumble dryers, etc (see Figure 5.28).

Also available, specially designed for outdoor use, is a weatherproof gas socket that allows for the easy connection and disconnection of appliances, leaving a neat, unobtrusive box when appliances are removed.

It is particularly suitable for gas leisure appliances such as barbecues.

A chained safety plug can be fitted to the shut-off valve when the hose has been disconnected, to prevent tampering or foreign matter entering the valve.

Pipework within builders' openings/fireplace recess

Soot and debris can be highly corrosive. All pipework and fittings installed within a builder's opening should therefore be either factory sheathed or wrapped/protected on site.

Pipe supports and fixings

All installation pipes shall be adequately supported in line with manufacture's instructions or in the absence of these the guidance in Table 5.6 can be followed.

The support shall be:

- designed to prevent the pipework from coming into contact with surfaces which are likely to cause corrosion (e.g. masonry, concrete, plaster)
- suitable for the environment and if used externally, corrosion and UV resistant.
- metallic in construction if used in protected areas where they need to be fire resistant.

Interrelation with other services

General

Install gas installation pipework so that it cannot cause danger by possible contact with electrical services. Make every effort to correctly space pipework from any electrical equipment – but if this is impracticable, the pipework should be sheathed with electrical insulating material rated at least 230 V ac or an electrical insulating panel inserted.

Pipework shall not be installed in ducts used for air conditioning or ventilation purposes.

Care is essential when you install gas pipework in buildings containing electrical damp-proofing protection systems, as this may cause accelerated pipework corrosion.

Do not bury gas installation pipework in floors where electrical under-floor heating is installed unless the heating system has been physically and permanently disconnected.

Do not install pipework or fittings where they may hinder the operation of any electrical item. e.g switch or socket outlet

Spacing

Space gas installation pipework and fittings as follows:

- at least 150mm away from electricity supply equipment, such as metering equipment, main service cut-outs or supplier (main) isolation switches, and distribution boards or consumer units; or
- at least 25mm away from electrical switches, sockets and electricity supply and distribution cables

Equipotential bonding – minimises hazards from electrical shock

All domestic gas installations should have a main protective bonding conductor on the gas installation pipework conforming to BS 7671: 2008 'Requirements for Electrical Installations. IET Wiring Regulations Seventeenth Edition'.

The purpose of protective bonding is to create a zone in which voltage differences through differing earth potentials are eliminated, and therefore hazards from electric shocks are minimized. This is achieved by connecting separate conductive components, such as the various metallic building services and structural steelwork, together with a main protective bonding conductor at the main earth terminal.

If an electrical fault occurs, it is possible for stray currents to be transmitted through the gas pipework. Therefore, to avoid electric shock or a spark which could ignite the gas, it is important to maintain electrical continuity in the pipework at all times.

Figure 5.29 Position of main protective bonding conductor

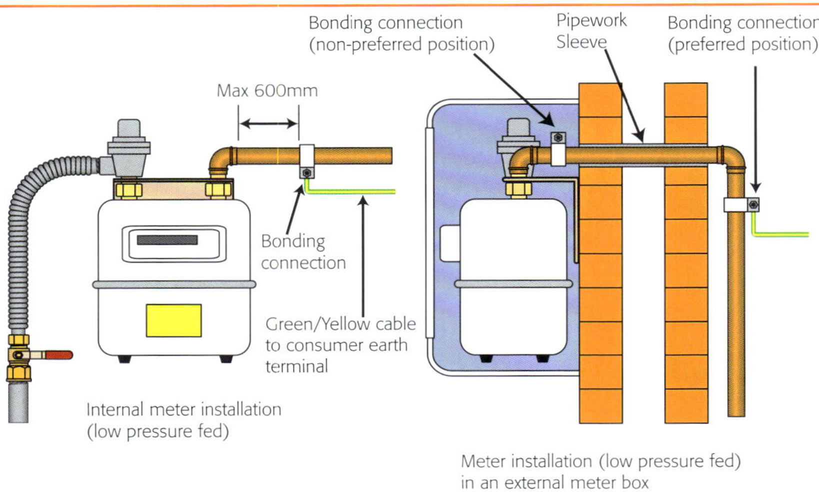

Figure 5.30 Main protective bonding conductor label (WL9)

ELECTRICAL BONDING

Most electrical Installations are required to be fitted with a main protective bonding conductor, which consists of the electrical connection of the internal gas and water pipes to the installation's main earthing terminal in accordance with the requirements of BS 7671, *Requirements for electrical instalations* (IET Wiring Regulations).

The gas installation pipework fitted in your premises does not appear to have the necessary main protective bonding conductor.

I am required by the Gas Safety (installation and Use) Regulations 1998 to tell you that any necessary main protective equipotential bonding should be carried out by a competent person.

I advise you to have this checked by a competent electrical operative.

If you are the tenant of this property, please bring this matter to the attention of the owner or their agent as soon as possible.

Order Ref: WL9
CORGI*direct* February 2016

The main protective bonding conductor (main equipotential bonding connection) shall be:

- connected to the installation pipework on the customer's side of the gas meter.
- connected as close as is practicable to the point of entry into a premises before any branch in the installation pipework.
- sited in a position where it can be readily seen, with a warning label stating 'Safety electrical connection. Do not remove' (see **Part 10 – Emergency notices, warning labels and reporting forms**).
- connected by a mechanically and electrically sound connection, which is not subject to corrosion (i.e. not exposed to the weather).
- for internally sited gas meters, fitted within 600mm of the meter outlet for verification purposes.
- for meters fitted in external meter boxes/compartments, fitted inside the building, within 600mm of the point of entry of the installation pipework (the preferred position). Alternatively, the connection may be within the meter box/compartment.

Note: It is essential that the bonding cable does not interfere with the integrity of the meter box/compartment or the sealing of any sleeve (see Figure 5.29).

For additional guidance on domestic gas meter installations, see Part 16 – Domestic gas meters.

What do you do when relocating a gas meter?

An existing main protective bonding conductor may be satisfactory, but you may need to either lengthen or shorten or, in some cases, completely renew the bond. The bond connection is satisfactory if the previously mentioned conditions are met.

Positioning of a main protective bonding conductor fitted to internal and external gas meter installations is shown in Figure 5.29.

See **Part 16 – Domestic gas meters** for further guidance, including required competency for the work activity

What do you do when gas installation pipework is connected to a primary meter that does not have a main protective bonding conductor fitted?

You should notify the responsible person (e.g. home owner, tenant, landlord/agent or builder in the case of new property) that a main protective bonding conductor should be installed and carried out by a competent person.

The same duty applies to you when inspecting primary gas meter installations in the course of carrying out gas work (e.g. gas safety inspections, routine maintenance or servicing etc.).

If you are competent to install a main protective bonding conductor, carry this out with the permission of the responsible person.

If you are not competent to carry out the work, provide written notification (e.g. a letter or card) to the responsible person (see Figure 5.30 and **Part 10 – Emergency notices, warning labels and reporting forms**).

Corrosion protection – a number of forms, including applying on site

Test assembled pipework and fittings for gas tightness (see **Part 6 – Testing for gas tightness (Natural gas)**) before any protection is applied and/or the pipework and fittings are buried.

Do not install gas installation pipework and fittings in positions where they are likely to be exposed to a corrosive environment. If this is unavoidable, the pipework and fittings need to be manufactured from a material that is resistant to corrosion, or you need to apply protection on site.

Ideally, use factory-finished, protected pipework and fittings that take the form of sheathing, wrapping, dipping, galvanising or painting which should be coloured yellow ochre in accordance with BS1710.

Take care not to damage the protection on the pipework when:

- storing
- jointing
- bending
- passing through holes and sleeves

Examine protection already applied for cuts or other defects and make good before use. Any flaw in the protection will provide a centre for accelerated corrosion.

Where jointing of pipework is required, you will need to cut away the protection to allow the joint to be made. Protection of the joints requires special attention, as exposed threads on steel pipe in particular are vulnerable to corrosion if left unprotected.

Site protection should be suitably selected for the application (for example, to stop the ingress of water) and can be achieved by the application of PVC wrapping tapes, self-bonding silicone tapes, grease-impregnated tape or bituminous paints applied in accordance with manufacturer's instructions where applicable.

Where it is not practicable to use or obtain factory-finished protected pipe, protection should be applied on site and in accordance with manufacturer's instructions where applicable.

Where wrapping tape is to be used, the pipework should be clean, dry and applied in accordance with the tape manufacturer's instructions. A minimum 55% overlap should be used to provide a layer of at least double thickness.

Some pre-insulation coatings, coverings, sheathings and wrapping tapes suffer from colour fade and surface cracking due to the effects of ozone or when exposed to direct sunlight or other ultra-violet (UV) sources. Therefore, care should be taken to ensure a suitable material is selected.

5 – Installation of pipework and fittings

Testing for gas tightness (Natural gas) – 6

6 – Testing for gas tightness (Natural gas)

Figures

Tables

Introduction

This part outlines the test procedures for domestic NG installations. For guidance on the testing of LPG installations, see **Part 18 – LPG – General requirements.**

These test procedures are based on the requirements of IGEM/UP/1B (Edition 3) 'Tightness testing and direct purging of small Liquefied Petroleum Gas/Air, Natural gas and Liquefied Petroleum Gas installations – Amendment October 2012' and apply to installations supplied with NG (see Note), where the installation pipework does not exceed 35mm copper or DN 32 (1¼ " BSP) steel tube.

The installation should operate at a pressure not exceeding 21mbar with or without a gas meter, to a maximum rating of 16m^3/hr with or without appliances fitted and a maximum installation volume not exceeding 0.035m^3 for an individual dwelling or non-domestic premises, see **Calculating installation volumes – to verify** in this Part.

This Part deals with NG only, but IGEM/UP/1B (Edition 3) covers LPG/Air and LPG installations also, which will be covered in parts of this manual and the other manuals within the Gas Installer Series.

The competence you need

When you carry out any work (see **Part 19 – Definitions**) relating to gas appliances and other gas fittings covered by this Part, you must be competent and hold a valid certificate of competence for each work activity that you wish to undertake (see **Part 1 – Gas safety legislation – How is your competence as a gas operative assessed?** for further guidance).

Why you must test for gas tightness – to prevent explosive mixture building up

When utilised under controlled conditions, NG is a safe, beneficial fuel. However uncontrolled mixtures of gas and air can be explosive.

Therefore strict mandatory regulations and procedures are in place to prevent gas escapes that may result in the build up of an explosive mixture (see **Part 3 – Characteristics of combustion**).

General requirements

IGEM/UP/1B (Edition 3) applies to:

1. Low pressure fed installations that are supplied with a gas pressure not exceeding 75mbar.
2. Medium pressure fed installations that incorporate a Meter Inlet Valve (MIV), positioned after the regulator, that are supplied with a gas pressure not exceeding 2bar.
3. Medium pressure fed installations that do not incorporate a MIV (see Note 2), positioned after the regulator, that are supplied with a gas pressure not exceeding 2bar.

Note 1: A MIV may also be known as a 'test valve'.

Note: 2: Medium pressure fed installations that DO NOT incorporate a MIV are outside the scope of IGEM/UP/1B (Edition 3) however, guidance on testing these installations is provided within the Appendices of the Standard and within this manual.

6 – Testing for gas tightness (Natural gas)

Gas installations with a pipework size in excess of 35mm copper or DN 32 (1¼" BSP) in diameter, or installations with a gas meter greater in volume than $16m^3/hr$ ($550ft^3/hr$) e.g. U16/G10, should be tested to:

- IGE/UP/1 (Edition 2) Strength testing, tightness testing and direct purging of industrial and commercial gas installations
- IGE/UP/1A (Edition 2) Strength testing, tightness testing and direct purging of small low pressure industrial and commercial natural gas installations
- IGEM/UP/1C Strength testing, tightness testing and direct purging of natural gas and LPG meter installations

– as appropriate.

For further guidance refer to the current Essential Gas Safety – Non-Domestic manual.

Low pressure fed installations

A low pressure fed NG meter installation is one that has a maximum operating pressure (MOP) not exceeding 75mbar at the inlet to the regulator with an outlet operating pressure of 21mbar – see Figure 6.1(a).

Medium pressure fed installations

A medium pressure fed NG meter installation is one where the operating pressure at the inlet to the regulator does not exceed 2bar, with an outlet operating pressure of 21mbar.

A medium pressure fed installation may or may not incorporate a MIV between the meter regulator and the meter itself – see Figure 6.1(b) and (c).

Report of a gas escape – urgent advice to be given to gas user(s)

1. When you are advised of a gas escape but are not on site, instruct the gas user(s):
   - to turn off the gas supply immediately at the ECV (see Figure 6.1)
   - to extinguish all sources of ignition
   - not to smoke
   - not to operate electrical light or power switches (on or off)
   - to ventilate the building(s) by opening doors and windows
   - to ensure that access to the premises can be gained
2. Also instruct the gas user(s) to report the gas escape to the appropriate gas emergency service call centre listed below:
   - in England, Scotland and Wales, contact the National Gas Emergency Service Call Centre, on 0800 111 999
   - in Northern Ireland contact Phoenix Natural Gas on 0800 002 001, or the LPG supplier
   - in the Isle of Man contact Manx Gas Ltd for all areas (including LPG) on 0808 1624 444
   - in the Channel Islands contact either Guernsey Gas Ltd on 01481 749000 or Jersey Gas Company Ltd. on 01534 755555, as appropriate
3. The gas supply should not be used until remedial action has been taken to correct the defect and the installation has been recommissioned by a competent person.

Figure 6.1 Typical domestic sized Natural gas meter installation

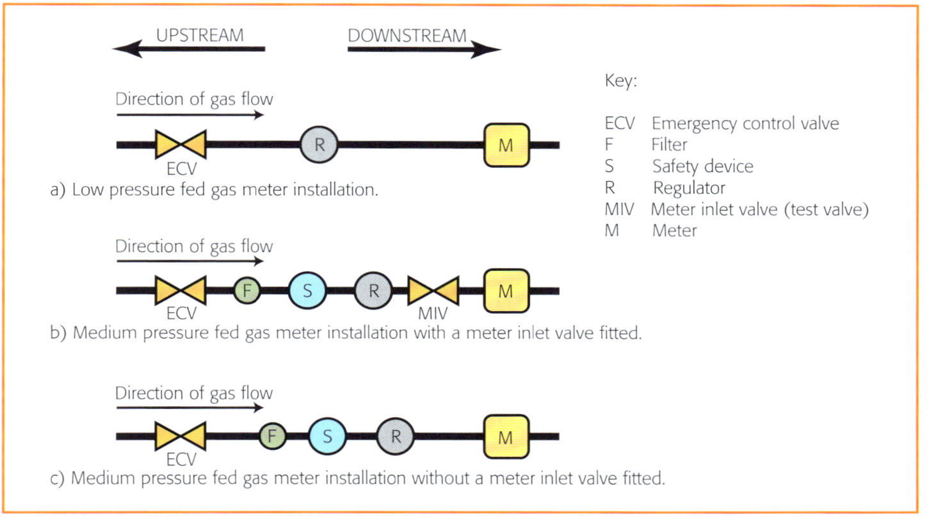

When you must test for gas tightness

- whenever a smell of gas is suspected, or reported
- on newly installed meters/pipework
- whenever work is carried out on a gas fitting which might affect its gas tightness (including pipework, appliances, meters and connections)
- before restoring the gas supply after work on an installation
- before working on an existing installation
- prior to the fitting of a gas meter on new or existing pipework or installation
- on the original installation prior to connecting any extension
- on completion of any extension or alteration
- before wrapping, painting, burying or concealing any installation pipework
- during routine testing of existing installation pipework
- whenever there has been an interruption of gas supply due to a gas supply failure
- immediately prior to purging the installation pipework

Note: Always remember – not all gas escapes will be identified by a pressure gauge test.

Gas pipework which is downstream of appliance control taps, flame supervision devices (FSD), etc. which cannot be subjected to a pressure gauge test (e.g. cooker grill supplies, boiler burner supplies) should be checked using a suitable non-corrosive leak detection fluid (LDF) after the tightness test has been completed and the gas supply restored. See also Part 19 – Definitions.

You could alternatively use a suitable portable electronic gas detector for this test.

Work following a reported gas escape

When you carry out a gas tightness test in any of the following emergencies:

1. When investigating a smell of gas or a reported gas escape.
2. When the installation has been isolated by the ESP (e.g. National Grid) following a reported escape of gas.
3. When on site, either you or the gas user(s) smell gas or suspect a gas escape.

There should be:

- no perceptible pressure loss (see **Perceptible pressure movement** in this Part) on the pressure gauge; and
- no smell of gas

Pressure gauges

General

Always use a suitable pressure gauge (water pressure gauge or electronic pressure gauge) and flexible tubing, which is in good condition (see Figure 6.2).

Periodically check the tube end connections with non-corrosive LDF to ensure these connections are gas tight.

For domestic tightness testing with NG, a water pressure gauge should have a minimum scale range of 0 to 30mbar.

How to use a water pressure gauge for testing

A water pressure gauge should be capable of being read to an accuracy of 0.5mbar, or better.

To use a water pressure gauge correctly, set the water levels in the 'U' tube limbs at zero. During the test period, the water pressure gauge should be in the vertical position. The vertical distance between the scale zero and liquid levels indicates the correct pressure reading and should be expressed in millibars (mbar) water gauge.

If the water gauge cannot be zeroed, measure the distance between the two meniscuses (water levels) and divide by two for the correct reading.

Note: It is important to remember that certain 'U' tube pressure gauges are designed for use with a fluid with an RD different to that of tap water. So do not use tap water with this type of gauge as you will get inaccurate readings. Contact the gauge manufacturer for further guidance.

How to use an electronic pressure gauge for testing

An electronic pressure gauge should be capable of being read to an accuracy of 0.1mbar or better.

When you use an electronic pressure gauge, consider the following points:

- the pressure gauge should be intrinsically safe if used in a gas emergency. In other circumstances carry out a risk assessment prior to testing

Note: This is not a requirement when testing pipework containing only air or inert gases.

- it is essential that the gauge is suitable for tightness testing and you operate it in accordance with the manufacturer's instructions (including ambient temperature requirements)

Figure 6.2 Water and electronic pressure gauges

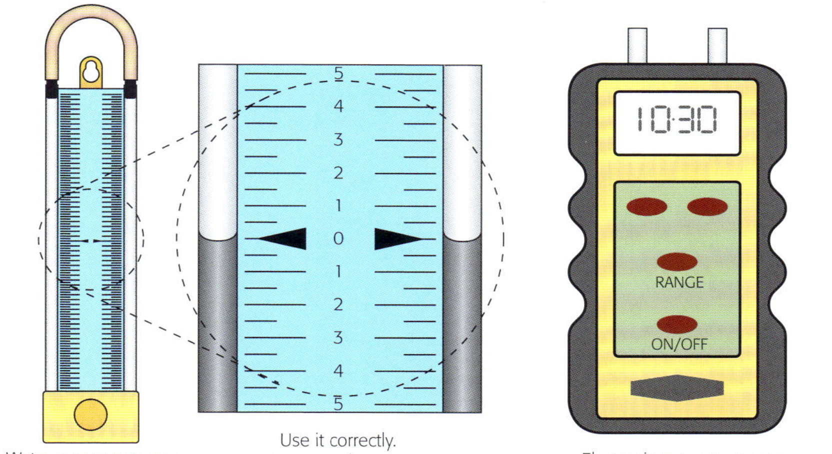

- test batteries prior to use
- stabilise the gauge at ambient temperature in accordance with manufacturer's instructions
- it should be zeroed at the start of each test and then zero checked at the end of each test
- unless otherwise stated by the pressure gauge manufacturer, the instrument should be tested, calibrated and certificated annually

Note: Retain copies of the calibration certificates for future reference.

Perceptible pressure movement

Certain tightness tests and let-by tests require that there is 'no perceptible pressure movement' on the pressure gauge.

No perceptible pressure movement would be less than 0.25mbar on a water pressure gauge and a maximum of 0.25mbar on an electronic pressure gauge, or 0.2mbar when the electronic gauge reads to one decimal place only.

'No perceptible pressure loss' is required where:

- an escape of gas has been reported, or there is a suspected smell of gas
- the installation has been isolated by the ESP (e.g. National Grid) following a reported escape of gas
- the installation is newly installed
- only the meter/pipework on an existing installation is being tested
- all appliances are isolated from the meter/pipework

Figure 6.3 Typical gas meters

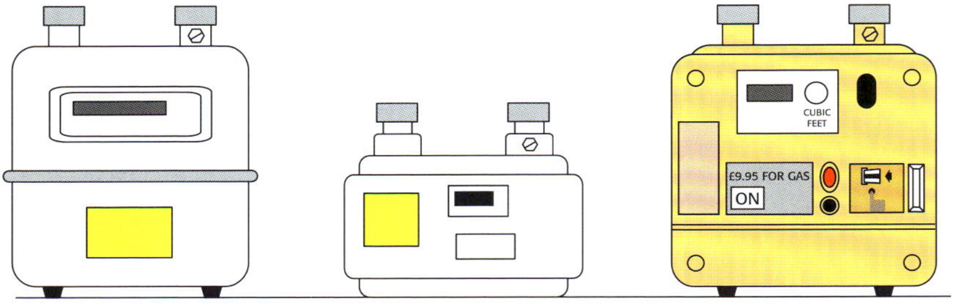

U6/G4 U16/G10 gas meter

E6 gas meter

Typical electronic token gas meter (ETM)

Identifying gas meter – helps you determine allowable pressure loss (if any)

To carry out a tightness test with the gas meter fitted, you need to identify the meter type and case size – to determine the permissible pressure loss (tolerance), if any (see Figure 6.3 and Table 6.1).

For additional guidance on gas meter installations, see **Part 16 – Domestic gas meters**.

The allowable pressure loss relates to the differing internal volumes of individual meter types and average lengths of installation pipework in domestic NG installations.

Note 1: The permissible pressure loss is determined by the total volume of the meter and the installation pipework.

The volume of an E6 meter is about half that of a U6 or G4 meter and therefore, the permissible pressure loss is doubled (see Figure 6.3 and Table 6.1).

Note 2: Electronic token meters (ETMs) and 'smart' meters are fitted with anti-tamper devices that, when activated, will prevent gas passing through the meter until it has been reset by the gas supplier or authorised agent. Therefore, take extreme care when removing or handling an ETM/smart meter as the anti-tamper device is easily triggered.

Table 6.1 Maximum permissible tightness test pressure losses

	Maximum permissible pressure loss (mbar)
Gas meter only (new or existing)	None*
Installation pipework only (new or existing)	None*
Gas meter and installation pipework only (new or existing)	None*
Gas meter, installation pipework (new or existing) and new appliance(s)	None*
Installations with existing appliances connected	**Maximum permissible pressure loss (mbar)**
No meter – AECV and pipework only ≤ 28mm	8.0mbar**
No meter – AECV and pipework only ≤ 35mm	4.0mbar**
Electronic meter (E6) ≤ 6m^3/h with pipework ≤ 28mm diameter	8.0mbar**
Electronic meter (E6) ≤ 6m^3/h with pipework ≤ 35mm diameter	4.0mbar**
Diaphragm meter (U6/G4) ≤ 6m^3/h with pipework ≤ 28mm diameter	4.0mbar**
Diaphragm meter (U6/G4) ≤ 6m^3/h with pipework ≤ 35mm diameter	2.5mbar**
Diaphragm meter (U16/G10) ≤ 16m^3/h with pipework ≤ 35mm diameter	1.0mbar**

* See **Perceptible pressure movement** in this Part.

** Where a tightness test is carried out due to a reported gas escape/smell of gas, the gas installation must be gas tight i.e. no permissible pressure loss (see Note).

Note: If a smell of gas persists with the gas turned off, there may be an escape of gas from upstream of the ECV/MIV. In this case, immediate contact should be made with the appropriate ESP to make the installation safe.

≤ less than or equal to

Table 6.2 Meter volumes (m^3)

E6 meter	$0.0024m^3$
G4/U6 meter	$0.008m^3$
G10/U16 meter	$0.025m^3$
Note: For any other gas meters, contact the meter manufacturer for the declared volume.	

Table 6.3 Pipework volumes (m^3)

Tubing sizes	Volume of 1m length of tube (m^3)
Steel, stainless steel or CSST*	
15mm $^1/_2$"	0.00024
20mm $^3/_4$"	0.00046
25mm 1"	0.00064
32mm $1^1/_4$"	0.0011
Copper tube	
15mm	0.00014
22mm	0.00032
28mm	0.00054
35mm	0.00084
PE pipe	
20mm	0.00019
25mm	0.00033
32mm	0.00053
*Corrugated stainless steel tube	

Calculate installation volumes – to verify procedure to use

Gas installations incorporating gas meters up to U16/G10 and pipework up to 35mm diameter will normally have an internal volume less than $0.035m^3$. If you have any doubt, calculate the volume of the installation to verify that you are using the correct tightness testing/purging procedures.

The total installation volume (m^3) will be the sum of:

$$\text{Total } m^3 = M_V + P_V + F_V$$

Where –

M_V – Meter internal volume (m^3), see Table 6.2

P_V – Pipework (m^3), see Table 6.3

F_V – Fittings, valves etc (m^3) if known, or add 10% of Pv

Example 1

A gas installation with a U6 gas meter, 10m of 35mm copper pipework, 5m of 22mm copper pipework and a further 5m of 15mm copper pipework. What would be the estimated internal volume of the complete installation?

Answer

Installation volume (m^3)

$= 0.008 + [(10 \times 0.00084) + (5 \times 0.00032) + (5 \times 0.00014)] + (Pv \div 10)$

$= 0.008 + [0.0084 + 0.0016 + 0.0007] + (Pv \div 10)$

$= 0.008 + 0.0107 + (0.0107 \div 10)$

$= 0.008 + 0.0107 + 0.00107$

$= 0.01977m^3$

Therefore the installation volume as calculated in Example 1 would be suitable for the tightness test procedures described in this Part.

Example 2

A gas installation with a U16 gas meter, 15m of 32mm steel pipework, 15m of 22mm copper pipework and a further 10m of 15mm copper pipework. What would be the estimated internal volume of the complete installation?

Installation volume (m^3)

= 0.025 + [(15 x 0.0011) + (15 x 0.00032) + (10 x 0.00014)] + (Pv ÷ 10)

= 0.025 + [0.0165 + 0.0048 + 0.0014] + (Pv ÷ 10)

= 0.025 + 0.0227 + (0.0227 ÷ 10)

= 0.025 + 0.0227 + 0.00227

= 0.04997m^3

The installation volume as calculated in Example 2 wouldn't be suitable for the described tightness test procedures in this Part (see Note), and would therefore require a tightness test to non-domestic procedures (IGE/UP/1 or 1A, as appropriate – see 'Essential Gas Safety – Non-domestic' manual for further guidance).

Note: The complete installation in Example 2 could be isolated into sections (each section needs to be less than 0.035m^3 in volume) and each section tightness tested separately. See also Part 15 – Re-establishing the gas supply and re-lighting appliances – Purging.

Tightness testing procedures

General

There is no requirement to carry out a strength test on any part of the installation which is subjected to a MOP of 75mbar or less. However, any component/assembly that will be subjected to a MOP exceeding 75mbar needs to be strength tested by the manufacturer or tested in accordance with IGEM/UP/1B (Edition 3) or IGE/UP/1 (Edition 2), as appropriate.

These test procedures apply to installations supplied with NG with pipework not exceeding 35mm in diameter, using fuel gas or air as the test medium. Where a gas meter is to be included in the tightness test, it should have a maximum capacity of 16m^3/hr (550ft^3/hr) (see **General requirements** in this Part for further details).

Attention: When you carry out a tightness test on a medium pressure fed installation that does not incorporate a MIV, refer to 'Tightness testing existing medium pressure fed installations without a MIV fitted' in this Part.

Note 1: The GSIUR require all newly made gas installation joints to be visually inspected to ensure they have been correctly made.

Note 2: For the purpose of this procedure, the ECV/additional emergency control valve (AECV) and MIV will be referred to as the 'appropriate isolation valve'.

Procedure for testing new pipework only using air (no gas meter connected)

What you need to do:

1. Prior to testing, where practicable, visually inspect all pipework joints to ensure that they have been assembled correctly.
2. Cap or plug all open ends on the installation with an appropriate fitting.
3. Confirm the pressure gauge and test hose to be used are in a sound working condition i.e. gauge is readable, in calibration (if electronic), test hose isn't showing signs of ageing/stress (cracking, brittleness, splits), etc.
4. Connect a suitably zeroed pressure gauge to the installation via one branch of a test tee (see Figure 6.4).

Figure 6.4 Typical test tee

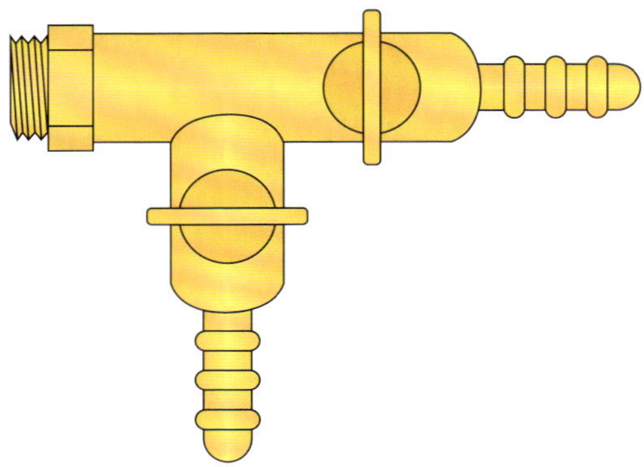

Note: The other end of the test tee is valved and used to introduce air into the installation pipework.

5. Slowly raise the pressure to between 20 – 21mbar.
6. Turn off the valved branch.
7. Allow one minute for temperature stabilisation, then where necessary re-adjust the pressure to between 20 – 21mbar.
8. Test the installation for a further two minutes. There should be no perceptible pressure loss (see **Perceptible pressure movement** in this Part).
9. Where a pressure loss is observed, trace and repair and re-test the installation, or make the installation safe.
10. Remove the test tee and plug or cap the point.
11. Complete all necessary documentation for example, CORGI*directs* 'Gas Testing and Purging – Domestic (NG)' form (see **Part 10 – Emergency notices, warning labels and forms**).

Note: Where the installation has passed the tightness test and you are not fitting a gas meter immediately, cap or plug the pipework with an appropriate fitting.

It would also be considered 'good practice' if you label the installation, stating that it has passed a tightness test using air only.

Procedure for testing new installations with or without appliances connected – and existing installations using gas as the test medium (low pressure and medium pressure fed incorporating a MIV)

What you need to do (including emergency procedures, as appropriate):

1. Turn off the gas supply at the appropriate isolation valve:
    - open all appliance isolating valves
    - turn off any pilot and burner controls
    - raise fold-down lids on hotplates so that any safety shut off valves on the supply to the hotplate taps, cooker grills, etc. are open
2. Confirm the pressure gauge and test hose to be used are in a sound working condition i.e. gauge is readable, in calibration (if electronic), test hose isn't showing signs of ageing/stress (cracking, brittleness, splits), etc.
3. Remove the appropriate pressure test point screw and using suitable tubing, connect a correctly zeroed pressure gauge to the test point.
4. Slowly turn on the gas supply at the appropriate isolation valve and raise the pressure to between 7 – 10mbar.
5. Turn off the gas supply at the appropriate isolation valve.
    - For medium pressure fed installations, hold open the release mechanism on the medium to low pressure regulator to balance the pressure on both sides of the regulator (see Figure 6.6)
6. Observe the pressure gauge for one minute. If there is a perceptible pressure rise (see **Perceptible pressure movement** in this Part), the means of isolation may be 'letting-by' (leaking). Disconnect the outlet union of the isolation valve and apply non-corrosive LDF to the barrel or ball of the valve. If the LDF bubbles then this will confirm let-by.

Note: Do not use household or 'washing up' liquid as LDF, as the high chloride content is corrosive to metals.

Attention: You should connect a temporary continuity bond to the installation pipework during this operation (see Part 5 – 'Installation of pipework and fittings – Safety precautions you must take – Temporary continuity bond – use for your own safety before you start work').

If you confirm let-by, make the installation safe (isolate gas supply and seal all open ends with an appropriate fitting); and:

- for isolation valves upstream of the primary gas meter, contact the ESP to arrange for the fault to be rectified (see **Report of a gas escape – urgent advice to be given to gas user(s)** in this Part for contact details)
- for isolation valves downstream of the primary gas meter, the valve should be repaired/replaced by a competent gas operative

Note: If you attribute the rise in pressure to a defective ECV or MIV, you must rectify the fault before proceeding with the tightness test.

When the ECV/MIV is part of the primary meter installation, make immediate contact with the appropriate ESP. See 'Report of a gas escape – urgent advice to be given to gas user(s)' in this Part for contact details.

7. After completing a successful let-by test, turn on the appropriate isolation valve and slowly raise the pressure to between 20 – 21mbar. Turn off the appropriate isolation valve.

Attention: It is important not to have a test pressure higher than 23mbar, as this may cause the meter regulator to 'lock up' – which may affect the results of the tightness test.

'Lock-up' is where at zero-flow conditions, the meter regulator closes off to limit the maximum pressure to an acceptable level at its outlet.

Should the pressure exceed 23mbar, drop the pressure back down to between 7 – 10mbar (this ensures the regulator isn't locked-up) before re-pressurising the installation to between 20 – 21mbar.

8. Allow one minute for temperature stabilisation then, where necessary, re-adjust the pressure to between 20 – 21mbar.
9. Test the installation for two minutes.

New installations

There should be:

- no perceptible pressure loss (see **Perceptible pressure movement** in this Part) on the gauge; and
- no smell of gas

Note: In the past, there has been confusion as to whether any pressure loss is allowable when testing new appliances, due to possible leakage from gas taps on cookers, etc. When new appliances are tested in the factory, the tightness test is carried out at 150mbar and any leakage recorded using a bubble leak indicator. As the domestic tightness test is only carried out at between 20 – 21mbar, there is no allowable leakage rate for new appliance installations.

- where the installation fails the test, trace and repair the leak(s) then re-test
- where the leak(s) cannot be traced, make the installation safe

Existing installations

For existing gas installations refer to Figure 6.5 and Table 6.1.

Note: If a gas meter is exchanged e.g. a U6/G4 for an E6 gas meter, then apply the appropriate permissible pressure loss value.

- record any pressure loss greater than that considered acceptable
- where possible, trace and repair the gas escape and re-test or make safe in accordance with the current Gas Industry Unsafe Situations Procedure (GIUSP)
- if a smell of gas persists with the gas turned off, there may be an escape of gas from upstream of the appropriate isolation valve (e.g. ECV). In this case, make immediate contact with the appropriate ESP to make the installation safe. See **Report of a gas escape – urgent advice to be given to gas user(s)** for contact details
- if the tightness test is satisfactory, remove the pressure gauge and re-seal the test point. Slowly turn on the gas supply and test the test point, ECV, regulator connections and where appropriate, the MIV connections with LDF (this is to ensure that you test equipment outlined above at the maximum operating pressure)
- record the test results (see **Part 10 – Emergency notices, warning labels and forms – Gas testing and purging – domestic (NG) (CP32)**) and where appropriate, inform the responsible person
- upon completion of a satisfactory tightness test the installation may need to be purged of air (needed for new or where air has entered the installation), see **Part 15 – Re-establishing gas supply/re-lighting appliances** in this manual

Figure 6.5 Dealing with gas escapes

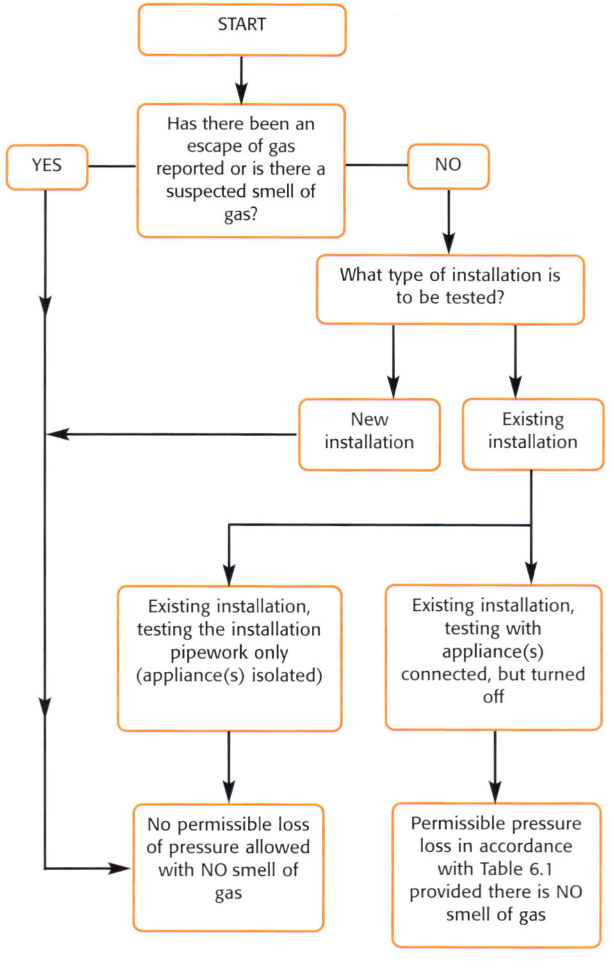

Tightness testing existing medium pressure fed installations without a MIV fitted

Test an existing installation for gas tightness before you commence any work using NG as the test gas.

Before you test, ensure that:

- there are no open ends on the installation
- all appliance isolation valves are open
- pilot and burner control valves are turned off
- fold-down lids on hotplates are raised so that any safety shut off valves on the supply to the hotplate taps are set in the open position

1. Turn off the gas supply at the ECV.
2. Confirm the pressure gauge and test hose to be used are in a sound working condition i.e. gauge is readable, in calibration (if electronic), test hose isn't showing signs of ageing/stress (cracking, brittleness, splits), etc.
3. Connect a suitable pressure gauge to an appropriate test point on the installation.
4. Carry out a let-by test of the ECV as follows:
    - reduce the pressure of the system to between 7 – 10mbar using a suitable point in the installation, (e.g. cooker hotplate burner) and turn off
    - hold open any release mechanism on the medium to low pressure regulator to balance the pressures on both sides of the regulator (see Figure 6.6)

Attention: Some medium to low pressure regulators require a different operation to the one outlined in Figure 6.6 or this may be carried out automatically through the regulator.

Where you find a medium to low pressure regulator that does not operate as outlined in Figure 6.6, contact the GT to follow the correct procedure. For contact details, see Part 21 – References – Licensed Gas Transporters.

- re-adjust pressure, were required, to between 7 – 10mbar
- observe the test gauge over the next 1 minute period for any perceptible movement
- at the end of the 1 minute test period, operate the release mechanism again (at these reduced pressures the UPSO or excess flow valve may have activated).
- note the final reading
- if over a period of one minute there is a perceptible pressure rise (see **Perceptible pressure movement** in this Part) the means of isolation may be 'letting-by' (leaking). Disconnect the outlet union of the isolation valve and apply non-corrosive LDF to the barrel or ball of the valve. If the LDF bubbles then this will confirm let-by. Make the installation safe and contact the appropriate ESP. See **Report of a gas escape – urgent advice to be given to gas user(s)** for contact details.

Suspend the test until a check has been made and any necessary repair carried out.

5. Following a successful let-by test of the ECV and while you operate any reset lever on the medium to low pressure regulator, very slowly open the ECV and raise the test pressure to between 18 – 19mbar.

Note: It is important that you avoid a test pressure higher than 21mbar to prevent the risk of the medium to low pressure regulator locking-up.

Figure 6.6 Typical medium to low pressure regulator

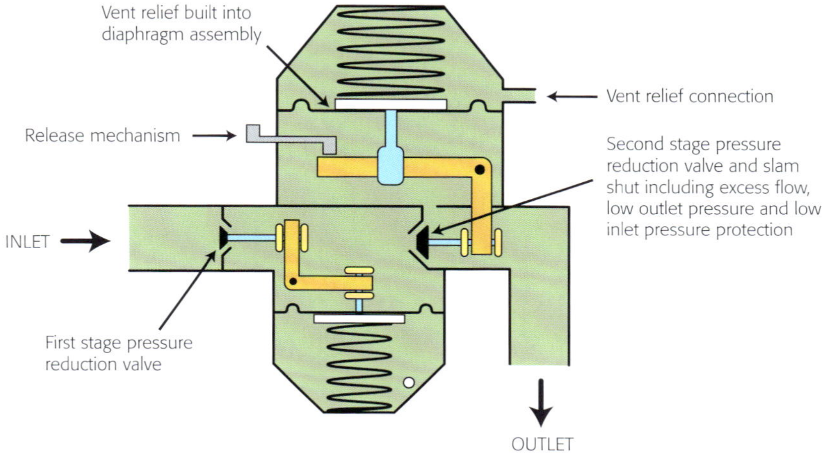

Should the pressure exceed 21mbar, drop the pressure back down to between 7 – 10mbar (this ensures the regulator isn't locked-up) before re-pressurising the installation to between 18 – 19mbar.

6. Turn off the gas supply at the ECV.
7. Allow one minute for temperature stabilisation.
8. Check for any pressure loss over the next two minute period ensuring that it does not exceed the values given in Table 6.1.
9. Where no appliances are connected to the installation or they are isolated, there should be no perceptible loss on the pressure gauge and no smell of gas.
10. If the installation fails the tightness test, trace and repair the escape(s) and re-test the installation until it passes the test. If you cannot trace and repair the escape(s) you must make the installation safe.
11. If a smell of gas persists with the gas turned off, there may be an escape of gas from upstream of the appropriate isolation valve (e.g. ECV). In this case, make immediate contact with the appropriate ESP to make the installation safe.
12. When the tightness test is satisfactory, remove the pressure gauge and re-seal the test point. Slowly turn on the gas supply and test the test point, ECV outlet connection and regulator connections with LDF.

Note: You could alternatively use a suitable portable electronic gas detector for this test.

13. Record the test results and where appropriate, inform the responsible person.
14. Purge the installation (see **Part 15 – Re-establishing the gas supply and re-lighting appliances**).

6 – Testing for gas tightness (Natural gas)

Connecting extensions

What you need to do:

1. Before you start any work, test the complete existing installation following the appropriate tightness testing procedure. Record any permissible pressure loss.
2. After connecting the extension, you must re-test the complete installation.
3. You must not accept any pressure loss greater than that recorded in 1. above.
4. Be aware that there should be no smell of gas.

Tracing and repairing gas escapes – indicators to help you

When you trace and repair gas escapes, always question the gas user. Try to pinpoint the area where the smell exists and:

- look for indicators such as old appliances with stiff or leaking control taps, old unprotected pipework passing through walls or under solid floors, etc.
- also use suitable portable electronic gas detectors to trace the source of gas escape

If you cannot pinpoint the escape straightaway:

- test the whole system in small sections
- then trace the escape using LDF following a process of elimination

Checking and/or setting regulators – 7

Figures

Figure 7.1 Typical low-pressure fed NG meter installation

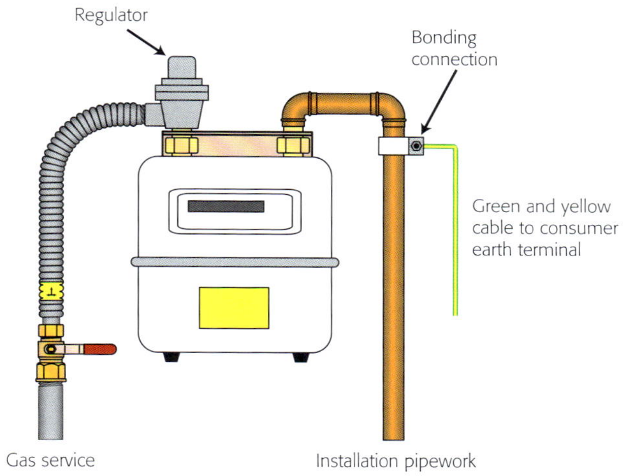

Gas supplied at differing pressures

NG is piped onshore and transported around the country in transmission and distribution pipelines operating at a variety of pressures. When the pressure is reduced in the distribution system to less than 7 bar a stenching agent (distinctive odour) is added to the gas.

The majority of domestic premises are supplied with pressures not exceeding 75mbar (low pressure). However, some premises may be supplied from mains operating at pressures up to 2 bar (medium pressure) and in a small number of cases 7 bar (high pressure). In all cases the gas will have been odorised.

On domestic NG meter installations the pressure is further reduced to provide a nominal operating pressure of 21mbar by a meter regulator (see Figures 7.1 and 7.3).

Maintaining the correct pressure: it is your responsibility to check that the appliance regulator is set correctly

Most gas appliances used in domestic and some non-domestic applications employ atmospheric burners (see Note).

A good illustration of the principle of operation for an atmospheric burner would be the humble Bunsen burner, where a stream of gas issuing from the injector enters the mixing tube and in the process, entrains primary air through the primary air ports. This gas/air mixture is then ignited at the burner head.

Note: Modern gas appliances, particularly boilers utilise forced draught burners which are controlled to more exacting standards in order to meet energy efficiency demands and given this, will become the dominant burner type in the not to distant future.

Figure 7.2 Typical multifunctional gas control valve

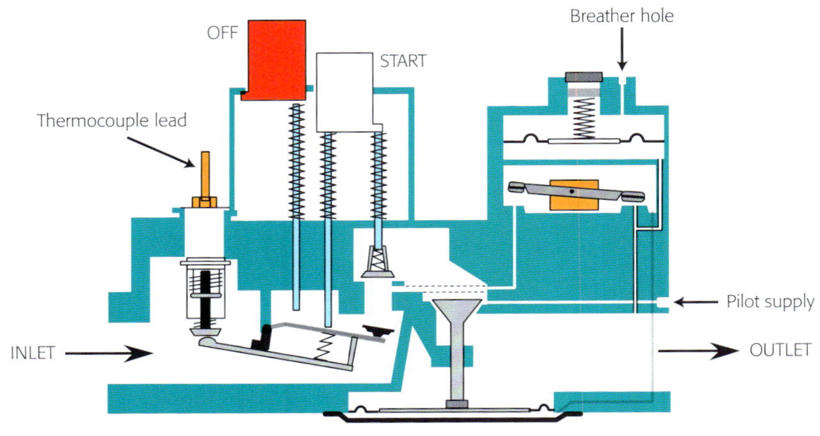

To ensure satisfactory combustion at a gas burner:

- it is important to maintain a constant pressure at the appliance burner injector

As the optimum pressure for NG to entrain the correct amount of primary air is 20mbar:

- gas should therefore be supplied to the appliance inlet gas connection at the appropriate pressure

Although it is your responsibility to check and ensure that the regulator pressure is set correctly, any adjustment necessary to the meter regulator must be carried out by the GT, or their approved agents.

The GSIUR require that the meter regulator is secured with a seal so as to prevent the setting of the regulator from being interfered with.

What happens where NG appliances are designed to operate on a range of gas rates and pressures?

These are therefore fitted with an individual appliance regulator, which may take the form of:

- a multifunctional gas control valve (see Figure 7.2), or
- a safety device at the inlet connection

Due to the inherent restrictions associated with these valves, you may record a pressure less than that indicated above at the appliance inlet test point.

In all cases where there is a reduced pressure:

- refer to the appliance manufacturer's installation instructions for an indication of the range of acceptable gas pressures

Figure 7.3 Essential parts of a NG meter regulator

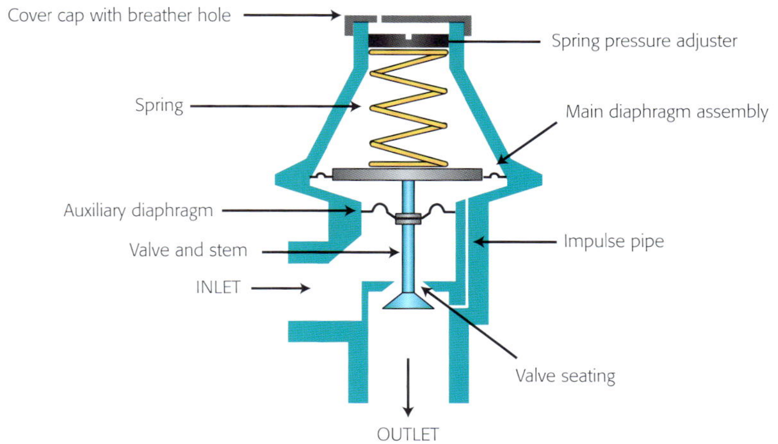

NG – the operating pressure you need to set

Because of frictional resistance in the gas pipework between the meter and appliance(s):

- the meter regulator (see Figure 7.3) is set to provide a pressure slightly above the optimum pressure
- the regulator is adjusted to provide a nominal operating pressure at the meter outlet of 21mbar

Note: It is the gas operative's responsibility to check/ensure that the regulator pressure is set correctly however any adjustment deemed necessary has to be carried out by the GT, or their approved agents.

The GSIUR require that the meter regulator, is secured with a seal to prevent the setting of the regulator from being interfered with.

Method of checking the operating pressure

1. Attach a suitable pressure gauge to the meter test point.
2. For a water pressure gauge ('U' gauge), ensure the pressure gauge is vertical and the water levels are zeroed i.e. water levels should have the same reading in each limb, or in the case of an electronic gauge, calibrate it in accordance with the manufacturer's instructions.

Note: With no gas in use, the pressure recorded will be the standing pressure, which normally exceeds 21mbar. Do not confuse this with the operating pressure.

Figure 7.4 Meter regulator pressure setting (NG)

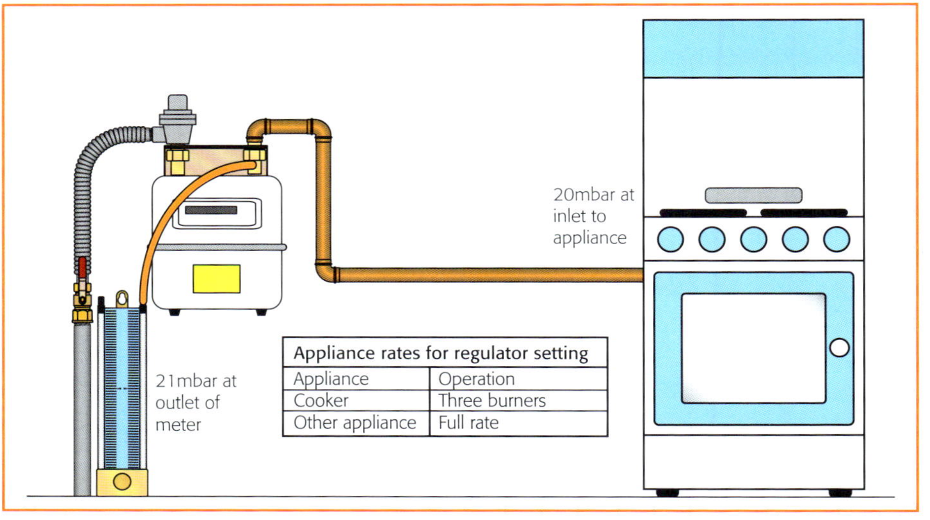

Appliance rates for regulator setting	
Appliance	Operation
Cooker	Three burners
Other appliance	Full rate

To establish the operating pressure, sufficient gas has to pass through the meter regulator.

You can achieve this by operating the largest single appliance connected to the installation (see also Figure 7.4):

- for a cooker – three hotplate burners
- any other appliance – full gas rate

Note: Meter regulators for NG are set at the factory to deliver an operating pressure of 21mbar with a flow rate equivalent to 3.5m³/hr of gas. Therefore when you check meter regulator settings on site, it is important that you obtain a sufficient flow rate.

How to obtain a sufficient flow rate:

- where connected, operate the largest single appliance to achieve this, or
- on new installations without any appliances connected, use a suitable industry meter regulator check device (see **Industry meter regulator check device (NG) – used where no gas appliances are connected** in this Part) to provide a minimum flow rate of $0.5m^3$/hr.

Meter regulators are designed to provide a constant gas pressure. However, at low through-puts (typically less than $0.5m^3$/hr) and high through-puts (typically greater than $3.5m^3$/hr), the outlet pressure provided by the meter regulator will be in the range of 19 to 23mbar.

Take immediate action as follows where you find readings outside those specified:

1. **GT** or their approved agents – adjust pressure and reseal regulator.
2. **Gas operative** contact the following as appropriate:
    - In England, Scotland and Wales, contact the National Gas Emergency Service Call Centre Telephone: 0800 111 999
    - in Northern Ireland contact Phoenix Natural Gas on 0800 002 001, or the LPG supplier
    - in the Isle of Man contact Manx Gas Ltd for all areas (including LPG) on 0808 1624 444
    - in the Channel Islands contact either Guernsey Gas Ltd on 01481 749000 or Jersey Gas Company Ltd. on 01534 755555 as appropriate

Industry meter regulator check device (NG) – used where no gas appliances are connected

This simple device comprises a fixed drilled orifice vented to atmosphere at one end and a union for attaching to the meter outlet at the other end.

The drilled orifice is sized to pass a known quantity of NG and to ensure that the meter regulator is operating correctly where no gas appliances are connected.

See **Part 16 – Domestic gas meters** for further guidance.

Gas Industry Unsafe Situations Procedure – 8

How you assess situations and what you must then do

When you carry out new installation work in a customer's premises, you are required to ensure that the appliance/installation is installed and fully commissioned in accordance with the GSIUR.

You must also pay due regard to the appropriate standards and the manufacturer's installation instructions, bearing in mind that:

- if this cannot be achieved, you must not leave the appliance/installation connected to the gas supply
- if an appliance cannot be fully commissioned, you must disconnect it from the gas supply, seal open ends (cap/plug the point of connection or installation) and label it accordingly

Two categories of risk: 'Immediately Dangerous' and 'At Risk'

Dependent upon the apparent risk, you can classify unsafe gas appliances/installations as:

1. 'Immediately Dangerous' (ID) appliances/installations

An 'ID' appliance/installation is one which, if operated or left connected to a gas supply, is an IMMEDIATE danger to life or property.

Broadly, these will be:

- installations that fail gas tightness tests
- appliances that fail spillage tests, or
- appliances which have serious flue and/or ventilation, or combustion defects when measured against manufacturer's instructions and/or BS 5440–1 and 2 or other relevant standards/guidance documents

2. 'At Risk' (AR) appliances/installations

An 'AR' appliance/installation is one which, if operated, may lead to a situation which could create risk to life or property.

The current Gas Industry Unsafe Situations Procedure – concentrates on gas safety matters only

Each revision following the first publication in1998 of the Gas Industry Unsafe Situations Procedure (GIUSP) provides installers with an up-to-date procedure for dealing with unsafe situations and installations in customer's premises.

The current procedure reflects changes in legislation and technical guidance arising from past gas incidents.

It adopts a wider industry consultation than previous versions of the Procedure, for a more comprehensive view. All sectors of the gas industry have input – including the Health and Safety Executive, large installer businesses, emergency service providers, the LPG industry, the non-domestic sector and Gas Safe Register.

In this procedure the 'Not to Current Standards' (NCS) classification has been removed as such situations are not unsafe.

Non-gas related safety issues are omitted – but still need addressing

The procedure concentrates on gas safety matters mainly, as it was felt that other trade sectors should be dealing with non-gas related safety issues through their respective guidance documents/procedures.

For this reason certain non-gas defects relating to water and electricity have been omitted. These exclusions do not minimise the importance of contraventions of other industry safety standards or legislation, but highlight the need for industry sectors to produce procedures for other fuels and utilities.

You must also be aware of other non-gas safety related defects which may exist on installations and your general duty of care under health and safety legislation. Engineers can still inform the gas user/responsible person via thier job reports of 'best advice' but must use thier judgement around what advice to give and the likelihood of it being feasible to implement.

The current procedure (7th Edition) provides guidance for competent gas operatives on how:

- you deal with unsafe situations, based on assessment of risk and the application of 'best practice'
- you must recognise that no procedure is able to identify every situation you may encounter and you need to use your competencies, skills and experience in assessing risks to ensure safe conditions are achieved

Major changes you need to note include the following:

- all warning notices and labels now are simplified to emphasise the words 'DANGER - DO NOT USE'
- a revised approach to 'At Risk' installations where turning off the gas WILL NOT remove a risk and a 'Danger Do Not Use' label is NOT required
- a revised approach to dealing with the lack of combustion air ventilation based on the latest research and analysis of carbon monoxide (CO) incidents. This is designed to provide a clearer 'stepped' approach for dealing with undersized ventilation issues and highlights the need for you to measure the actual ventilation free area provided, before you classify the risk presented to the end user. See **Part 4 – Ventilation – Maintenance of air vents**
- the re-classification of domestic wall adjacent open-flued termination positions to 'AR' situations, recognising that these installations offer an unacceptable risk to safe flue operation. See **Part 13 – Chimney standards – Wall-faced/wall-adjacent termination**
- the inclusion of many more examples of unsafe gas situations in the Tabled section of the procedure. These include additional information on non-domestic, LPG and commercial catering.

- specific guidance to operatives, on their obligations when applying the requirements of the procedure, whilst recognising the limits to which their assessed competency extends
- expanded glossary of terms and definitions.

You will find this document a significant tool in your armoury. Take care to fully understand it so you apply its guidance and content. Doing this will continue to improve gas safety and raise competency standards throughout the industry.

Note: You can obtain a printed version of the current Gas Industry Unsafe Situations Procedure from CORGI*direct* – Tel: 0800 915 0490, or visit www.corgi-direct.com

Alternatively as a registered engineer, you can download the Technical Bulletin (TB) from Gas Safe Register, TB 001 at https://engineers.gassaferegister.co.uk

Emergency controls and isolation valves – 9

Figures

Introduction

This part covers NG installations only.

For the operation and positioning of emergency controls and isolation valves for LPG installations, refer to **Part 18 – LPG – General requirements**.

These essential controls are for use by operatives and gas users

Gas installations must be able to be turned off (isolated) from the fuel supply, both:

- in an emergency; and
- to facilitate service/maintenance or the removal of appliances

ECVs are an essential part of a gas installation and are legislated for in the GSIUR. They must be accessible and clearly labelled and must be in good working condition.

Types and uses of valves

The three types of valves are:

- **emergency control valve (ECV)**: to shut the gas supply off in an emergency

Note: There can be more than one ECV, termed 'Additional Emergency Control Valve' (AECV) – see 'ECVs remote from the gas meter' in this Part.

- **meter inlet valve (or 'test valve')**: a valve located between the outlet of a medium to low pressure regulator and the gas meter used for gas tightness testing purposes
- **isolation valve**: all gas appliances should have a means of isolation. This valve is not primarily for use by gas users, but is to allow service/maintenance to be carried out

You must ensure that the valve is suitable for the gas family being used, that it conforms to British Standard (BS) 1552 or BS EN 331and that it is not a water fitting.

Some of these valves will be an integral part of an appliance, whereas others will only be suitable for the type of appliance for which they have been designed e.g. cooker bayonet connection.

The installation pipework material may be the limiting factor, in terms of the jointing method used. However, compression, threaded or capillary connections are acceptable on the valve.

Note: The positioning and labelling of ECV/AECV, MIV and appliance isolation valves is important. It needs to be related to the location of the gas meter, the type of installation and the appliance. See Part 10 – Emergency notices, warning labels and reporting forms for further guidance.

NG meter installations must have an ECV

The ECV will normally be the property of the gas supplier, who has a duty to maintain and label it:

- In England, Scotland and Wales, contact the National Gas Emergency Service Call Centre Telephone: 0800 111 999
- in Northern Ireland contact Phoenix Natural Gas on 0800 002 001, or the LPG supplier
- in the Isle of Man contact Manx Gas Ltd for all areas (including LPG) on 0808 1624 444
- in the Channel Islands contact either Guernsey Gas Ltd on 01481 749000 or Jersey Gas Company Ltd. on 01534 755555

to carry out any necessary repairs or alterations to the valve.

9 – Emergency controls and isolation valves

Figure 9.1 Typical gas meter installation

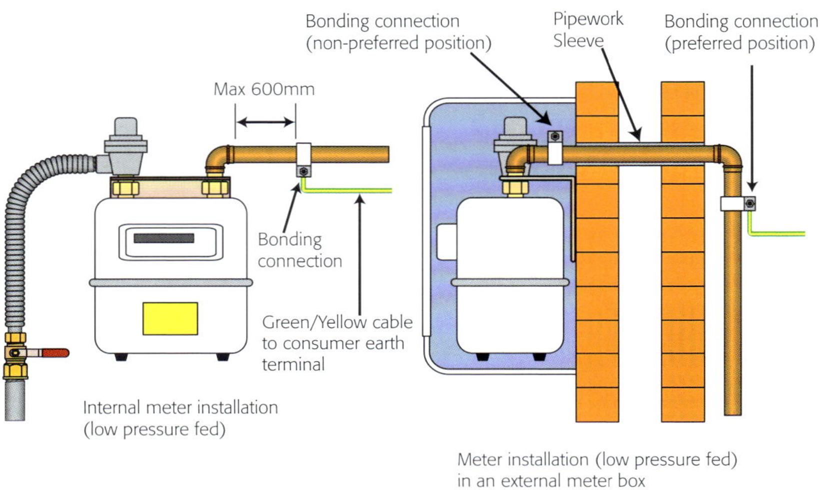

Figure 9.1 shows a typical gas meter installation.

Meter located internally or externally

Situate the ECV:

- as near as is practicable to the point where the gas supply enters the building, either internally or externally
- in a readily accessible position
- so that the lever (handle) of the valve is parallel to the pipe to which it is connected when it is in the open position (see Figure 9.2)
- so the lever (handle) only moves downwards one quarter of a turn to the off position (see Figure 9.2)
- so the lever (handle) is securely attached to the valve spindle
- so it is labelled with the on/off position of the valve and the gas escape emergency procedure (see **Part 10 – Emergency notices, warning labels and reporting forms**)

ECVs remote from the gas meter

Additional ECVs (AECVs) are required:

- at the point where the gas enters a dwelling when the gas meter installation is located remote from the dwelling it supplies.
- where the gas supply leaves the main dwelling and enters a separate building or area on the same property, that is not under the control of one responsible person
- where a gas supply enters separate dwellings within a building e.g. within a group of flats

See also Figure 9.3 for further guidance.

Figure 9.2 Emergency control valve NG

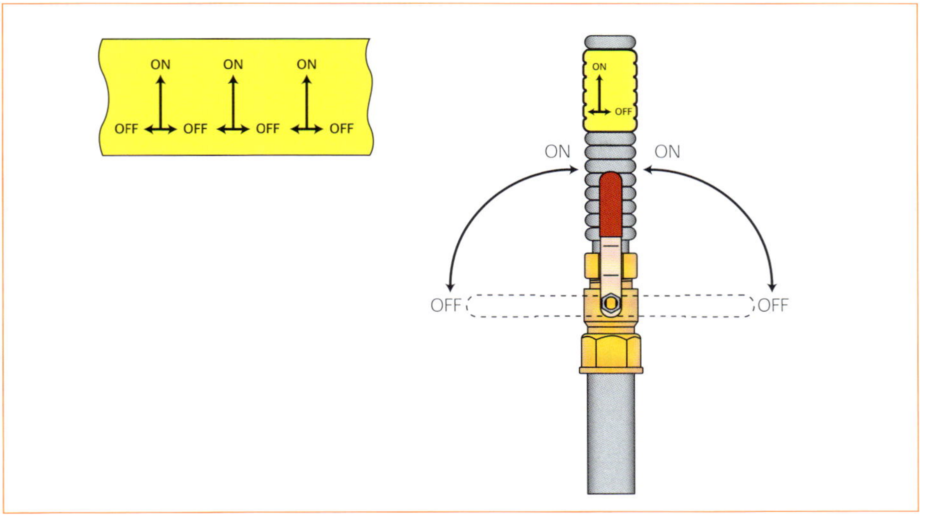

Figure 9.3 Emergency control valve positions

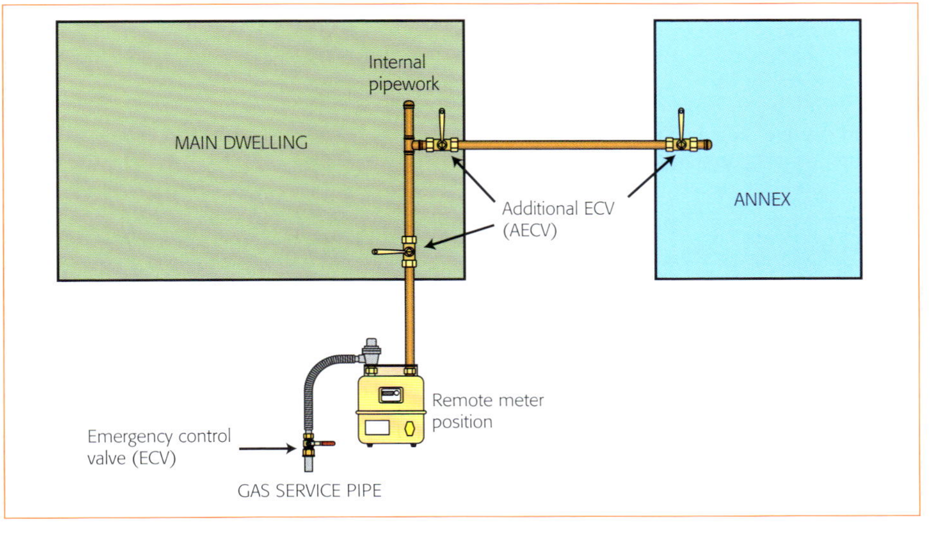

Note 1: It is also required that you fit a pressure test point within 300mm downstream of the AECV, unless it is provided for within the valve assembly.

Note 2: You must ensure that a notice is displayed on or near the meter indicating the position of the ECV where:

- a gas meter is installed more than 2m from the ECV; or
- is installed out of sight of the ECV altogether

See **Part 10 – Emergency notices, warning labels and reporting forms**.

Secondary gas meters need to have an ECV installed adjacent to them

The owner of the property is responsible for the maintenance and repair of secondary gas meters and associated emergency controls.

It is your responsibility when you install the secondary meter to ensure you correctly position and label the ECV. See **Part 16 – Domestic gas meters** and **Part 10 – Emergency notices, warning labels and reporting forms**.

Note: Ensure that either on or adjacent to the primary meter, the location of all secondary meters is clearly shown. There should be a line diagram showing the configuration of all meters, installation pipework and ECV/AECVs.

Multi-occupancy installations

Responsibility for ECVs depends on positioning

If there are independent metered gas supplies to separate dwellings in a multi-occupancy complex, ensure each one has a separate ECV as follows:

- where primary meters are located in each dwelling, the ECV is part of the associated meter fittings and is the property of the gas supplier
- where secondary meters are located in each dwelling, the owner of the property is responsible for the maintenance and repair of associated ECVs
- where the gas meter's location is remote from the dwelling and the ECV cannot be used by the occupiers as their sole means of turning off the gas supply, each dwelling should have their own AECV

  This valve is normally the property of the dwelling owner. If you are installing the AECV, you will be responsible for correct positioning and labelling. See **Part 10 – Emergency notices, warning labels and reporting forms**

Figure 9.4 Emergency control valve positions inside dwellings

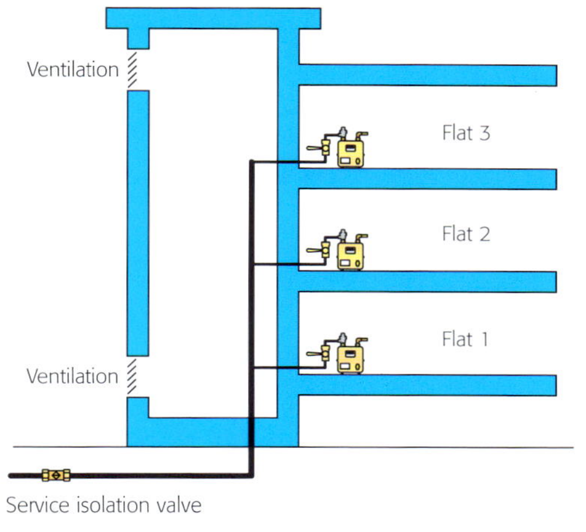

1. Each individual gas meter installation incorporates an emergency control valve (ECV).
2. Gas pipework fire stopped as it passes through the wall into each individual flat.
3. The service into the property will have a service isolation valve (SIV).
4. Air vents communicating with outside air.

Gas meters located inside each dwelling – what you must do

If a gas meter is located inside each dwelling, situate the ECV as shown in Figure 9.4 ensuring:

- it is as near as is practicable to the point where the gas supply enters the individual dwelling, either internally or externally
- it is in a readily accessible position
- it is fitted so that the lever (handle) of the valve is parallel to the pipe to which it is connected when it is in the open position – and only moves downwards one quarter of a turn (90°) to the off position
- the lever is securely attached to the valve spindle
- The meter installation shall incorporate a pressure test point fitted in the meter outlet boss or meter outlet liner
- it is labelled with the on/off position of the valve and gas escape emergency procedure (see **Part 10 – Emergency notices, warning labels and reporting forms**)

Figure 9.5 Position of emergency control valves

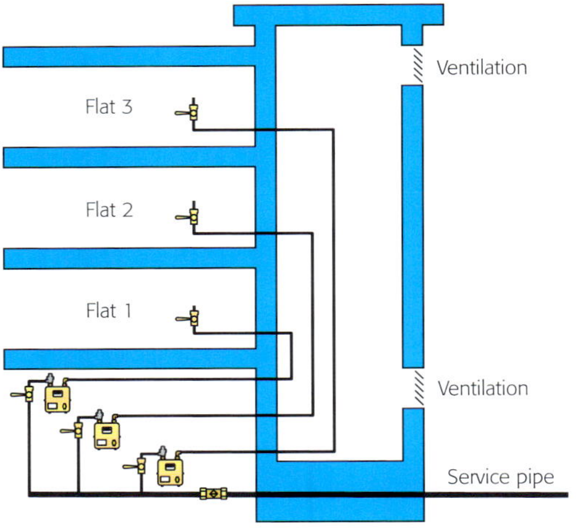

1. Each flat will have an additional emergency control valve (AECV) and a test point.
2. Each meter installation will incorporate an emergency control valve (ECV).
3. The service into the property will have a service isolation valve (SIV).
4. Pipes to be fire stopped where they pass into and out of the protected shaft.

Gas meters located remotely from dwellings: how to site both an ECV and an AECV

For gas meters located remotely from dwellings, site an ECV at the meter and an AECV should be situated as shown in Figure 9.5 ensuring:

- it is as near as is practicable to the point where the gas supply enters the individual dwelling, either internally or externally
- it is in a readily accessible position

Note: An appropriate height for an AECV is between 450mm and 1450mm above floor height.

- be easy to operate
- the lever (handle) of the valve is parallel to the pipe to which it is connected when it is in the open position
- the lever should only move downwards one quarter of a turn (90°) to the off position
- the lever is securely attached to the valve spindle
- there is a test point fitted downstream of the AECV (within 300mm)
- it is labelled with the on/off position of the valve
- it is labelled with a suitable notice giving advice on what to do in an emergency

See **Part 10 – Emergency notices, warning labels and reporting forms**.

Medium pressure fed installations: how to site additional meter inlet valve (MIV)

On medium pressure fed installations, you will normally install an additional valve downstream of the ECV, situated between the outlet of the medium to low pressure regulator and the inlet connection of the gas meter.

This is known as a meter inlet valve (MIV) or 'test valve.' It is not normally fitted with a lever, as it is designed for you to use when you test for gas tightness.

Note: This valve is not intended for use by the gas user.

Appliance isolation: different valves required for different appliances

The GSIUR require you to fit an isolating valve when installing a gas appliance. This valve allows the appliance to be isolated from the gas supply during maintenance and servicing, without turning off the whole gas supply.

The isolating valve depends on the appliance.

Gas fires and space heaters

A fire restrictor elbow or thumb cock is fitted into the inlet pipework adjacent to the appliance.

Gas cookers and tumble dryers

A bayonet connection is normally used. This is a self-sealing connector, which allows the gas user to disconnect the appliance for cleaning purposes. It is usually fitted adjacent to the appliance and positioned so that its outlet connection is facing downwards. This connection also acts as an appropriate fitting to seal a gas supply.

A rigid pipe connection to a hotplate/oven/grill should have an isolation valve and means of disconnection (via a union type joint located between the isolation valve and the appliance).

Central heating boilers, water heaters and warm air heaters

Isolating valves are fitted adjacent to the appliance, usually as part of the appliance associated controls.

Leisure appliances

Most leisure appliances are connected using a bayonet connection. These may be installed in wall-mounted micro-point boxes. See **Part 5 – Installation of pipework and fittings**.

For guidance on the operation and positioning of ECVs and isolation valves on LPG installations see **Part 18 – LPG – General requirements**.

Emergency notices, warning labels and forms – 10

10 – Emergency notices, warning labels and forms

Figures

Introduction

During the course of your work, you may encounter appliances/installations that require you to attach a Warning label or issue a Notice or Reporting Form. We explain here how these are used both to provide guidance for gas users and help you fulfil your mandatory duties.

If the appliance/installation is:

- its continued use would classify it to be 'Immediately Dangerous' (ID), or
- 'At Risk' (AR)

the appropriate notices/labels you can use best inform the responsible person about this. See also **Part 8 – Gas Industry Unsafe Situations Procedure.**

To fulfil your mandatory duties you must:

- inform the responsible person
- disconnect the appliance, or make safe a gas installation/appliance as appropriate
- use the relevant Warning label, Notice or Reporting Form

Labels, notices or forms can help you to demonstrate that you have fulfilled your mandatory duties.

It is not enough for you just to identify an 'ID' or 'AR' situation and verbally inform the responsible person; They may ignore your advice, or other gas users may be unaware of the danger.

CORGI*direct* warning notices, labels and forms – help users and operatives

It can be difficult and expensive for individuals, or small registered businesses, to provide warning notices, labels and forms. An additional concern is that the wording and safety checks identified may not cover current legislation.

To overcome these problems, CORGI*direct* has a range of inexpensive labels (in self-adhesive or tie-on form) and also notices and forms (available in pads and printed on 'no carbon required' (NCR) material) to meet these requirements (see **Part 22 – CORGI*direct* Publications**).

Use them to highlight the defect(s) and notify the responsible person. They also have a section for the responsible person to sign, accepting notification of work carried out, or of a defect(s).

The added advantage is you can issue the notice or form whilst still on-the-job, allowing you to leave a copy with the responsible person whilst retaining a copy for the office files (see Note).

Note: You may use any system that delivers the same level of safety, e.g. the use of a combined 'Do Not Use' label/Warning Notice. It may be particularly relevant if you use electronic recording systems.

Labels and notices

Warning 'Danger Do Not Use' label and notices– what you must do

For 'ID' and 'AR' situations, attach a warning label (WL1) to the appliance/installation, warning against further use, see Figures 10.1 at the end of this Part.

The gas industry agreed to standardise the design of warning labels, used by gas operatives when dealing with 'ID' and 'AR' situations in accordance with the current GIUSP.

The agreed format relates to the actual size and shape of a single ' Danger Do Not Use' label (see Figure 10.1)

It is believed that having a standardised approach for warning labels across industry will assist gas operatives, GTs, ESPs and customers to easily identify that a problem exists with a gas appliance/system and for 'ID' situations, provides additional guidance to an operative whereby an appliance maybe subject to a Health and Safety Executive (HSE) investigation.

To further assist gas operatives, CORGI*direct* designed the labels to be either tie-on or affixed (stuck to), by removing the backing, to an appliance/system.

Back this action up by completing a Warning Notice 'Danger Do Not Use' – see Figure 10.10 at the end of this Part – listing the fault(s) and identifying the responsible person.

Invite the responsible person to sign the notice, to accept responsibility for the faulty installation.

Retain a copy of the notice for the office file (see **Warning Notice – your duty to notify** at the end of this Part).

Gas meter labels – give emergency/other details

Primary meters – display notice as follows:

The GSIUR require all primary gas meters to have a suitably worded notice in permanent form prominently displayed on or near them, indicating:

1. What to do in the event of a gas escape.
2. The National Gas Emergency Service Call Centre:
   - in England, Scotland and Wales, contact the National Gas Emergency Service Call Centre, on telephone number 0800 111 999.

     In the case of Liquefied Petroleum Gas (LPG), contact the gas supplier, whose details can be found on the bulk storage vessel or cylinder (if no label can be found, information can be found under GAS in the local telephone directory)
   - in Northern Ireland contact Phoenix Natural Gas on 0800 002 001, or the LPG supplier
   - in the Isle of Man contact Manx Gas Ltd for all areas (including LPG) on 0808 1624 444
   - in the Channel Islands contact either Guernsey Gas Ltd on 01481 749000 or Jersey Gas Company Ltd. on 01534 755555 as appropriate

Note: A typical Notice or Label commonly used is shown in Figure 10.2 in this Part. See also Part 16 – Domestic gas meters.

Secondary meters – display notice as follows:

1. When you install a secondary meter (see **Part 16 – Domestic gas meters**), GSIUR require you to label it 'Secondary meter' and prominently display a notice, in permanent form, on or near the primary meter, indicating:
    - the number and location of all secondary meters (see Figure 10.3 at the end of this Part)
    - what to do in the event of a gas escape (see Figure 10.2 at the end of this Part)
    - the National Gas Emergency Call Centre (see information under **Primary meters – display notice as follows:** and Figure 10.2 at the end of this Part)
2. The person supplying or permitting the supply of gas through a primary meter to a secondary meter is also responsible for ensuring that a line diagram in permanent form is prominently displayed:
    - on or near the primary meter or gas storage vessel, and
    - on or near all emergency controls connected to the primary meter. This should show the configuration of all meters, installation pipework and ECV/AECVs

Any person who changes the configuration of any meter, installation pipework or ECV/AECVs needs to ensure the line diagram is amended to show this alteration.

ECV – for emergency/maintenance work

Every gas installation must have an ECV to allow the gas installation to be turned off in an emergency or for maintenance work.

- where the valve is installed adjacent to a gas meter it is known as the ECV
- where the primary meter is remote from the premises, as found in some flats, you need to install an AECV where the gas supply pipe enters the property

LPG bulk storage vessel and multiple cylinder installations should be fitted with an ECV, located externally where the supply enters the premises.

Attach a label clearly identifying the valve and marked to show its open and closed positions and its direction of closure (see Figure 10.4 at the end of this Part).

You must identify ECVs – so that their function is obvious to the user

Site a label, in permanent form, in a prominent position adjacent to the valve (see Figure 10.5 at the end of this Part) Indicating:

1. What to do in the event of a gas escape.
2. The National Gas Emergency Call Centre (see information under **Primary meters – display notice as follows:** in this Part).
3. Your name, your registration number and the date.

The GSIUR also require that where you install an ECV:

- you clearly mark the means of operating the lever or prominently display a notice in permanent form nearby, to indicate when the control is open and when the control is shut (see Figure 10.4 at the end of this Part)
- when it is not adjacent to the primary meter installation, you must prominently display (on or near the ECV) a suitably worded notice, in permanent form, indicating what to do in the event of a gas escape (see Figure 10.5 at the end of this Part)

Where a meter is installed in any premises at a distance of more than 2m from, or out of sight of, the nearest upstream ECV in the premises:

- you must not supply or provide gas for the first time through the meter unless you ensure that a suitably worded notice, in permanent form, is prominently displayed on or near the meter indicating the position of the ECV (see Figure 10.6 at the end of this Part)

Compartment ventilation label (WL8): warns about use/risks from storage

When you install a gas appliance(s) in a compartment, there is a risk that the compartment may later be used as a storage cupboard.

To avoid this potential fire hazard and to avoid the possibility of the ventilation grilles becoming restricted or blocked, label the compartment: to warn against such use.
(see Figure 10.7 at the end of this Part)

Balanced compartments: door must be closed (except for access)

For a balanced compartment to function correctly, it is important that all doors in the compartment walls are kept closed at all times. The door should be self-closing and fit tightly to its frame which should incorporate a draught sealing strip.

Attach a notice to the door or access cover stating that the door needs to be kept closed except for access.

Electrical bonding label (WL9)

Some types of electrical installations are fitted with Equipotential Bonding, which is the connection of the internal gas and water pipes to the electrical installation's earth terminal. By law, installations with Protective Multiple Earthing (PME) must be fitted with this. See **Part 5 – Gas installation pipework and fittings** and **Part 16 – Domestic gas meters**.

The GSIUR require any person who connects any installation pipe to a primary meter installation where Equipotential Bonding is necessary, to inform the responsible person that this needs to be carried out by a competent person. The WL9 label meets this requirement (see Figure 10.8 at the end of this Part).

If you are competent to carry out this work, make the connection on the customer's side of the meter – in a position where it can be easily observed, up to a maximum distance of 600mm from the meter outlet connection and before any branch in the installation pipework (see **Part 5 – Installation of pipework and fittings** and **Part 16 – Domestic gas meters**).

The connection to the gas pipe should have a warning label stating 'Safety electrical connection. Do not remove'.

Note: These requirements apply to both Natural gas and LPG installations.

Chimney/Hearth Notice Plate (CP3PLATE)

The Chimney/Hearth notice plate (See Figure 10.9 at the end of this Part) has been developed in conjunction with the CP3 form (see Figure 10.17 at the end of this Part) and conforms with the same standards.

In addition to the CP3 form mentioned above you must:

- provide a notice plate – detailing the performance capabilities and what that particular chimney system has been installed/designed for
- securely attach the notice plate to the fabric of the building in an unobtrusive yet obvious location – normally adjacent to the electricity consumer unit, mains water supply stop-valve or the specific chimney it relates to
- add the appropriate details to the notice plate in a permanent form

Note: Similar regulatory requirements apply also in other geographical areas of the UK.

See also **Chimney, Fireplace and Hearth Commissioning Record (CP3FORM)** at the end of this Part.

Forms

Warning Notice -Danger Do Not Use (CP14): your duty to notify

The principal warning notice for gas operatives is shown at the end of this Part (see Figure 10.10).

Use this notice to allow you to carry out your duties:

- to bring any defect(s) to the attention of the responsible person when you identify an 'ID' or 'AR' appliance/installation
- to back up this verbal advice with written confirmation that can be kept for possible future reference
- to ask the responsible person to sign the completed 'Warning Notice - Danger Do Not Use' as a record of receipt.

Landlord/Home Owner Gas Safety Record (CP12): to meet regulations

Use this form to provide a landlord with a record as required by Regulation 36 'Duty of Landlords' of the GSIUR. It covers:

- the date of inspection
- details of the appliance(s) checked for safety
- faults identified
- what remedial action, if any, is required to correct the fault(s) and make the appliance/installation safe for further use

See Figure 10.11 at the end of this Part.

You can also use this form as a safety record for home owners who may want a safety check carried out on the gas appliances within their home.

Note: How to complete this form is dealt with in Part 17 – Landlord's Gas Safety Record.

Leisure Industry Gas Safety Record (CP2): specifically for Leisure Accommodation Vehicles (LAVs) and Boats rented out to the public

This form fulfills a similar role to that of the Landlord's Gas Safety Record (CP12), but is specifically designed, both:

- to meet the needs of the leisure industry and the GSIUR; and
- to help you when carrying out Landlord's gas safety checks on LAVs and boats which are rented out to the public (see Figure 10.12 at the end of this Part)

Note: How to complete this form is dealt with in Part 17 – Landlord's Gas Safety Record.

Gas Safety Inspection form (CP4): for reports on condition of appliances/installations to a private householder/house surveyor

This is similar to the Landlord's Gas Safety Record, but is designed for gas operatives to provide a report on the condition of the appliance/installation to private householder or house surveyor (See Figure 10.13 at the end of this Part).

Note: Do not use this form as a substitute for a Landlord's Gas Safety Record, as it does not hold enough information to fulfil the requirements of the GSIUR.

Service/Maintenance Check List (CP6) – helps you check systematically

This is designed to record customer/appliance details and helps when you carry out all necessary checks when servicing/maintaining a gas appliance (see Figure 10.14 at the end of this Part).

The form is suitable for use with the "Benchmark" scheme for service and maintenance (see **Part 19 – Definitions**).

Gas Safety Record (CP1) – multi-purpose record you can use

You can use this multi-purpose CP1 (see Figure 10.15) instead of the CP4 or CP6.

Note: Although the form has similar information to that of the Landlord's Gas Safety Record (CP12), it doesn't include everything that the CP12 does. Therefore, where a landlord gas safety inspection is being undertaken you are recommended to use a CP12 instead of a CP1.

Visual Risk Assessment of Gas Appliances (CP9)

The forms purpose is to record, on a basic level, the often automatic 'checklist' of safety points when working on other gas appliances within a property.

For example, you may pass a gas fire on your way to work on a boiler and note signs of staining around the fire.

You haven't worked on the fire, but under the GSIUR you have a 'duty of care' which will require you to visually inspect the fire for any obvious safety defects centred around: flue, ventilation, signs of distress, stable and secure and location.

Although these checks are visual only and you are not required to record them, the form provides a mechanism for doing just that (see Figure 10.16 at the end of this Part).

Chimney, Fireplace and Hearth Commissioning Record (CP3FORM)

CORGI*direct* has developed The Chimney, Fireplace and Hearth Commissioning Record (see Figure 10.17 at the end of this Part) to enable gas installing businesses to conform to the requirements of Building Regulations/Standards.

The Building Regulations in England and Wales – Approved Document J (ADJ) requires that when:

- you install new open-flues, fireplaces or hearths; or
- modify existing chimney systems –

    you provide the client/gas user with a commissioning record (checklist) detailing the work carried out.

The commissioning record can also be used by the client/gas user as a means of confirming to the local Building Control Body that the work carried out conforms to Building Regulations requirements.

The checklist on the commissioning record is also a useful aid for you to ensure that you have undertaken all necessary checks (see also **Chimney/Hearth Notice plate (CP3 Plate)** in this Part).

Note: Similar building regulatory requirements also apply in other geographical areas of the United Kingdom.

Gas testing and purging – domestic (NG) (CP32)

The form enables you to record your findings in relation to the tightness testing of domestic gas installations. See **Part 6 – Testing for gas tightness (Natural gas), Part 15 – Re-establishing gas supply/re-lighting appliances – Purging** and Figure 10.18 at the end of this Part.

Fumes investigation report (CP26)

This report form will help support the suite of British Standards – BS 7967: 2015 in regards to the use of portable Electronic Combustion Gas Analysers (ECGAs) and the recording of findings where a customer reports a suspected escape of fumes.

See **Part 3 – Characteristics of combustion – Industry Standards for using flue gas analysers** and Figure 10.20a and 10.20b at the end of this Part.

Additionally, refer to the CORGI*direct* manual entitled 'Using Portable Electronic Combustion Gas Analysers for Investigating Reports of Fumes' – see **Part 22 – CORGI*direct* Publications**.

Existing concealed room-sealed fanned-draught boiler chimney assessment (CP43)

This form, which has built upon previous work by industry has been further developed by Gas Safe Register and the wider construction industry in response to the problem of chimney systems being installed within voids without sufficient access for inspection being installed.

The form, which is a risk assessment of the installation in question needs to be completed in conjunction with and only after reading Gas Safe Register's TB 008 (Edition 3), which is freely available to registered gas businesses/installers at https://engineers.gassaferegister.co.uk

See **Part 13 – Chimney standards – Room-sealed fanned draught chimney systems concealed within voids** and Figure 10.19 at the end of this Part.

Note: The third edition of this bulletin was updated in April 2013, which replaced the 'model' form with a process flowchart. CORGI*direct* has therefore, altered its form to reflect the process flowchart approach.

Figure 10.1 Danger ' Do Not Use' Label (WL1)

Figure 10.2 Typical meter label

IN THE EVENT OF AN ESCAPE
Turn off the supply at the control valve. Open windows.
Do NOT search with naked light and if gas escape persists
IMMEDIATELY CONTACT THE NATIONAL GAS EMERGENCY SERVICE CALL CENTRE THE TELEPHONE NUMBER FOR GAS ESCAPES CAN BE FOUND UNDER GAS IN YOUR LOCAL BRITISH TELECOM TELEPHONE DIRECTORY
Do NOT reopen the supply until remedial action has been taken by a competent person to prevent gas escaping.

Date................

Figure 10.3 Secondary meter label

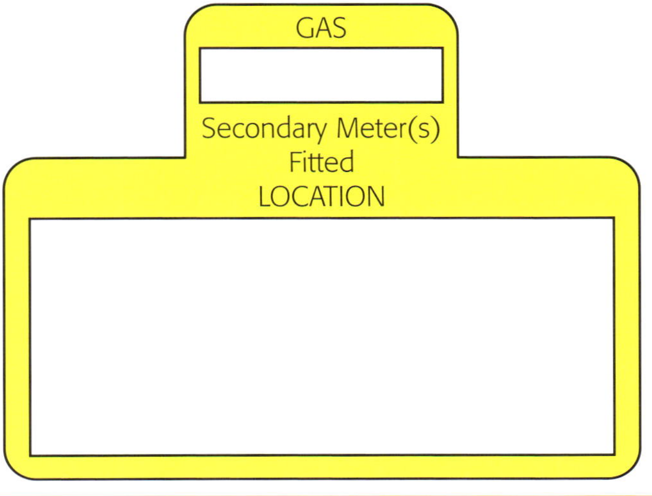

Figure 10.4 Emergency Control Valve for both NG and LPG

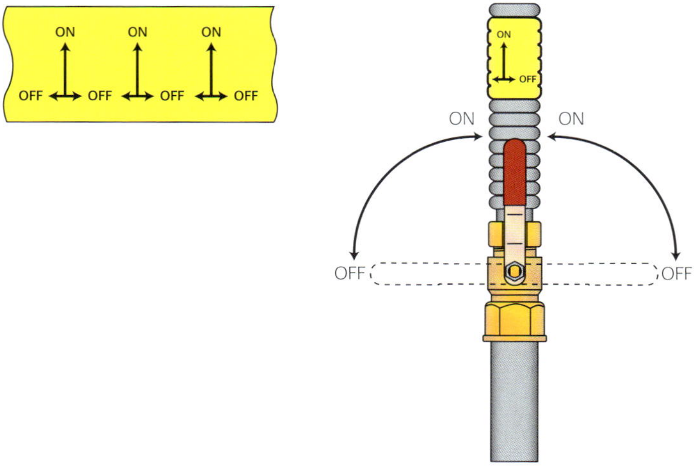

Figure 10.5 Emergency Control Valve label (WL5)

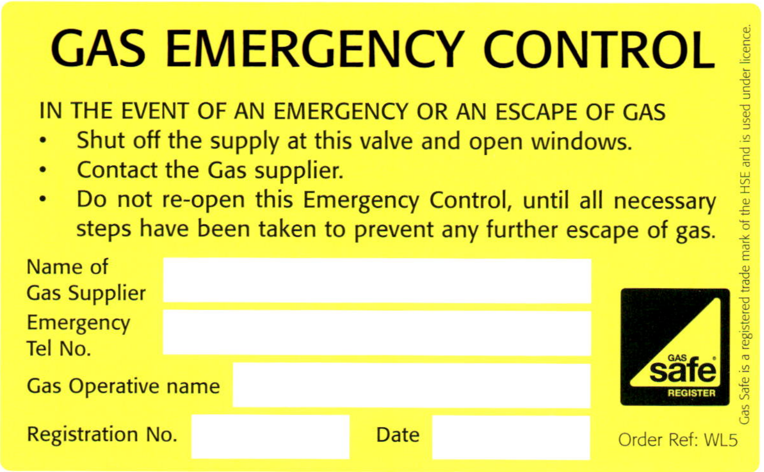

GAS EMERGENCY CONTROL

IN THE EVENT OF AN EMERGENCY OR AN ESCAPE OF GAS

- Shut off the supply at this valve and open windows.
- Contact the Gas supplier.
- Do not re-open this Emergency Control, until all necessary steps have been taken to prevent any further escape of gas.

Name of Gas Supplier

Emergency Tel No.

Gas Operative name

Registration No. Date

GAS safe REGISTER

Order Ref: WL5

Gas Safe is a registered trade mark of the HSE and is used under licence.

Figure 10.6 Gas Emergency Control valve (ECV) label

THE GAS
EMERGENCY CONTROL
IS LOCATED

Figure 10.7 Ventilation label (WL8)

IMPORTANT

FOR YOUR SAFETY

- Keep the compartment door CLOSED at all times, except for access
- DO NOT block or restrict any air grilles or louvres in the walls, door, floor or ceiling of this compartment
- DO NOT use this compartment for storage, airing or drying, unless there is a purpose-provided partition separating the appliance and/or chimney from stored articles

Gas Operative's name:

Registration Number:

Order Ref: WL8

Gas Safe is a registered trade mark of the HSE and is used under licence.

Figure 10.8 Electrical bonding label (WL9)

ELECTRICAL BONDING

Most electrical Installations are required to be fitted with a main protective bonding conductor, which consists of the electrical connection of the internal gas and water pipes to the installation's main earthing terminal in accordance with the requirements of BS 7671, *Requirements for electrical instalations* (IET Wiring Regulations).

The gas installation pipework fitted in your premises does not appear to have the necessary main protective bonding conductor.

I am required by the Gas Safety (installation and Use) Regulations 1998 to tell you that any necessary main protective equipotential bonding should be carried out by a competent person.

I advise you to have this checked by a competent electrical operative.

If you are the tenant of this property, please bring this matter to the attention of the owner or their agent as soon as possible.

Order Ref: WL9
CORGI*direct* February 2016

Gas Safe Register is a registered trade mark of the HSE and is used under licence.

Figure 10.9 Chimney/Hearth Notice plate (CP3PLATE)

CHIMNEY/HEARTH NOTICE PLATE

Property address:	
Chimney/hearth installed in the:	
Is suitable for:	
Chimney liner (type/diameter):	
Suitable for condensing mode:	
Installed date:	
Installer name/Registration No.	

Ref. CP3PLATE

IMPORTANT SAFETY INFORMATION
THIS DATA PLATE MUST NOT BE REMOVED OR COVERED

Gas Safe is a registered trade mark of the HSE and is used under licence.

Figure 10.10 Warning Notice - Danger Do Not Use (CP14)

To confirm the validity of the gas operative please contact the Gas Safe Register on Tel: 0800 408 5500

Serial No.

WARNING NOTICE - DANGER DO NOT USE

This form should be completed in accordance with the requirements of the current Gas Industry Unsafe Situations Procedure

Registered Business Details	Job Address	Client / Landlord's Details (if different)
Reg No:	Name:	Name:
Company:	Address:	Address:
Address:		
Postcode:	Postcode:	Postcode:
Tel No:	Tel No:	Tel No:

1. THE GAS APPLIANCE / GAS INSTALLATION (delete as appropriate, see also boxes 2 to 5 for details)

Make: ______ Model: ______ Type: ______ Serial No: ______

Installation Details: (e.g. gas pipework, gas meter) ______

Fixed in the room / location: (if known) ______

2. IS UNSAFE AND HAS BEEN CATEGORISED IMMEDIATELY DANGEROUS (ID) because:

☐ A gas escape has been detected on the appliance / installation (delete as necessary if location of the gas escape is known) or

☐ ______ and

☐ with your permission it has been disconnected from the gas supply and a DANGER DO NOT USE label attached, or

☐ as you have refused to allow it to be made safe, a DANGER DO NOT USE label has been attached *

Gas Emergency Contact Centre Reference: ______

3. IS UNSAFE AND HAS BEEN CATEGORISED AT RISK (AR) because:

☐ ______

☐ with your permission it has been TURNED OFF and a DANGER DO NOT USE label attached or,

☐ as you have refused to allow it to be made safe, it has a DANGER DO NOT USE label attached

☐ turning off will not reduce the risk, please contact the following to undertake further investigation

Name: ______ Tel No: ______

4. RIDDOR:

☐ The unsafe situation(s) identified is reportable to the Health and Safety Executive (HSE) under RIDDOR

** PLEASE NOTE: Registered gas operatives are required by law to report cases where they are refused permission to disconnect an IMMEDIATELY DANGEROUS gas installation to the Gas Emergency Service Call Centre for Natural Gas, or for Liquefied Petroleum Gas (LPG), the supplier. All Gas Transporters (GTs) operate a gas emergency service and have powers under the Gas Safety (Rights of Entry) Regulations to visit properties and disconnect unsafe gas appliances / installations.*

5. REMEDIAL ACTION REQUIRED:

FOR YOUR OWN SAFETY 'IMMEDIATELY DANGEROUS' & 'AT RISK' GAS APPLIANCES / INSTALLATIONS SHOULD NOT BE USED

Gas User / Responsible Person Declaration: I confirm that I have received this Warning Notice concerning the safety of the gas installation. I understand that the use of the appliance / installation in the case of an **IMMEDIATELY DANGEROUS** or **AT RISK** installation, could present a danger and could place me in breach of the Gas Safety (Installation and Use) Regulations

Signed: ______ Print Name: ______

Date: ______

The gas user was not present at the time of the visit, the installation has been made safe and this notice left on the premises ☐

Gas Operative Declaration: I confirm that the situation(s) recorded above, has been identified and brought to the attention of the Gas User / Responsible Person in accordance with the Gas Safety (Installation and Use) Regulations (GSIUR), Industry Standards and Procedures.

Licence No: ______

Signed: ______ Print Name: ______

Issue Date: ______ Issue Time: ______

See overleaf for further guidance & definitions

Key: Top Copy – Gas User / Responsible Person Green Copy – Client / Landlord Yellow Copy – Gas Operative

To re-order quote Ref. CP14

Figure 10.11 Landlord/Home Owner Gas Safety Record (CP12)

To confirm the validity of the gas operative please contact Gas Safe Register on Tel: 0800 408 5500

Serial No. B 123456

LANDLORD/HOME OWNER GAS SAFETY RECORD

Gas Safe Register

This form allows the recording of the results of the required checks as defined by the Gas Safety (Installation and Use) Regulations.
The information recorded on this form does not confirm that the installation was installed by a Registered Installer or that the installation complies with any relevant Building Regulations.
Chimney systems were inspected visually and checked for satisfactory evacuation of products of combustion, a detailed internal inspection of the chimney system has not been carried out.

Registered Business Details REG NO ______	Job Address	Landlord (or where appropriate their agent)
Gas operative ______ (Print name)	Name (Mr/Mrs/Miss/Ms) ______	Name (Mr/Mrs/Miss/Ms) ______
Operative licence No. ______	Address ______	Address ______
Company ______		
Address ______	Postcode ______	Postcode ______ Tel No. ______
Postcode ______ Tel No. ______	Tel No. ______	Number of appliances tested ☐

APPLIANCE DETAILS

	Location	Appliance type	Make	Model	Landlord's appliance Yes/No/NA	Appliance inspected Yes/No	Flue type OF/RS/FL
1							
2							
3							
4							

INSPECTION DETAILS / **AUDIBLE CO ALARM**

	Operating pressure in mbar or heat input in kW	Initial combustion analyser reading (if applicable)	Final combustion analyser reading (if applicable)	Safety device(s) correct operation Yes/No/NA	Ventilation provision satisfactory Yes/No	Visual condition of chimney/termination satisfactory Yes/No/NA	Flue performance checks Pass/Fail/NA	Appliance serviced Yes/No	**Appliance safe to use Yes/No**	Approved CO alarm fitted Yes/No/NA	Is CO alarm in date Yes/No/NA	Testing of CO alarm satisfactory Yes/No/NA
1												
2												
3												
4												

DEFECT(S) IDENTIFIED	If Warning/Advice Notice issued insert serial No.*
1	
2	
3	
4	

Gas installation pipework satisfactory visual inspection Yes/No ☐
Emergency Control Valve (ECV) accessible Yes/No ☐
Satisfactory gas tightness test Yes/No/NA ☐
Protective equipotential bonding satisfactory Yes/No ☐

NEXT SAFETY CHECK DUE WITHIN 12 MONTHS

REMEDIAL ACTION TAKEN
1
2
3
4

This Safety Record issued by: Signed ______
Print Name: ______
Received by: Signed ______ Tenant/Landlord/Agent/Home Owner
Date appliance(s)/chimney(s) checked: ______

Key: Top Copy – Landlord/Managing Agent/Home Owner Green Copy – Tenant Yellow Copy – Registered Business * Refer to separate Warning/Advice Notice To re-order quote Ref. CP12

Figure 10.12 Leisure Industry Landlord's Gas Safety Record (CP2)

To confirm the validity of the gas operative please contact Gas Safe Register on Tel: 0800 408 5500

Serial No. 123456

LEISURE INDUSTRY LANDLORD'S GAS SAFETY RECORD

NEXT SAFETY CHECK DUE WITHIN 12 MONTHS

This inspection is for gas safety purposes only in accordance with The Gas Safety (Installation and Use) Regulations. Flues were inspected visually and checked for satisfactory evacuation of products of combustion. A detailed internal inspection of the flue integrity, construction and lining has not been carried out.

Registered Business Details REG NO ____

Gas operative ____ (Print name)

Operative licence No. ____

Company ____

Address ____

Postcode ____ Tel No. ____

Job Address

LAV serial No./Boat Name ____

Pitch/Mooring Number ____

Park/Marina ____

Address ____

Postcode ____ Tel No. ____

Landlord (or where appropriate their agent)

Name (Mr/Mrs/Miss/Ms) ____

Address ____

Postcode ____ Tel No. ____

No. of appliances tested ☐

APPLIANCE DETAILS

	Location	Appliance type	Make	Model	Landlord's appliance Yes/No/NA	Appliance inspected Yes/No	Flue type OF/RS/FL
1							
2							
3							
4							

INSPECTION DETAILS / **AUDIBLE CO ALARM**

	Operating pressure in mbar or heat input in kW	Initial combustion analyser reading (if applicable)	Final combustion analyser reading (if applicable)	Safety device(s) correct operation Yes/No/NA	Ventilation provision satisfactory Yes/No	Visual condition of chimney/termination satisfactory Yes/No/NA	Flue performance checks Pass/Fail/NA	Appliance serviced Yes/No	**Appliance safe to use Yes/No**	Approved CO alarm fitted Yes/No/NA	Is CO alarm in date Yes/No/NA	Testing of CO alarm satisfactory Yes/No/NA
1												
2												
3												
4												

	DEFECT(S) IDENTIFIED	**If Warning/Advice Notice issued insert serial No.***
1		
2		
3		
4		

	REMEDIAL ACTION TAKEN
1	
2	
3	
4	

Cylinder/final connection hoses to LAV/boat satisfactory Yes/No ☐
Gas installation pipework (visual inspection) satisfactory Yes/No ☐
Gas tightness test satisfactory Yes/No/NA ☐
ECV accessible and operable Yes/No ☐
LPG regulator operating pressure (mbar) ☐
LPG regulator lock-up pressure (mbar) ☐

Safety Record issued by: ____

Print Name: ____

Received by: ____ Tenant/Landlord/Agent

Date appliance(s)/flue(s) checked: ____

Key: Top Copy – Landlord/Managing Agent **Green Copy** – Tenant **Yellow Copy** – Registered Business

* Refer to separate Warning/Advice Notice

To re-order quote Ref. **CP2**

Figure 10.13 Gas Safety Inspection form (CP4)

To confirm the validity of the gas operative please contact Gas Safe Register on Tel: 0800 408 5500

Job No:

Invoice no:

GAS SAFETY INSPECTION

This form is not to be used as a Landlord's Gas Safety Record and the details recorded below do not confirm that the installation was installed by a Registered Installer or that the installation complies with Building Regulations.

Registered Business Details (REG NO)

Company

Gas Operative Licence No.

Address

Postcode Tel No.

Job Address

Name (Mr/Mrs/Miss/Ms)

Address

Postcode Tel No.

Rented accommodation state Yes or No

Gas Installation

	Yes/No	DETAILS
Is the installation gas tight?		
Have the correct materials been used in the installation?		
Is the installation pipework correctly sized?		
Where appropriate, has protective electrical bonding been carried out?		

Emergency Control(s)

	Yes/No	DETAILS
Is the emergency control valve(s) correctly positioned/accessible?		
Is the emergency control valve(s) labelled?		

Appliance Details

Type	Make	Model	Flue type OF/RS/FL	Chimney condition and termination satisfactory Yes/No/NA	Flue performance checks Pass/Fail/NA	Is ventilation satisfactory? Yes/No	Operating pressure in mbars or heat input kW or Btu/h	Combustion analyser reading (if applicable)	**Appliance safe to use Yes/No**

Findings

	Yes/No		Yes/No
Is the gas installation safe for use?		Have warning labels been affixed?	
If No, issue a Warning/Advice Notice (insert serial No.)		Is any remedial work required?	

Details of remedial work required

Gas user signature

Print name

Date

Gas operative signature

Print name

Date

Key: Top Copy – Gas User. Green Copy – Gas Operative

To re-order quote Ref. CP4

Figure 10.14 Service/Maintenance Check List (CP6)

To confirm the validity of the gas operative please contact Gas Safe Register on Tel: 0800 408 5500

Job No:

Invoice no:

SERVICE/MAINTENANCE CHECK LIST

Gas Safe Register

This form should not be used as a Landlord's Gas Safety Record as it does not hold sufficient information

Registered Business Details ____________ (REG NO)
Company ____________
Address ____________
Postcode ________ Tel No. ________
Gas tightness test carried out Yes or No
Gas tightness test result Pass or Fail

Job Address
Name (Mr/Mrs/Miss/Ms) ____________
Address ____________
Postcode ________ Tel No. ________
Rented accommodation Yes or No
Type of work: Service or Maintenance

APPLIANCE DETAILS		COMMENTS
Type		
Make		
Model		
Location		

COMBUSTION GAS ANALYSIS INFORMATION

Initial combustion analyser reading, where applicable | Final combustion analyser reading, where applicable

In relation to the initial combustion analyser reading above, has a full strip and clean service been undertaken? YES ☐ NO ☐

SAFETY CHECKS (see overleaf)	YES	NO	N/A	REMEDIAL ACTION/NATURE OF DEFECT
Ventilation satisfactory				
Chimney termination satisfactory				
Flue flow test satisfactory				
Spillage test satisfactory				
Safety device(s) satisfactory				
Operating pressure (mbar) or heat input (kW or Btu/h)				

APPLIANCE CHECKS (satisfactory)	YES	NO	N/A	REMEDIAL ACTION/NATURE OF DEFECT (see overleaf)
Burner/injectors				
Heat exchanger				
Ignition				
Electrics				
Controls				
Gas/water leaks				
Gas connections				
Seals (appliance case etc.)				
Gas pipework				
Fan(s)				
Fireplace opening/void				
Closure plate				
Flame picture				
Location				
Stability				
Return air/plenum				

FINDINGS	YES	NO
Is the Appliance/Installation safe for use?		
If No, issue a Warning/Advice Notice (record Serial No. of form) and attach a Warning Label; Insert Serial No.		
The following remedial work is required (see 'Findings' overleaf):		

Gas User signature ____________
Print name ____________
Date ____________

Gas operative signature ____________
Print name ____________ Date ________
Operative licence No. ____________

Key: **Top Copy** – Gas User **Green Copy** – Gas Operative

To re-order quote Ref. CP6

Figure 10.15 Gas Safety Record (CP1)

To confirm the validity of the gas operative please contact Gas Safe Register on Tel: 0800 408 5500

Customer Ref No.

GAS SAFETY RECORD

GAS safe REGISTER

This form allows the recording of the results of the required checks as defined by the Gas Safety (Installation and Use) Regulations.
The information recorded on this form does not confirm that the installation was installed by a Registered Installer or that the installation complies with any relevant Building Regulations.
Chimneys/flues were inspected visually and checked for satisfactory evacuation of products of combustion. A detailed internal inspection of the chimney/flue integrity, construction and lining has not been carried out.

Registered Business Details REG NO ______
Gas operative ______ (Print name)
Operative licence No. ______
Company ______
Address ______
Postcode ______ Tel No. ______

Job Address
Name (Mr/Mrs/Miss/Ms) ______
Address ______
Postcode ______
Tel No. ______

Client Details (if different)
Name (Mr/Mrs/Miss/Ms) ______
Address ______
Postcode ______ Tel No. ______
Rented accommodation Yes / No ☐
No. of appliances tested ☐

APPLIANCE DETAILS

	Location	Appliance type	Make	Model	Landlord's appliance Yes/No/NA	Appliance inspected Yes/No	Flue type OF/RS/FL
1							
2							
3							
4							

INSPECTION DETAILS

	Operating pressure in mbar or heat input in kW	Initial combustion analyser reading (if applicable)	Final combustion analyser reading (if applicable)	Safety device(s) correct operation Yes/No/NA	Ventilation provision satisfactory Yes/No	Visual condition of chimney/termination satisfactory Yes/No/NA	Flue performance checks Pass/Fail/NA	Appliance serviced Yes/No	**Appliance safe to use Yes/No**
1									
2									
3									
4									

	DEFECT(S) IDENTIFIED	If Warning/Advice Notice issued insert serial No.*
1		
2		
3		
4		

	REMEDIAL ACTION TAKEN
1	
2	
3	
4	

Details of work carried out (e.g. Gas safety check/record, Service etc.)

Gas installation pipework satisfactory visual inspection Yes/No ☐
Emergency Control Valve (ECV) accessible Yes/No ☐
Satisfactory gas tightness test Yes/No/NA ☐
Protective equipotential bonding satisfactory Yes/No ☐

Safety Record issued by: ______
Print Name: ______
Received by: ______ Client/Agent/Home Owner
Date appliance(s)/chimney(s) checked: ______

Key: Top Copy – Gas User/Responsible Person Second Copy – Client Third Copy – Gas Operative

* Refer to separate Warning/Advice Notice

To re-order quote Ref. CP1

Figure 10.16 Visual Risk Assessment of Gas Appliances (CP9)

To confirm the validity of the gas operative please contact Gas Safe Register on Tel: 0800 408 5500

VISUAL RISK ASSESSMENT OF GAS APPLIANCES

Gas safe REGISTER

The purpose of this form is to provide registered gas operatives a mechanism for the recording of visual risk assessment ONLY of gas appliances that they come in to contact with whilst engaged in other gas related work. The appliances documented HAVE NOT been tested/serviced or otherwise worked on, but have been visually assessed in accordance with Regulation 26 of The Gas Safety (Installation and Use) Regulations (GSIUR).

Registered Business Details ____________ (REG NO)	Landlord/Managing Agent (where appropriate)
Company:	Name/Company:
Gas Operative Licence No:	Address:
Address:	
Postcode: Tel No:	Postcode: Tel No:

Job Address	Description of Gas Work Undertaken –
Name (Mr/Mrs/Miss/Ms):	
Address:	
Postcode: Tel No:	

Gas Appliances Installed

Whilst engaged in gas work described above, how many other gas appliances installed in the property have you encountered/required to visually examine for safety as per Regulation 26 of The Gas Safety (Installation & Use) Regulations: ☐ (complete Appliance Details below)

Appliance Details

	Type	Make	Mode	Flue Type (OF/RS/FL)
1.				
2.				
3.				
4.				

Visual Inspection Details

	Flueing Satisfactory? (Yes/No/NA)	Ventilation Satisfactory? (Yes/No)	Signs of Distress? (Yes/No)	Stable & Secure? (Yes/No)	Location Satisfactory? (Yes/No)	Flame Picture Satisfactory? (Yes/No/NA)
1.						
2.						
3.						
4.						

Observations/Comments (as and where appropriate)		If Warning/Advice Notice issued insert serial No.*
	1.	
	2.	
	3.	
	4.	

I confirm that during the course of other gas works described on this form that I have visually inspected **ONLY** those gas appliances documented for safety as required by the GSIUR.	Print name: Gas operative signature: Date:

Key: Top Copy – Gas User (PTO). Green Copy – Gas Operative

To re-order quote Ref. CP9

Figure 10.17 Chimney, Fireplace and Hearth Commissioning Record (CP3FORM)

To confirm the validity of the gas operative please contact the gas registration body.

Customer Ref:

CHIMNEY, FIREPLACE AND HEARTH COMMISSIONING RECORD

GAS safe REGISTER

This form complies with the requirements of relevant Building Regulations/Standards

Registered Business Details REG NO ____________
Gas operative ____________ (Print name)
Operative licence No. ____________
Company ____________
Address ____________
____
Postcode ____________ Tel No. ____________

Building/Site Address
Name (Mr/Mrs/Miss/Ms) ____________
Address ____________
____
____
Postcode ____________ Tel No. ____________
Customer's signature ____________

Client/Landlord Details (if different)
Name (Mr/Mrs/Miss/Ms) ____________
Address ____________
____
____
____
Postcode ____________ Tel No. ____________

General chimney, fireplace and hearth details	
Location of chimney, fireplace or hearth	
Application capability (gas/oil/solid fuel/all)	
Intended type of appliance (state type or make where possible)	
If open fire give finished fireplace dimensions	
Recommended ventilation requirements	
State type of ventilation and free area required	
Hearth construction (new or existing)	
Chimney inspection/testing	
Satisfactory visual inspection Yes/No	
Chimney system swept Yes/No	
Satisfactory coring ball check Yes/No	
Satisfactory smoke test Yes/No	
Satisfactory appliance spillage test Yes/No* ** When necessary*	
Chimney, hearth notice plate fitted Yes/No	
Chimney, hearth notice plate location State	

Chimney construction	
Type/make	
New/existing	
Give internal flue dimensions (for natural draught appliances, provide the equivalent flue height where calculated)	
For clay/concrete chimney liners/blocks, confirm that all joints are socket end up and suitable jointing material is used	
For refurbished chimneys state type and make of liner used	
Details of chimney outlet including compliance criteria	
Details of any bends incorporated, together with their angles	
Details of provision for cleaning and frequency necessary	

I confirm that the work described in this record has been satisfactorily completed in accordance with Building Regulations/Standards, the current Gas Safety (Installation and Use) Regulations (GSIUR), appropriate standards and relevant manufacturer's/industry requirements.

Gas operative's signature ____________ **Date** ____________

Key: Top Copy – Customer/client Green Copy – Gas Operative

To re-order quote Ref. CP3FORM

Figure 10.18 Gas Testing and Purging - Domestic (NG) (CP32)

To confirm the validity of the gas operative please contact the gas registration body.

Customer Ref.

GAS TESTING AND PURGING – DOMESTIC (NG)

This form should be completed in accordance with the requirements of IGE/UP/1B (Edition 2)

Registered Business Details	Job Address	Client/Landlord's Details (if different)
Registered Business Details ______ REG NO	Job Address	Client/Landlord's Details (if different)
Gas operative ______ (Print name)	Name (Mr/Mrs/Miss/Ms) ______	Name (Mr/Mrs/Miss/Ms) ______
Operative licence No. ______	Address ______	Address ______
Company ______		
Address ______		
Postcode ______ Tel No. ______	Postcode ______	Postcode ______
	Tel No. ______	Tel No. ______

Installation details

Indicate new or existing installation:

Meter size/type e.g. U6, E6, G4, U16 or G10:

Medium pressure fed meter installation Yes/No ______ If Yes is a meter inlet valve (MIV) fitted:

Record maximum installation pipework diameter installed:

Calculate installation volume (where the installation volume is believed to be greater than 0.035m³)

Meter volume (m³) (see Table 2 overleaf)

Installation pipework volume (m³) (see Table 3 overleaf)

Total volume (m³) m³

Where the installation volume is greater than 0.035m³ the installation needs to be tightness tested in accordance with either IGE/UP/1A or IGE/UP/1

Tightness test(s) carried out

Reason for tightness test 1. ______
Let-by test carried out Yes/No ______ If Yes Pass/Fail ______
Indicate what is included in the tightness test: Meter ☐ Installation pipework ☐ Appliance(s) ☐
Permissible pressure loss ______mbar* Actual pressure loss ______mbar Tightness test Pass/Fail ______

Reason for tightness test 2. ______
Let-by test carried out Yes/No ______ If Yes Pass/Fail ______
Indicate what is included in the tightness test: Meter ☐ Installation pipework ☐ Appliance(s) ☐
Permissible pressure loss ______mbar* Actual pressure loss ______mbar Tightness test Pass/Fail ______

Reason for tightness test 3. ______
Let-by test carried out Yes/No ______ If Yes Pass/Fail ______
Indicate what is included in the tightness test: Meter ☐ Installation pipework ☐ Appliance(s) ☐
Permissible pressure loss ______mbar* Actual pressure loss ______mbar Tightness test Pass/Fail ______

Reason for tightness test 4. ______
Let-by test carried out Yes/No ______ If Yes Pass/Fail ______
Indicate what is included in the tightness test: Meter ☐ Installation pipework ☐ Appliance(s) ☐
Permissible pressure loss ______mbar* Actual pressure loss ______mbar Tightness test Pass/Fail ______

* *See Table 1.* ***Maximum permissible tightness test pressure losses*** *overleaf.*

Does the installation need to be purged Yes/No ______ If Yes proceed to **Purging details**

Purging details

Is a purge of the installation required? Yes/No

Calculate the installation purge volume below m³

For an E6/G4/U6 gas meter and installation pipework ≤ 28mm use 0.01m³ (0.35ft³)

For installations with E6/G4/U6 gas meters and installation pipework ≤ 35mm or G10/U16 meters with or without installation pipework, the purge volume needs to be calculated (see below)

A. Gas meter purge volume m³ (see Table 4. overleaf)

B. Installation pipework volume m³ x 1.5 (see Table 3. overleaf)

C. Total installation purge volume m³ = A + B (see Note) m³

Has purge been completed satisfactorily? Yes/No

Note: Where the calculated purge volume is greater than 0.02m³ the purged mixture must be ignited at a burner as soon as possible either at an installed appliance or a temporarily installed burner.

Installation general

Record the operating pressure at the outlet of the gas meter (mbar), where applicable

Is gas installation pipework satisfactory? (visual inspection) Yes/No

Is Emergency Control Valve (ECV)/Additional Emergency Control Valve (AECV) accessible? Yes/No

Is equipotential bonding satisfactory? Yes/No

Has installation/appliance(s) been commissioned/put back into operation as appropriate? Yes/No

Is gas installation safe for use? Yes/No

If No, has a Warning/Advice Notice been issued? Yes/No

If Yes give Warning/Advice Notice Serial No.

Check any disturbed joints e.g. pressure test point with a suitable Leak Detection Fluid (LDF) or gas leak detector.

This Gas Testing and Purging form is issued by: Signed ______

Received by: Signed ______ Tenant/Landlord/Agent/Home Owner

Date ______ The gas user was not present at the time of the visit ☐

Key: Top Copy – Gas User/Responsible Person Second Copy – Client/Landlord Third Copy – Gas Operative

To re-order quote Ref. CP32

Figure 10.19 Existing Concealed Room-Sealed Fanned-Draught Boiler Chimney Assessment (CP43)

Customer Ref.

EXISTING CONCEALED ROOM-SEALED FANNED-DRAUGHT BOILER CHIMNEY ASSESSMENT

This form should be completed in conjunction with Gas Safe Registers Technical Bulletin 008 (Edition 3):
Existing concealed room-sealed fanned-draught boiler chimney/flue systems in domestic premises - visit www.gassaferegister.co.uk

Registered Business Details (REG NO) ______

Gas operative ______ (Print name)
Operative licence No. ______
Company ______
Address ______
______
Postcode ______ Tel No. ______

Job Address

Name (Mr/Mrs/Miss/Ms) ______
Address ______
______
______
Postcode ______
Tel No. ______

Client/Landlord's Details (if different)

Name (Mr/Mrs/Miss/Ms) ______
Address ______
______
______
Postcode ______
Tel No. ______

Positively record in the boxes provided either a 'Y' (Yes) or 'N' (No) to the questions posed

Q1.
Is the chimney system routed through adjacent property? ☐ Y - Go to **Q2** N - Go to **Q3.**

Q2.
Can access be gained to the adjacent property? ☐ Y - Go to **Q3** N - Go to **Q4.**

Q3.
Are there acceptable means to examine the entire chimney system, i.e. Inspection hatches & CO alarms installed? ☐ Y - Go to ‡ N - Go to **Q5.**

‡ Subject to appropriate and satisfactory operational safety checks in accordance with manufacturer's instructions and GSIUR 26(9), the boiler can be deemed safe and left operational.

FINISH - sign and date form.

‡‡ Provided that; from what can be determined from the parts of the installation that can be examined and where there is NO evidence that the concealed chimney system is otherwise unsafe the boiler can be reasonably considered safe and left operational.

Subject to appropriate and satisfactory operational safety checks in accordance with manufacturer's instructions and GSIUR 26(9).

FINISH - sign and date form.

Q4.
Take all 'reasonable steps' to verify overall system integrity, e.g. by;
- Attempting to gain access.
- Arranging access with the Responsible Person for the building.

Where access cannot be gained provide relevant address information to Gas Safe's 'Flues in Voids' database - https://engineers.gassaferegister.co.uk/FluesInVoids.aspx

Does evidence exist indicating any historical issues relating to concealed chimney within the building/development and with no evidence of subsequent remedial action to correct the issue (answer in a.)?

a. ☐ Y - Does evidence constitute an ID (answer in b.)? N - Go to ‡‡

b. ☐ Y - Go to § N - Go to *

Q5.
Is a Carbon Monoxide Safety Shut-Off Void Monitoring (COSSVM) system installed – see Gas Safe's TB 008, Version 3 – (answer in a.)?

a. ☐ Y - Is the COSSVM installed and operating correctly (answer in b.)? N - Does the ceiling/enclosure indicate any signs of distress along the route of the chimney system which cannot be attributed to other causes, e.g. water leaks, etc. (answer in c.)?

b. ☐ Y - Go to ‡ N - Go to *

c. ☐ Y - Go to § N - Go to *

*** Classify as 'AR' in accordance with the current GIUSP. Carry out appropriate operational safety checks including combustion performance & confirm boiler is operating correctly. Affix appropriate Warning label, complete a Warning/Advice Notice and with the permission of the responsible person turn off the boiler. Recommend the installation of CO alarms/protection equipment compliant with BS EN 50291 and check the operation of any existing CO alarms.**

§ Classify as 'ID' in accordance with the current GIUSP, affix appropriate Warning Label and complete a Warning/Advice Notice. With the permission of the responsible person, disconnect and seal the gas supply to the boiler with an appropriate fitting. Complete an online RIDDOR 6(2) report at www.hse.gov.uk/riddor

I confirm that the assessment has been undertaken and the results have been brought to the attention of the Responsible Person in accordance with the Gas Safety (Installation and Use) Regulations, Industry Standards and Procedures.

Gas operative's signature ______
Date of issue ______

See overleaf for definitions & additional guidance

Key: Top Copy – Gas User/Responsible Person Green Copy – Client/Landlord Yellow Copy – Gas Operative

To re-order quote Ref. CP43

Figure 10.20a Fumes Investigation Report – Part 1 (CP26)

To confirm the validity of the gas operative please contact Gas Safe Register on Tel: 0800 408 5500

Customer Ref.

FUMES INVESTIGATION REPORT PART 1

This form should be completed in accordance with the current requirements of BS 7967

Registered Business Details
Reg No: ____
Company: ____
Address: ____
Postcode: ____
Tel No: ____

Job Address
Name: ____
Address: ____
Postcode: ____
Tel No: ____

Client / Landlord's Details (if different)
Name: ____
Address: ____
Postcode: ____
Tel No: ____

Gas users reason for concern

What is the gas user's reason for concern?			
When does it occur/is there a pattern?			
Is it occurring now?			
Which gas appliance(s) was in operation at the time?			
Was there any other fuel burning appliance(s) in operation at the same time?	Yes	No	
If Yes to above provide details			
Was anyone suffering from illness?	Yes	No	
If Yes to above provide details of individuals affected?			
If anyone was suffering from illness, what were the symptoms?			
Were there any abnormal weather conditions at the time?	Yes	No	Not Known
If Yes to above provide details			

Observations

A. What was the ambient CO reading (ppm) before entering the property?	ppm
B. What was the initial CO reading (ppm) upon entering the property?	ppm
C. If CO was recorded in B what was the highest CO reading (ppm) anywhere in the property	ppm
How many appliances were in operation on arrival at the property?	

List all fuel burning appliances in the property.	No.	Location	Type	Fuel type	Checked Yes/No	In operation on arrival at property?
• List gas appliances first.	1					
• *See Part 2 of this form to*	2					
record checks carried	3					
out on gas appliances.	4					

If alarm has been activated but CO not found in the property have ALL external sources been checked?	Yes	No
Gas tightness test of installation *(see separate Gas Testing and Purging form for details)?*	*Pass*	*Fail*

Property type – House ☐ Bungalow ☐ Flat ☐ Maisonette ☐ Detached ☐ Semi-detached ☐ Terraced ☐

Any other comments:

Conclusions

1. Problems were identified with the installation/appliance(s)?	Yes	No	2. If the answer to 1. is Yes, has a separate **Warning Notice** been raised?	Yes	No
3. Has rectification work been completed?	Yes	No	4. Is the installation safe to use with no defects identified?	Yes	No

Gas User / Responsible Person Declaration: I confirm that I have received this Fumes Investigation Report concerning the safety of the gas installation.

Signed: ____ Print Name: ____

Date: ____

The gas user was not present at the time of the visit and where appropriate (an IMMEDIATELY DANGEROUS (ID) or AT RISK (AR) situation), the installation has been made safe and a **Warning Notice** completed, which along with this Report, is left on the premises: ☐

Gas Operative Declaration: I confirm that the investigation has been undertaken and the results have been recorded above and brought to the attention of the Gas User / Responsible Person in accordance with the Gas Safety (Installation and Use) Regulations (GSIUR), Industry Standards and Procedures.

Licence No: ____

Signed: ____ Print Name: ____

Issue Date: ____ Issue Time: ____

To re-order quote Ref. CP26

Key: Top Copy – Gas User/Responsible Person Second Copy – Client/Landlord Third Copy – Gas Operative

Part 1 of 2

Figure 10.20b Fumes Investigation Report – Part 2 (CP26)

To confirm the validity of the gas operative please contact Gas Safe Register on Tel: 0800 408 5500

Customer Ref.

FUMES INVESTIGATION REPORT PART 2

This form should be completed in accordance with the current requirements of BS 7967

Gas Appliance Investigation Report – Appliance No. 1	
Type	
Was appliance checked (Yes/No)?	
If No, State why	
Gas type	
Location	
Make	
Model	
Serial No.	
Flue type	
What is visual condition of appliance?	
Does chimney comply with current standards?	
Flue flow test (Pass/Fail/NA)?	
Appliance spillage test (Pass/Fail/NA)?	
Ventilation satisfactory (Yes/No)?	
Burner operating pressure (mbar)	
Gas rate (kW/Btu)	
Weather conditions during test	
Ambient & Combustion Test Readings	
Ambient CO (ppm) reading – Initial	
Appliance sweep test CO (ppm)	
Combustion test CO (ppm)	
Combustion test CO_2 (%)	
Combustion test CO/CO_2 ratio	
Ambient CO (ppm) reading – Final	
Satisfactory after detailed investigation?	
Faults identified:	
Faults rectified:	
Is appliance safe for use Yes/No*?	
** If No see separate Warning Notice*	

Gas Appliance Investigation Report – Appliance No. 2	
Type	
Was appliance checked (Yes/No)?	
If No, State why	
Gas type	
Location	
Make	
Model	
Serial No.	
Flue type	
What is visual condition of appliance?	
Does chimney comply with current standards?	
Flue flow test (Pass/Fail/NA)?	
Appliance spillage test (Pass/Fail/NA)?	
Ventilation satisfactory (Yes/No)?	
Burner operating pressure (mbar)	
Gas rate (kW/Btu)	
Weather conditions during test	
Ambient & Combustion Test Readings	
Ambient CO (ppm) reading – Initial	
Appliance sweep test CO (ppm)	
Combustion test CO (ppm)	
Combustion test CO_2 (%)	
Combustion test CO/CO_2 ratio	
Ambient CO (ppm) reading – Final	
Satisfactory after detailed investigation?	
Faults identified:	
Faults rectified:	
Is appliance safe for use Yes/No*?	
** If No see separate Warning Notice*	

Gas Appliance Investigation Report – Appliance No. 3	
Type	
Was appliance checked (Yes/No)?	
If No, State why	
Gas type	
Location	
Make	
Model	
Serial No.	
Flue type	
What is visual condition of appliance?	
Does chimney comply with current standards?	
Flue flow test (Pass/Fail/NA)?	
Appliance spillage test (Pass/Fail/NA)?	
Ventilation satisfactory (Yes/No)?	
Burner operating pressure (mbar)	
Gas rate (kW/Btu)	
Weather conditions during test	
Ambient & Combustion Test Readings	
Ambient CO (ppm) reading – Initial	
Appliance sweep test CO (ppm)	
Combustion test CO (ppm)	
Combustion test CO_2 (%)	
Combustion test CO/CO_2 ratio	
Ambient CO (ppm) reading – Final	
Satisfactory after detailed investigation?	
Faults identified:	
Faults rectified:	
Is appliance safe for use Yes/No*?	
** If No see separate Warning Notice*	

Table A - Ambient Readings For Flueless Appliances

CO(ppm) Readings	Appliance No	Appliance No	Appliance No
Initial Reading			
1			
2			
3			
4			
5			
6			
7			
8			
9			
10			
11			
12			
13			
14			
15			
16			
17			
18			
19			
20			
21			
22			
23			
24			
25			
26			
27			
28			
29			
30			
Final Reading (Ambient)			

To re-order quote Ref. CP26

Key: Top Copy – Gas User/Responsible Person Second Copy – Client/Landlord Third Copy – Gas Operative

Part 1 of 2

Checking/setting burner pressures/gas rates – 11

11 – Checking/setting burner pressures/gas rates

Figures

Tables

Checking/setting burner pressures/gas rates – 11

11 – Checking/setting burner pressures/gas rates

Figures

Tables

Introduction

We explain here how these procedures are essential to achieve the right heat output/efficiency and how you must follow them as an essential safety requirement under the GSIUR.

NG and LPG must be supplied to appliance burners at the pressure specified by the appliance manufacturer to ensure the expected heat output and efficiency.

So you must be able:

- to carry out tests to verify correct operating pressures and gas rates (volume/usage)
- to adjust appliance burner pressures when necessary

As many existing appliances are in circulation, you must know how to convert rates.

You must be able to convert and adjust:

- gas pressures in millibars (mbar) and inches water gauge (ins w.g.)
- the gas rate to kilowatts (kW) and British Thermal Units (Btu)

Gas rate and burner pressure tests – important in commissioning and fault finding

Where appropriate, you can further reduce NG supplied to an appliance at 20mbar via pressure regulators in the appliance to provide a range of pressures at the burner. This allows flexible heat outputs.

Note: Domestic NG installations should be designed to ensure that a pressure drop across the installation does not exceed 1mbar (between the outlet of the gas meter and the inlet connection of each appliance), when the installation is subjected to the maximum load. See Part 5 – 'Installation of pipework and fittings' for further guidance.

LPG appliances are normally limited to a burner pressure of approximately 37mbar for Propane and 28mbar for Butane.

Note: In the case of LPG installations, the pipework should be of sufficient size to prevent a pressure drop greater than 2.5mbar, between the final pressure regulator to the property and all appliances, when the installation is subjected to the anticipated maximum load. For further guidance see Part 18 – 'LPG – General requirements'.

The gas rating test determines the volume (amount) of gas being burned by an appliance. The test converts the time it takes to pass a known volume of gas through a gas meter to an appliance into the heat input. You can then compare this figure with the appliance's data plate.

In some cases the permissible variation from the 'nominal' gas rate will be specified by the appliance manufacturer. Where this is the case the manufacturer's instructions should be followed. However, if the relevant information is not available or cannot be obtained from the manufacturer, the gas rate of a domestic boiler can be considered acceptable if this is measured to be within plus 5% or minus 10% of the rate related to the appliance heat input. This slight additional tolerance over approval/production requirements will normally be sufficient to allow for variances and inaccuracies which can occur in the field.

You also need to know the CV (see **Part 19 – Definitions**) of the gas: to use as a constant figure in the calculation. It is a test you can carry out in a matter of minutes with the use of a stopwatch and a calculator or a pre-calculated gas rate chart.

Burner pressure alone does not confirm the correct volume of gas is being burned.

During commissioning or servicing, a burner would have been 'set up' in accordance with the manufacturer's data plate or written instructions.

Although a wrongly sized or blocked injector is not likely to change the gas pressure at the burner, it could significantly affect the rated heat input/output of the appliance.

The average CVs are:

- NG 38.76MJ/m^3 (gross) 1040btu/ft^3
- Propane 93.1MJ/m^3 (gross) 2496btu/ft^3
- Butane 121.8 MJ/m^3 (gross) 3265btu/ft^3

The difference between 'Gross' and 'Net' Calorific Values

The heat input of a gas appliance in the UK used to be given in terms of gross CV. This changed to net CV in 2000 – as part of the 'Europeanisation' of appliance safety standards.

So do make sure you know whether the manufacturer provides details of the heat input on the appliance data plate or appliance installation instructions, on the basis of:

- gross CV or net CV

The CV of a fuel is basically the amount of heat contained in a given quantity of that fuel. When fuels that contain hydrogen in their chemical make-up are burned, the hydrogen is converted to water vapour.

The water vapour holds some of the heat produced as 'latent heat' (in the case of NG, this is about 10% of the total heat).

It helps you to know that this latent heat:

- is included in the figure given for the gross CV of a fuel
- is excluded from the figure given for the net CV

A gas rate check carried out using a gas meter will normally give you a reading in gross terms. You may then need to convert this reading to net terms depending on the information supplied by the appliance manufacturer.

The ratio of gross to net CV heat input is dependent on the fuel type and is approximately as follows:

- NG 1.11:1
- Propane 1.09:1
- Butane 1.08:1

Therefore, the following calculation needs to be applied.

Example 1 An open-flued appliance with a gross CV heat input of 27kW using NG would have a net CV heat input of 24.32kW, based on the following calculation:

$27 \div 1.11 = 24.32$kW (net).

Example 2 An open-flued appliance with a gross CV heat input of 27kW using Propane would have a net CV heat input of 24.77kW, based on the following calculation:

$27 \div 1.09 = 24.77$kW (net).

Example 3 An open-flued appliance with a gross CV heat input of 27kW using Butane would have a net CV heat input of 25.00kW, based on the following calculation:

$27 \div 1.08 = 25.00$kW (net).

Note: The determination of a net CV for gas rating purposes is different to that used to determine ventilation requirements for gas appliances. See Part 4 – 'Ventilation – Gross or net calorific values are used to quote heat input'.

Gas meter types

The most common domestic gas meter in present use is the unit construction diaphragm type.

This will be either a U6/U16 meter which measures gas in cubic feet (ft^3) or the G4/G10 meter which measures gas in cubic metres (m^3).

There are E6 electronic type meters as well, which also measure gas in cubic metres, have a smaller case size, pass an equivalent volume of gas and give a more accurate reading than the 'U6' type meter.

Rating – what you must do

Method – cubic metres (m^3): Electronic or metric diaphragm meter

Because E6/G4/G10 meters do not have a test dial (see Figure 11.1), the gas rating test and calculation is different from that of the U6/U16 meter test.

E6/G4/G10 meters measure gas in cubic metres (m^3), not cubic feet (ft^3). The figures on the meter index display move significantly more quickly than those on the U6 meter and update the gas flow measurement approximately every two seconds.

To compensate, time the gas flow volume over a two minute period.

Procedure:

1. Make sure that all other gas appliances are turned off.
2. Operate the gas appliance that is under check, allowing sufficient time for it to warm up (normally 10 minutes) and making sure that it will continue to run throughout the check period, e.g. check that any thermostat/controls are set accordingly.
3. Record the meter reading.
4. After 2 minutes, take another reading. Turn off the appliance and subtract the first reading from the second reading.

Note: To be as accurate as possible it is recommended you use a stopwatch.

5. At this point determine whether the appliance heat input rating is in gross or net CV terms.

Note: Table 11.1 is a chart showing the amount of gas passed in a 2 minute test period in cubic metres (m^3).

Figure 11.1 Typical electronic E6 and G4 gas meter index displays reading m^3

E6 gas meter index display

G4 gas meter index display

You can also calculate the appliance heat input using the following formula:

$$kW = \frac{a \times b \times c}{d}$$

Where -

a = number of seconds in one hour

b = cubic metres (m^3)

c = number of kW/m^3 (NG)

d = 120 seconds (plus number of seconds till the next digit appears on the test drum or electronic display)

If the heat input is still not to the manufacturer's requirements, carry out further checks on the appliance and the installation to establish the cause. Rectify any faults.

Note: To determine the number of kW in 1m^3 of NG, Propane or Butane, divide the CV of the gas (in MJ/m^3) by 3.6.

Therefore:

NG = 10.76

Propane = 25.86

Butane = 33.83

For NG the calculation looks like this:

$$kW = \frac{3600 \times m^3 \times 10.76}{\text{time in seconds}}$$

Example for NG:

- first reading = 99999.010m^3
- second reading = 99999.052m^3

difference between the two readings

= 0.042m^3

$$= \frac{3600 \times 0.042 \times 10.76}{123} = \frac{1626.9}{123}$$

= 13.23kW (gross)

Table 11.1 Gas rating chart for metric gas meters (up to 6m³/hr) which measure NG in cubic metres (m³) – averaged CV = 38.76 MJ/m³

2 min gas flow m³	Heat input kW (gross)	Heat input kW (net)	2 min gas flow m³	Heat input kW (gross)	Heat input kW (net)	2 min gas flow m³	Heat input kW (gross)	Heat input kW (net)	2 min gas flow m³	Heat input kW (gross)	Heat input kW (net)
0.002	0.65	0.58	0.052	16.80	15.13	0.102	32.95	29.68	0.152	49.10	44.23
0.004	1.29	1.16	0.054	17.44	15.71	0.104	33.59	30.26	0.154	49.74	44.81
0.006	1.94	1.75	0.056	18.09	16.30	0.106	34.24	30.85	0.156	50.39	45.39
0.008	2.58	2.33	0.058	18.73	16.88	0.108	34.88	31.43	0.158	51.03	45.98
0.010	3.23	2.91	0.060	19.38	17.46	0.110	35.53	32.01	0.160	51.68	46.56
0.012	3.88	3.49	0.062	20.03	18.04	0.112	36.18	32.59	0.162	52.33	47.14
0.014	4.52	4.07	0.064	20.67	18.62	0.114	36.82	33.17	0.164	52.97	47.72
0.016	5.17	4.66	0.066	21.32	19.21	0.116	37.47	33.75	0.166	53.62	48.30
0.018	5.81	5.24	0.068	21.96	19.79	0.118	38.11	34.34	0.168	54.26	48.89
0.020	6.46	5.82	0.070	22.61	20.37	0.120	38.76	34.92	0.170	54.91	49.47
0.022	7.11	6.40	0.072	23.26	20.95	0.122	39.41	35.50	0.172	55.56	50.05
0.024	7.75	6.98	0.074	23.90	21.53	0.124	40.05	36.08	0.174	56.20	50.63
0.026	8.40	7.57	0.076	24.55	22.12	0.126	40.70	36.66	0.176	56.85	51.21
0.028	9.04	8.15	0.078	25.19	22.70	0.128	41.34	37.25	0.178	57.49	51.80
0.030	9.69	8.73	0.080	25.84	23.28	0.130	41.99	37.83	0.180	58.14	52.38
0.032	10.34	9.31	0.082	26.49	23.86	0.132	42.64	38.41	0.182	58.79	52.96
0.034	10.98	9.89	0.084	27.13	24.44	0.134	43.28	38.99	0.184	59.43	53.54
0.036	11.63	10.48	0.086	27.78	25.03	0.136	43.93	39.57	0.186	60.08	54.12
0.038	12.27	11.06	0.088	28.42	25.61	0.138	44.57	40.16	0.188	60.72	54.71
0.040	12.92	11.64	0.090	29.07	26.19	0.140	45.22	40.74	0.190	61.37	55.29
0.042	13.57	12.22	0.092	29.72	26.77	0.142	45.87	41.32	0.192	62.02	55.87
0.044	14.21	12.80	0.094	30.36	27.35	0.144	46.51	41.90	0.194	62.66	56.45
0.046	14.86	13.39	0.096	31.01	27.94	0.146	47.16	42.48	0.196	63.31	57.03
0.048	15.50	13.97	0.098	31.65	28.52	0.148	47.80	43.07	0.198	63.95	57.62
0.050	16.15	14.55	0.100	32.30	29.10	0.150	48.45	43.65	0.200	64.60	58.20

Figure 11.2 Meter index test dial reading cubic feet (ft^3)

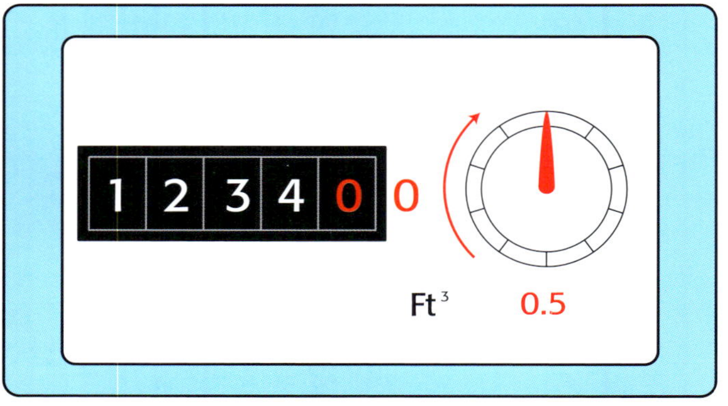

Method – cubic feet (ft^3): U6 diaphragm meter

1. Make sure that all other gas appliances are turned off.
2. Operate the gas appliance that is under check, allowing sufficient time for it to warm up (normally 10 minutes) and making sure that it will continue to run throughout the check period, e.g. check that any thermostat/controls are set accordingly.
3. Observe the meter index test dial and record the time (in seconds) taken to pass one cubic foot of gas. This is normally one complete revolution of the test dial, (see Figure 11.2). Turn off the appliance.

Note: To be as accurate as possible it is recommended you use a stopwatch.

4. At this point determine whether the appliance heat input rating is in gross or net CV terms.
5. Table 11.2 will provide a direct reading for both kW (gross) and kW (net) based on the time taken for the meter to pass one cubic foot ($1ft^3$) of NG.

You can also calculate the appliance heat input using the following formula:

$$\frac{\text{seconds in 1hr (3600) x the CV of the gas}}{\text{one revolution of test dial in seconds}}$$

Note: As the meter is imperial (ft^3) the averaged CV for NG of 38.76Mj/m^3 needs to be converted to Btu/ft^3; divide by 0.0373 to convert to Btu/ft^3, i.e. 38.76 ÷ 0.0373 = 1040Btu/ft^3.

If the heat input is still not to the manufacturer's requirements, carry out further checks on the appliance and the installation pipework to establish the cause. Then rectify any faults.

Table 11.2 Gas rating chart for imperial gas meters (up to 212ft³/hr) which measure NG. Averaged CV = 38.76MJ/m³ = 1040btu/ft³

Time (in secs) to burn 1ft³ at the appliance	Heat input in kW (gross)	Heat input in kW (net)	Time (in secs) to burn 1ft³ at the appliance	Heat input in kW (gross)	Heat input in kW (net)
17	64.57	58.17	68	16.14	14.54
18	60.98	54.94	70	15.68	14.13
19	57.77	52.05	74	14.83	13.36
20	54.89	49.45	78	14.07	12.68
21	52.27	47.09	82	13.39	12.06
22	49.90	44.95	86	12.76	11.50
23	47.73	43.00	90	12.20	10.99
24	45.74	41.21	94	11.68	10.52
25	43.91	39.56	98	11.20	10.09
26	42.22	38.04	102	10.76	9.70
27	40.66	36.63	106	10.36	9.33
28	39.20	35.32	110	9.98	8.99
29	37.85	34.10	114	9.63	8.67
30	36.59	32.96	120	9.15	8.24
31	35.41	31.90	130	8.44	7.61
32	34.30	30.90	140	7.84	7.06
33	33.26	29.97	150	7.32	6.59
34	32.29	29.09	160	6.86	6.18
35	31.36	28.25	170	6.46	5.82
36	30.49	27.47	180	6.10	5.49
37	29.67	26.73	190	5.78	5.20
38	28.89	26.02	200	5.49	4.94
39	28.15	25.36	210	5.23	4.71
40	27.44	24.72	220	4.99	4.50
41	26.77	24.12	230	4.77	4.30
42	26.14	23.55	240	4.57	4.12
43	25.53	23.00	250	4.39	3.96
44	24.95	22.48	260	4.22	3.80
45	24.39	21.98	270	4.07	3.66
46	23.86	21.50	280	3.92	3.53
47	23.36	21.04	290	3.79	3.41
48	22.87	20.60	300	3.66	3.30
49	22.40	20.18	310	3.54	3.19
50	21.95	19.78	320	3.43	3.09
52	21.11	19.02	330	3.33	3.00
54	20.33	18.31	340	3.23	2.91
56	19.60	17.66	350	3.14	2.83
58	18.93	17.05	360	3.05	2.75
60	18.30	16.48	370	2.97	2.67
62	17.70	15.95	380	2.89	2.60
64	17.15	15.45	390	2.81	2.54
66	16.63	14.98	400	2.74	2.47

Example: A gas operative is to check the heat input of a central heating boiler burning NG. The boiler's data plate reveals a maximum rated heat input rating of 15.24kW (gross). The time for one complete revolution of the test dial (1ft^3) is 72 seconds. The calculation is as follows:

Note: To convert a gross CV to a net CV, divide the gross figure by 1.11 (NG), 1.09 (Propane) and 1.08 (Butane) – see 'Gross and net CVs' in this Part.

$$\frac{3600 \text{ (N}^{\text{o}}\text{. seconds) x 1040 (CV of gas)}}{72 \text{ (N}^{\text{o}}\text{. of seconds)}}$$

$$= \frac{3744000}{72}$$

$$= 52000 \text{ btu/hr input}$$

To then convert btus to kW:

$$= \frac{52000}{3412^{*}}$$

$$= 15.24\text{kW (gross)}$$

*Number of btus per kW.

For larger gas meters, up to U16, the following calculation applies:

$$\frac{\text{seconds in 1hr x the CV of the gas x test dial}}{\text{one revolution of test dial in seconds}}$$

Therefore for a U16 gas meter with a 5ft^3 test dial, the calculation would be:

$$= \frac{3600 \times 1040 \times 5}{360}$$

$$= \frac{18720000}{360}$$

$$= 52000 \text{ btu/hr input}$$

To then convert btus to kW:

$$= \frac{52000}{3412}$$

$$= 15.24\text{kW (gross)}$$

The same principles of gas rating apply to LPG installations.

Where gas meters are installed, the formulas for both gas meters reading ft^3 and m^3 are the same as described for NG. The differences will be in the particular CV for each gas type.

Testing appliance burner pressures – ensure compliance with manufacturer's instructions

- this must be carried out during appliance commissioning or service/maintenance work
- check the burner pressure for compliance with the manufacturer's instructions by carrying out a test at the appliance with a suitable pressure gauge (water 'U' gauge or electronic gauge)

Note: Air/gas ratio valves are found on many new condensing boilers. These valves operate differently from conventional gas valves fitted to standard efficiency boilers over recent years. Air/gas ratio valves are also set up very differently from conventional gas valves and care must be taken to ensure the valves are set and operated correctly to avoid the risk of poor combustion performance and the production of high levels of CO. Generally burner pressure is NOT measured and gas rating is required to confirm safe operation along with combustion analysis.

Range rating – when and how to adjust pressure regulator

Most NG burning central heating boilers are designed to work correctly over a range of gas pressures. This allows manufacturers to produce one appliance for use for a number of heat output applications.

Many appliances, especially combination boilers, have modulating burners. You can

Figure 11.3 Water 'U' gauge and appliance

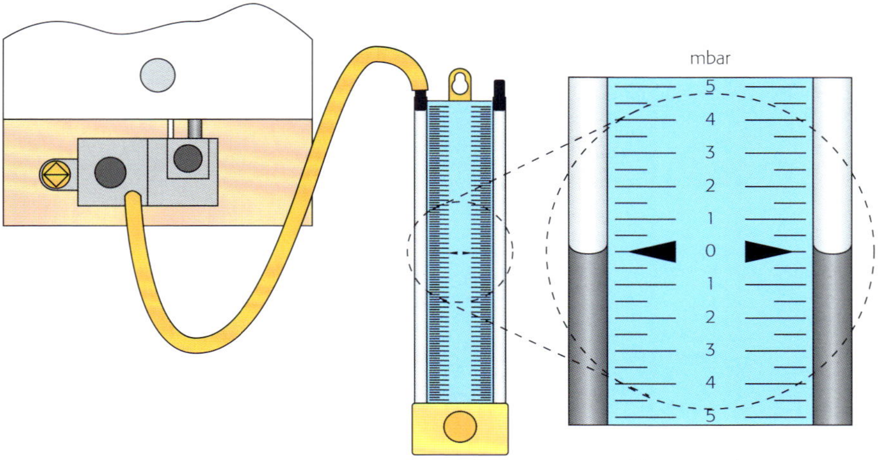

adjust a pressure regulator integral to the appliance to a range of pressures. The appliance data plate or commissioning instructions will provide this information.

What to do for a non-modulation burner

1. Check that the water levels in a 'U' gauge are at zero or, in the case of an electronic gauge, that it has been calibrated in accordance with its manufacturer's instructions (see Figure 11.3).
2. Turn off the appliance under test.
3. Identify the correct burner pressure test point (refer to the manufacturer's instructions).
4. Remove or loosen the test point screw and connect the pressure gauge via suitable tubing (see Figure 11.3).
5. Turn on the appliance and allow it to reach operating temperature (usually 10 minutes, or the manufacturer's recommended time period).
6. Record the pressure gauge reading and compare it with the manufacturer's data.
7. Alter the burner pressure, if necessary, by adjusting the burner pressure regulator.
8. Turn off the appliance, remove the pressure gauge and replace or tighten the test point screw.
9. Turn on the appliance and check the test point for gas tightness with non-corrosive leak detection fluid (LDF).

Figure 11.4 Modulating valve

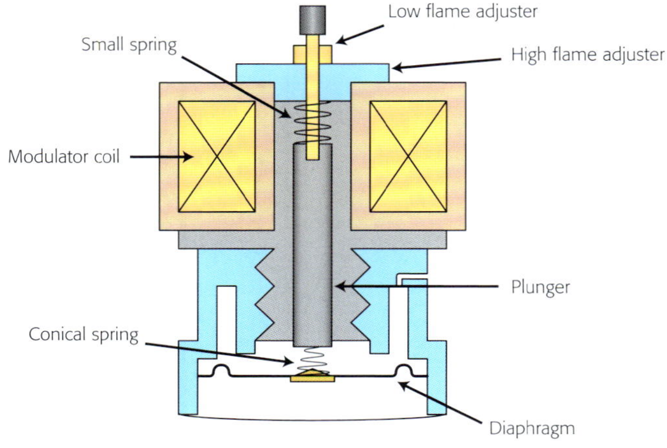

What to do for a modulation burner

Only adjust the regulators serving this type of burner (see Figure 11.4) when the manufacturer's instructions allow.

1. Check that the water levels in a 'U' gauge are at zero or, in the case of an electronic gauge, that it has been calibrated in accordance with its manufacturer's instructions (see Figure 11.3).
2. Turn off the appliance under test.
3. Identify the correct burner pressure test point (see manufacturer's instructions).
4. Identify the mode being checked where appropriate (hot water or central heating).
5. Remove or loosen the test point screw and connect the pressure gauge via suitable tubing.
6. Turn on the appliance and allow it to reach operating temperature (usually 10 minutes or manufacturer's recommended time period).
7. Record the pressure gauge reading and compare it with the manufacturer's data.
8. Adjust burner modulation pressure if necessary, by adjusting the burner pressure modulation regulator.
9. Turn off the appliance, remove pressure gauge and replace or tighten test point screw.
10. Turn on the appliance and check the test point for gas tightness with non-corrosive LDF.

Gas pressure fault finding – an outline to help

On occasion you may receive a customer complaint about lack of heat or poor performance from a gas appliance.

Your initial inspection may reveal the appliance is working and, when you check, it may have the correct or near-correct gas burner pressure. This does not necessarily mean that the appliance is performing satisfactorily.

The following list may help identify other causes, but in all cases, check the gas rate:

Poor/insufficient heat output

Look for:

- blocked or partially blocked burner injector
- incorrect injector size
- incorrect appliance or burner inlet pressure

Burner flame lift-off

Look for:

- excessive burner pressure
- over or under-sized burner injector
- appliance not suitable for gas type

Incorrect pressure at appliance

Look for:

- undersized gas supply pipework from the gas meter to the appliance
- meter regulator incorrectly set
- blockage in supply pipe to appliance

Note: This list is not exhaustive.

Checking of appliance gas safety devices/controls – 12

12 – Checking of appliance gas safety devices/controls

Checking of appliance gas safety devices/controls – 12

12 – Checking of appliance gas safety devices/controls

Figures

Tables

Why NG and LPG need controls/safety devices

NG and LPG need to be controllable when utilised for heating, cooking or lighting:

1. For safety reasons.
2. To regulate temperature.
3. To ensure that the varied gas heat input rates of appliances are achieved.

A wide variety of controls and safety devices are available, which can be either electronically or manually operated. They range from a simple quarter turn brass tap to sophisticated electronic flame rectification systems and their main functions are:

- for gas isolation
- to stop gas flow to the installation or appliance when a fault or hazard occurs
- to ensure safe ignition
- to control temperature, time or function
- to regulate pressure

The GSIUR stipulate:

- gas supplies must be able to be controlled
- appliances must be able to be isolated
- safety devices must always be operative (and never rendered inoperative for any reason)
- non-room-sealed appliances in sleeping areas must incorporate a safety device designed to shut down the appliance before there is a build up of a dangerous quantity of the products of combustion in the room concerned, or when there are adverse flue conditions such as down draft

Note: When you change a component, you must ensure that it is suitable for the gas type being used. Some types of valve seals and lubricating greases, for example, will be suitable for NG, but not LPG.

Pressure regulation (gas) – protects against high or low pressure faults

NG and LPG can be supplied and stored at greater pressures than required at an appliance burner.

Service pipe (NG), storage vessel and cylinder (LPG) supply pressures need to be reduced and regulated to ensure a constant gas pressure, flow and volume at the inlet connection of an appliance.

NG is very versatile. With the correct air/gas mixture, gas appliances can achieve complete combustion over a range of different pressures.

Pressure regulation devices are available to help achieve this and protect the appliance and installation against high or low pressure faults.

Constant pressure regulator – a simple device to help

Operation and use:

This ensures a constant working pressure is achieved in a gas installation. Gas flows from the inlet to the outlet of the regulator via a valve connected to a weighted or spring-loaded diaphragm.

The higher inlet pressure is reduced by the position of the valve in relation to the valve seating, thereby maintaining a lower outlet pressure.

The device maintains this constant outlet pressure by opening or closing the valve, depending on the demand.

Single diaphragm and valve models are normally used for domestic sized appliances and installations (see **Part 7 – Checking and/or setting regulators**).

Figure 12.1 shows a typical constant pressure meter regulator.

Figure 12.1 Essential parts of a constant pressure meter regulator

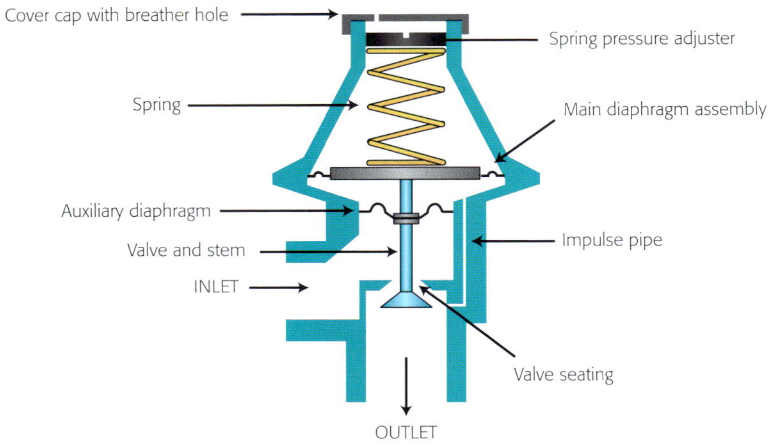

Checks:

- in some older appliances that have the diaphragm manufactured from leather, you may need to maintain (oil) it, using a suitable vegetable oil such as Neats foot or almond oil, or replace it
- check and adjust the outlet pressure periodically

Note: The gas meter regulator fitted to low pressure fed installations (up to 75mbar) is also of the constant pressure type and should only be adjusted and sealed by gas operatives or persons authorised to do so by the gas supplier.

Medium/low pressure regulators (NG) – include shut off valves to avoid over-pressurisation

Operation and use:

These regulators are normally used to reduce a medium gas pressure of up to 2bar to a low gas pressure of 21mbar, through a single reduction.

The pressure reduction is achieved by the restriction of the regulator's small valve seating orifice.

Due to the possible danger of the higher inlet pressure affecting the downstream pipework and equipment, an over pressure safety shut off valve is built into this type of regulator.

In the event of over-pressurisation:

- rectify the fault; then
- manually re-set the valve

Checks:

- check the outlet pressure

You will see regulators of this type on medium pressure fed meter installations.

Note: As with meter regulators fitted to low pressure fed installations, this type of regulator should only be adjusted and sealed by gas operatives or persons authorised to do so by the gas supplier.

For further information relating to LPG pressure regulation, also see **Part 18 – LPG – General requirements**.

Gas controls must be in place – for use both by gas users and operatives

1. All gas installations must be able to be controlled both by gas users and gas operatives, especially in an emergency.
2. Every gas appliance should have an isolation valve fitted to its inlet gas supply pipework to allow work to be carried out or for it to be disconnected, as necessary.
3. All appliance isolation control devices must be accessible for operation and service/maintenance when necessary.

Note: Do not consider the appliance isolation valve as an emergency control valve. See Part 9 – 'Operation/positioning of emergency controls/isolation valves'.

Gas appliance operating valve – used for lower pressures/smaller supplies

Operation and use:

This device is generally manufactured as an integral part of a gas appliance. To isolate or control gas flow you turn a tapered plug or barrel, or a ball or a disk.

Some valves will have multiple orifices controlling gas flow to effect different gas rates or temperatures. Operate these manually, turning the valve to pre-set positions by use of a lever, thumb piece or knob.

Most will be spring-loaded with an initial 'push in' action required, which acts as a safety device to prevent accidental turning of the valve to the 'on' position.

Checks:

- grease the valves with high temperature grease (or similar) to assist smooth operation and prevent gas escapes as the grease forms a gas-tight seal (the valves are prone to drying out, especially when used as a cooker hotplate valve)
- they should not be stiff in operation or travel past pre-set positions
- they should turn the gas fully off in the closed position and allow no let-by
- take care to ensure excess grease is not allowed to block orifices that control turn down rates, as this could cause the gas flame to be extinguished when the tap is turned down

Note: Valves suitable for LPG need to be spring-loaded.

Gas valve (manual) – used for higher pressures/larger supplies

Operation and use:

These valves can be used for higher pressures and larger supplies than gas taps or cocks. The gas flow is controlled by the movement of a gate, flap or disc.

Checks:

- check the valve for gas leakage around the valve glands or spindles
- levers, handles or wheels should not be stiff to turn and should have full travel to their on/open – off/closed positions
- it is essential that gas valves turn the gas supply fully off and that they allow no let-by

Gas valve (electronic) – a control device that operates a motor

Operation and use:

This acts by electrical current flow operating a motor, which moves or holds a valve away from its seating or orifice, thus opening the valve.

The return action to the closed position may be operated by a spring or operated by the electric motor turning a spindle/stem through 180°.

Checks:

- ensure the area around the valve head or motor is kept clear to allow air flow and prevent over heating
- check smooth action of the valve by manually operating it (if applicable)

Cooker bayonet (self-sealing) connection – appropriate to seal a gas supply

Operation and use:

This device allows the disconnection of a free standing gas cooker or gas tumble dryer. It allows a gas user to clean the cooker/dryer and it enables you to carry out any necessary maintenance.

The flexible connector is isolated from the gas supply by means of a spring loaded, 'push in' and turn, manually-operated, bayonet type valve.

When the flexible connector is inserted into the bayonet fitting, it pushes a valve off its seating, allowing gas to flow.

When the connector is then fully turned to its 'lock' position, a gas tight seal is made between the connector and bayonet body. Removing the bayonet fitting allows the spring loaded valve to re-seat, stopping the gas flow.

Note: This type of fitting is recognised as an 'appropriate fitting' to seal a gas supply within the GSIUR.

Checks:

- always use non-corrosive LDF to ensure that the self-sealing connector turns off completely when disconnected
- the valve body should be firmly secured to the wall i.e. the fabric of the building
- the entry point for the bayonet fitting should face downwards (variations will be encountered however)
- the flexible hose should not be strained, and should always hang down in a 'U' shape
- the flexible hose should not be in contact with the floor

Gas fire restrictor elbow – aesthetically pleasing fitting

Operation and use:

This gas control device will allow the connection, disconnection or isolation of a gas fire from its gas supply by an aesthetically pleasing fitting.

It incorporates a compression type gas connection with a manual screw-in isolation valve. The valve is accessed via a screw-in cap on the top of the fitting.

The outlet gas supply will normally be 8mm in diameter, although you may find older imperial sizes. The fitting is available with a pedestal base, which can be screwed to the floor for safety.

The fitting should have its outlet connection sealed with a suitable plug if the fire is disconnected.

Checks:

- always ensure that the olive on the compression connection is in good condition and is correctly positioned on the pipe before you connect it into the fitting
- ensure the top cap and washer are secure and gas tight
- be aware that this cap and washer combination is a common cause of gas escapes

Cooker hotplate lid control

Operation and use:

This safety device is fitted onto many gas hotplates, particularly those with glass drop down lids and is operated by opening/closing the lid.

A gas valve is fitted into the supply pipe to the hotplate burners normally at the rear of the appliance, which will turn off the gas supply to the hotplate if the lid is accidentally closed when the burners are in use (conversely, the lid needs to be raised fully before gas will be permitted to flow to the hotplate burners).

The valve is operated by a push-down spindle connected to the hotplate lid, closing the valve as the lid drops.

Checks:

- the valve should fully stop the gas flow to the hotplate when operated

Electronic solenoid valve

Operation and use:

A solenoid is an electrically operated device (see Figure 12.2). An integral electrical coil within the valve generates a magnetic field that holds a gas valve in the open position. When the electrical supply to the valve is stopped, the valve will close.

Solenoids will operate on either an a.c. or d.c. electric current and units are available in various sizes. Normally they are small devices, which are integral to a multi-functional gas control valve.

Checks:

- the solenoid valve can be prone to nuisance valve 'buzz' on a.c. systems
- they should fully close when in the closed position and allow no let-by
- do not box in or cover over remote solenoid units – as they may become subject to overheating and subsequent failure

Figure 12.2 Electronic solenoid valve

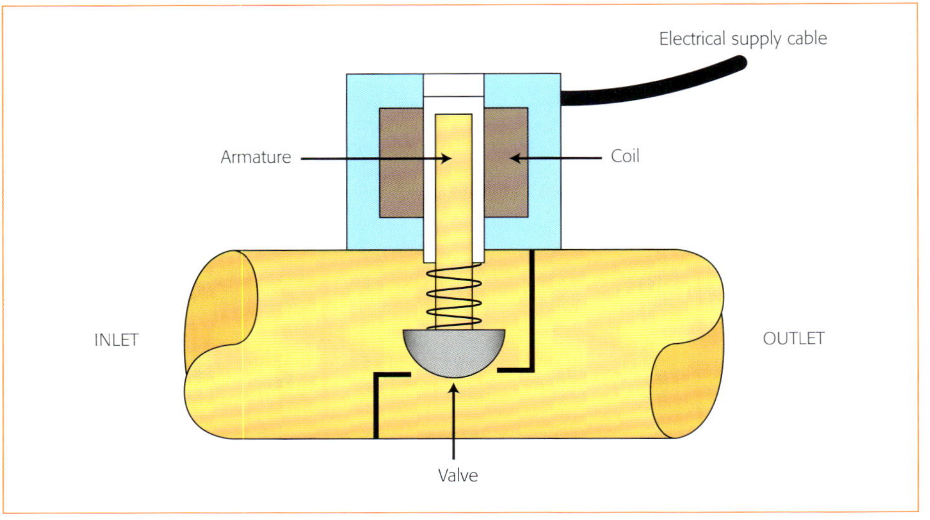

Thermal cut-off – protects vulnerable areas of a building

Operation and use:

This safety device is fitted to some gas meters to protect vulnerable areas of a building, such as a fire escape route.

The device consists of a spring-loaded gas valve held in the open position by solder with a low temperature melting point. In the event of a fire, the solder melts and the valve shuts off the gas supply.

Checks:

- the valve should fail-safe at temperatures in excess of 95°C

Low pressure cut-off – for installations prone to interruption of supply

Operation and use:

You can fit this device into a gas installation that is prone to risk by interruption of the gas supply, e.g. an empty LPG cylinder or a pre-payment meter that has run out of credit.

A valve attached to a spring-loaded diaphragm closes when the gas pressure on the valve fails or reaches a pre-set minimum pressure. The valve should only be a manual re-set type.

Checks:

The safety device should:

- not allow full rate gas flow to appliances after the fault condition has occurred
- fail at a pre-set pressure

Table 12.1 Flame supervision device maximum fail safe period

Appliance		Fail safe period
Central heating boiler (max. heat input 70kW net)		60 seconds
Cooker (see Note)	Oven	60 seconds
	Hotplate, griddle,etc	90 seconds
Space heater		180 seconds
Decorative fuel-effect gas appliance		60 seconds
Warm air heater (max. input 60kW gross)		60 seconds
Storage water heater	(≤35kW(net))	60 seconds
	(35< to ≤70kW(net))	45 seconds
Combination boiler and multi-point water heater		60 seconds
Note: For cookers, the gas valve should shut down within 60 seconds if the burner is situated in a compartment (oven) and within 90 seconds in the case of an open burner, covered burner or griddle.		

Relay valve – works with remote control devices

Operation and use:

This valve is designed to control or operate a gas supply via a remote control device, e.g. a solenoid valve, or a thermostat via a weep tube.

It operates using a valve attached to a diaphragm, which opens or closes when pressure is removed from, or applied to it by the opening or closing of the remote device.

Gas is normally supplied to the remote device via a small orifice in the diaphragm.

Checks:

- when you clean or check the diaphragm, the orifice should be clear and not enlarged
- neither the weep tube nor the remote control device should allow gas to escape or the valve may stay in the open position
- if the diaphragm is made of leather, you may need to maintain (oil) it using a suitable vegetable oil such as Neats foot or almond oil, or you may need to replace it
- you also need to check and clean the valve seating

Flame protection controls: stop gas flow when risk occurs – and ensure safe ignition

These safety devices, known as flame supervision devices (FSD) are used:

- to stop gas flow to burners when a fault or risk occurs
- to ensure the safe ignition of a gas burner by preventing quantities of unignited gas from building up

They need to 'fail safe' in a maximum time measured in seconds (see Table 12.1 or follow manufacturer's specifications).

The flame heating the device's probe/sensor normally also ignites the main burner, via a cross lighting or ladder pilot.

Bimetallic strips – work by metal expanding on heat

Operation and use:

A bimetallic device acts on the principle of metal expansion caused by a flame's heat (usually a pilot).

The strip is made from two dissimilar metals joined together which, when heated, expand at different rates, causing the strip to bend.

To utilise this bending effect, a valve spindle, electric contact or micro switch is attached to the strip. When the strip is heated it bends, opening or activating the attached gas valve, electric contact or switch.

When the device cools, the action is reversed, which returns the valve, electronic contact or switch to its closed/off position.

Checks:

- you should ensure that any valve spindle is not sticking or misshapen. Bimetallic strips can be prone to metal fatigue due to constant flame contact
- these devices cannot be repaired: you need to replace them when they fail

Vapour pressure device – locate for full flame contact

Operation and use:

This device usually consists of a mercury liquid-filled probe (sensor) that is in contact with a flame (usually a pilot light or burner's by-pass rate).

The probe is connected, via a capillary tube, to a set of flexible bellows, to which is attached a valve in the gas supply line to the burner.

When heat is applied to the probe, it causes the liquid within the device to vaporise and expand. This causes the valve to lift off its seating, allowing full gas rate flow to the burner.

Whilst heat is applied to the probe the valve remains open. Should the heat to the probe fail or be interrupted, the liquid/vapour will cool, causing the bellows to contract. This closes the valve and stops the gas flow to the burner (see Figure 12.3).

Checks:

- the probe needs to be correctly located for full flame contact
- you may need to clean or grease the valve spindle to move easily
- these devices are not normally designed for repair: you must replace them with a new unit when faulty
- test vapour pressure devices in accordance with the manufacturer's instructions

Thermoelectric device (thermocouple)

Operation and use:

This device uses an electric current generated by the thermocouple to hold open an electromagnetic gas valve.

A thermocouple is a thin copper tube encompassing a core of dissimilar metals. The two metals are joined together at one end to form a tip. When the tip is heated by a gas flame it generates a small electrical current. The other end of the lead is connected to an electromagnetic valve.

This electrical current energises the electromagnet, which holds a valve off its seating. If the pilot flame/heat source is extinguished, the electrical current fails, causing the valve to shut.

Figure 12.3 Typical vapour pressure device

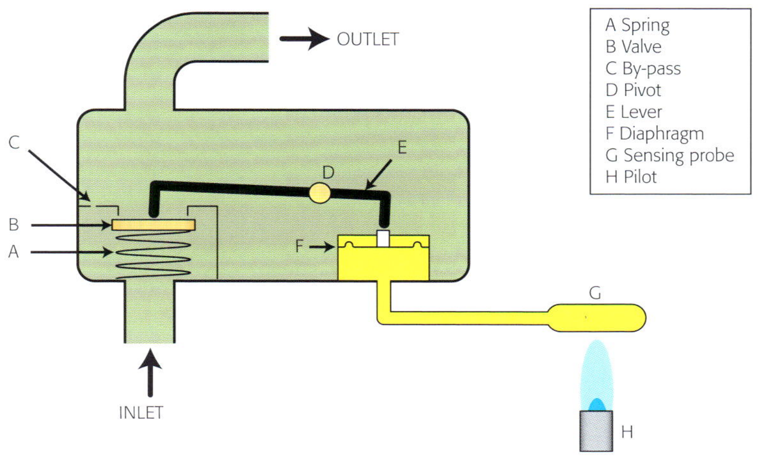

Checks:

- correctly position the thermocouple tip for full flame contact
- ensure the end connected to the electromagnet, which is part of the control valve, is secure
- check that the tip is clean and that the flame is burning correctly and also inspect the lead for damage
- the electrical current generated should be between 10mV and 30mV
- constant flame contact will cause a slow deterioration of the probe tip. As you cannot repair thermocouples, renew them when burnt out (many universal types are available, with a range of screw-in fittings)

Note: When you use universal thermocouples, take care to use the correct threaded nut.

Interrupter device

Some thermocouples also incorporate an interrupter, which is connected to an additional safety control (e.g. an overheat thermostat, fusible link or spillage monitoring device).

The interrupter can either be part of the thermocouple or an additional adaptor that is fitted 'in line' with the thermocouple (see Figure 12.4).

The interrupter works by cutting the electrical current generated by the thermocouple, which in turn causes the electromagnet to de-energise, allowing the valve to close.

Figure 12.4 Thermocouple interrupter

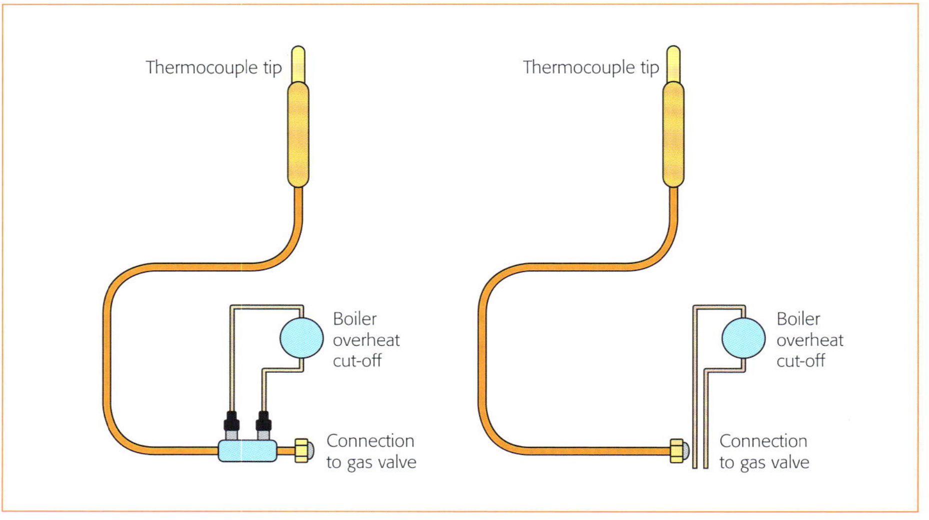

Atmosphere sensing device (ASD) – provides additional protection only

Operation and use:

Many modern gas appliances, including flueless appliances such as LPG cabinet heaters, incorporate an atmosphere sensing device (ASD). Note that these devices provide an additional level of protection only.

The device is normally a two port aerated pilot burner. The pilot flame responds to the presence of an ever-increasing concentration of combustion products (vitiation) in the room or space in which the appliance is fitted, by progressively causing the flame to lift off of the thermocouple. The thermocouple operates an electromagnetic gas valve (see Figure 12.5).

When operated, the device will shut off the gas supply to the burner. This occurs as the concentration of carbon dioxide (CO_2) in the combustion air reaches 1.5 – 2.0% and as the corresponding carbon monoxide (CO) concentration reaches a level less than 200 parts per million (ppm) – which is within safety limits.

The dimensional relationship between the pilot burner and the thermocouple is critical and is factory set.

Note: The GSIUR require that non-room-sealed gas fires and other space heaters installed in sleeping accommodation after 1st January 1996 – and non-room-sealed water heaters installed in any location after 31st October 1998 – must incorporate a safety control designed to shut down the appliance before there is a dangerous quantity of products of combustion in the room concerned. An ASD is considered suitable for this purpose.

For further information, see the current CORGI*direct* manuals entitled – 'Central Heating – Wet and Dry', 'Gas Fires and Space Heaters' and 'Water Heaters' from the Gas Installer Manual Series – Domestic.

Figure 12.5 Atmosphere sensing device (oxygen depletion system)

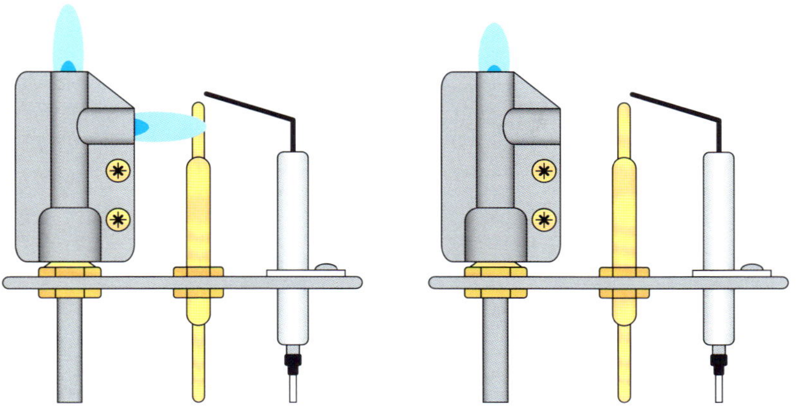

Flame heats thermocouple lead under normal conditions

Under poor atmospheric conditions (vitiation) flame heating the thermocouple is extinguished

Checks:

- ensure the safety device is kept dust/lint free around its air port(s) and injector
- correctly position the thermocouple tip for full flame contact
- ensure the device is cleaned regularly otherwise it could be prone to nuisance drop-out
- be aware it is an offence under GSIUR to interfere with or alter (bridge out) a gas appliance's safety device. ASDs are not normally designed to be repaired by a gas operative. Some manufacturers recommend that they are replaced at five yearly intervals

Spillage monitoring device

This device is often referred to as a TTB (an acronym of the Dutch words 'Thermische Terugslag Beveiliging'). It is located in a flange or bracket at or near the edge of the appliance draught diverter. The sensors are pre-set by the manufacturer and are calibrated to avoid nuisance shut down while maintaining safe tolerances.

Most TTB sensors look like circular overheat thermostats on boiler heating system flow pipes. Two leads connect to the sensor, which is generally linked to a thermocouple using a typical interrupter principle. See also **Thermoelectric device (thermocouple)** in this Part.

When adverse flue conditions occur, the heat generated by the spilling products of combustion passes across the sensor and activates it. This causes pilot shut down through interruption.

Checks:

- check the sensor is located in accordance with the manufacturer's instructions
- check the connections to the thermocouple interrupter for tightness

Flame conduction and rectification device

Operation and use:

This device acts on the principle that a gas flame is able to conduct an electrical current.

When a flame burns, it creates a flow of electrically charged ions. If this flame is located between two electrodes, the small electrical current can flow between the electrodes.

This is utilised via suitable electronic controls to operate a gas control valve. The advantage of this device is that it operates (i.e. shuts down) as soon as the flow of electricity fails.

Checks:

The electrodes need to be in good, clean condition and correctly located for full flame effect .

- regularly clean the conducting flame burner

Photoelectric device – senses the flame picture

Operation and use:

Some metals are not able to conduct electricity in the dark, but when a light is sensed, an electrical current can be passed through the metal.

Some metals will also only conduct electricity when subject to ultra-violet or infra-red radiation. This radiation can be generated by a gas flame and sensed by a photoelectric cell, causing an electrical current to pass through the cell and control a gas valve.

Devices using this method will only sense the flame picture and will not be subject to heat from the flame or other heat sources.

Checks:

- clean the sensing flame burner regularly to ensure a correct flame picture
- locate the photoelectric cell correctly to ensure full exposure to the flame profile
- clean it of any soot deposits

Thermostats – various non-electric types, designed to control temperature

Thermostats control the temperature of objects or areas. They can be a component part of an appliance or a remote sensor.

Some types will gradually open or close, others will snap open or shut (fast action). Simple types will manually control gas flow via an integral valve; others will operate via the electrical control of a remote flow valve.

It may be possible for you to re-calibrate the thermostat. However, many will be factory sealed and the manufacturer must replace or repair them, if faulty.

There are 4 main types of thermostat operation:

- metal expansion (manual control)
- liquid expansion (manual control)
- vapour pressure (manual control)
- electrical (remote control)

Bimetallic (rod and strip type)

Operation and use:

The action of the device is mainly produced by heat in or around the thermostat's location.

- the 'rod' is formed by a tube of two dissimilar metals with different expansion rates. One metal type forms the tube, enclosing the other metal type as a core. The expansion of the rod's core controls the gas flow to a burner. When the rod is connected directly to the valve, the action will be gradual. For fast action, the rod is connected to a lever, which quickens it
- bimetallic 'strip' types are usually formed in the shape of a coil or spring to increase the action movement and they will normally activate electrical switches or control a gas relay valve. Thermostat settings are normally factory set and sealed for these types

Checks:

- if the rod type thermostat is controlling an oven, ensure it operates on its bypass rate (if applicable). You can check temperature differentials against the manufacturer's specification with a suitable thermometer
- check that any control tap/knob stem is not stiff to turn and that it is not allowing gas to escape from its gland or stuffing box
- always ensure, as far as possible, that rods are not contaminated by grease or scale

Liquid expansion thermostat

Operation and use:

This type of device houses the valve remotely from the point of heat that is being sensed.

The movement of this thermostat is achieved by liquid expanding in a sensing phial that is connected by a capillary tube to a set of flexible bellows or a diaphragm that activates a gas valve. The phial and other parts containing the liquid are completely full to ensure maximum movement.

Locate the phial to detect heat generated by a burner flame so that it will control the gas flow to it.

Checks:

- always ensure that the phial is in its correct location
- when possible ensure that it is not contaminated by dust, grease or scale
- be aware that if the unit is subject to excessive heat, the temperature control may be erratic
- if there is any damage caused to liquid-containing parts, you need to fit a complete new unit

Vapour pressure – similar to liquid expansion type

Operation and use:

In this device, the liquid containing parts are not completely full. When subject to heat, the liquid will boil and vaporise, quickly increasing in volume and causing movement of a bellows or diaphragm.

Checks:

- as for liquid expansion type (previously)

Figure 12.6 Typical multifunctional gas control valve

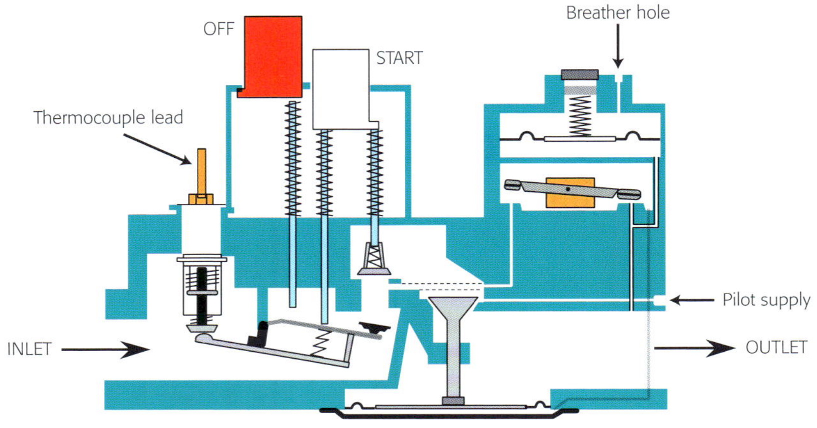

Multi-functional controls in a single unit

A multi-functional device (see Figure 12.6) can incorporate many, if not all, of the controls necessary to operate a gas appliance in a single compact unit. This will normally include:

- gas flow control
- gas pressure regulation
- flame protection
- temperature control; and
- burner ignition

They will normally be a component part of an appliance, but you may be able to change separate parts of the valve, as necessary.

Gas modulation can also be a function of this type of valve, which is common on most appliances.

Operation and use:

The valve is designed:

- for use by the gas user to start the appliance; and
- for your use to carry out work/adjust the gas appliance

Different makes will have manual or electronic gas control via a tap, knob, push button or remote electrical device.

Main burner flame protection will usually be via a thermocouple or flame rectification. You can set gas operating pressure regulation on many types of NG models within a pressure range specified by the appliance manufacturer. A thermostat, bimetallic or electric contact or solenoid valve may control the main gas supply.

Operate gas modulation by using a thermistor to sense the temperature generated by the appliance. The variable current provided by the thermistor can open or close a solenoid by various degrees, allowing different volumes of gas to pass to the burner.

Checks:

- when you work on the appliance, always ensure that any flame protection device operates within its maximum time period (see **Flame protection controls: stop gas flow when risk occurs – and ensures safe ignition** in this Part)
- check that any gas control tap operates smoothly
- 'buzzing' may indicate a faulty solenoid
- if there are any gas carrying joints that are disturbed, test for gas tightness, using non-corrosive LDF

Multi-functional controls with integral zero regulators – complex to maintain

These controls are now commonly fitted to a range of high efficiency boilers and use the working principals of zero regulators, where the negative pressure created by the combustion fan of the appliance controls the gas rate.

Typically, they feature two 24V (rectified a.c.) inlet and servo solenoid valves and a primary air-sensitive pressure regulator.

Servo solenoid valves are closed by spring pressure in the normal shut down position and can only be opened when gas pressure is sufficient to overcome the spring force.

This safety feature ensures that the main valve will automatically close down in the event of a power or gas supply failure.

When a demand for heat is detected, both solenoid valves operate together – the inlet valve feeding the servo valve to pass gas to the main valve-opening diaphragm.

As the main valve opens and the outlet (burner) pressure rises, it is sensed under the pressure regulator diaphragm against the opposing force set by the primary air pressure and the (factory-set) regulator spring setting. When the two forces are equal, the burner pressure is stable.

As the heat demand nears satisfaction, the control printed circuit board (PCB) reduces the fan speed in response to the thermistor signal. This unbalances the force on the regulator diaphragm to allow servo pressure to 'spill' to the control outlet, thus permitting a reduction in the main valve opening to reduce the pressure.

On/off control is managed by the room and/or boiler temperature control thermostats.

- multi-functional gas controls tend to be very complex and they often have very small components
- they are often difficult to maintain and other than exchanging the solenoid operator, you need to replace them as a complete unit

Figure 12.7 Room thermostat

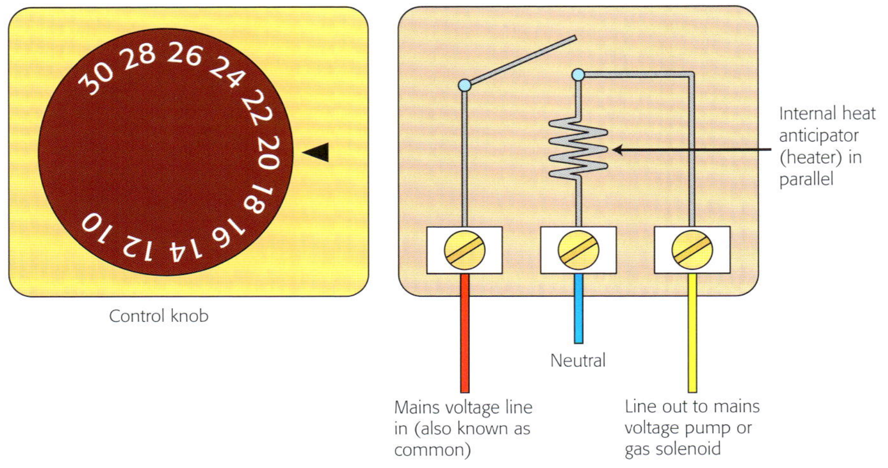

Thermostats electrical – also used to control temperature

Room thermostat – senses air temperature where located

Operation and use:

This temperature sensing device (see Figure 12.7) controls the heat of a room or area by sensing the air temperature in the area in which it is located.

Operation is normally via a bimetallic coil or spring, which has an electrical contact fitted to it. This contact makes or breaks with another contact, controlling (switching) a gas valve or similar gas flow device.

On some models, the contacts are located in a closed phial containing a small amount of mercury.

When the bimetallic coil or spring moves (due to room temperature) the phial moves, causing the mercury to cover the contacts and make the switch.

Some types incorporate an internal heat anticipator (heater) that will cause the device to operate slightly before the pre-set room temperature is reached. This is to reduce the temperature differentials between 'on' and 'off' and to prevent overheating of the room.

Checks:

- keep a wall mounted unit containing the bimetallic parts free of dust/lint – to allow a circulation of air to reach the sensing device
- ideally site the thermostat so it is not affected by other heat sources, e.g. a radiator, solar gain, or by cold draughts. These affect its accuracy

- you must install the thermostat in accordance with manufacturer's instructions so it operates correctly
- when the thermostat incorporates an internal heat anticipator in addition to the live, switched live and earth wires (if applicable), you must connect a neutral wire to the internal heat anticipator. If you fail to wire the internal heat anticipator correctly, this will result in large temperature differentials between 'on' and 'off'

Frost thermostat – protects an appliance/area from freezing

Operation and use:

This device (similar in design to a room thermostat) protects an appliance or area from the danger of freezing. The majority are set to operate at 4°C, when they will override any pre-timed controller and turn on a heating appliance or central heating zone valve.

Locate the thermostat in the area to be protected (normally between the flow and return pipework of the appliance). Connect strap-on types directly to the heating pipes.

Strap-on types (often known as 'pipe stats') when used with wall mounted frost thermostats, also help to prevent nuisance problems, i.e. continual running of boiler/controls.

A strap-on type is normally wired in series with a frost thermostat. This:

- allows adequate protection to be afforded to the appliance/system; and
- also guards against nuisance calls and ultimately wasted energy resources

Checks:

As for room thermostat. If it is a strap-on type it should:

- be in contact with the pipework; and
- not be fitted on top of any pipe lagging

Thermistors – extremely accurate devices

Operation and use:

A thermistor is an extremely accurate, non-metallic heat-sensing device that alters its electrical resistance when subjected to temperature changes: i.e. the higher the temperature it detects, the greater the flow of electricity it will pass. This variable resistance can be used to control a modulating gas valve or fan.

Thermistors are also found in modern, programmable room thermostats.

Checks:

As for **Room thermostat – senses air temperature where located.**

Cylinder thermostat – controls temperature of stored hot water

Operation and use:

There are two main types:

1. Liquid expansion (manual) cylinder thermostats control a water flow valve, which is fitted to either the primary flow or return pipe of a hot water storage vessel.
2. Bimetallic (electric) types are in contact with the vessel body, which remotely controls a gas flow or system control valve.

They are normally set at 60°C, which reduces scale build-up in the domestic hot water system and to guard against legionella.

Checks:

As for **Room thermostat – senses air temperature where located.** Also:

- ideally locate models that are in direct contact with the hot water storage vessel at a third of the vessel's height from its base
- inset them into pre-lagged vessels and restrain with the manufacturers' supplied straps/fittings to ensure direct/continual contact

Air/gas ratio valves – minimise excess air to maximise performance

Important: A number of appliance manufacturers will indicate that the air/gas ratio valves fitted to their appliances are not intended to be adjusted. If this is the case then do not alter any of the valve settings.

If you adjust an air/gas ratio valve without following the specific manufacturer's instructions, it may result in the boiler combustion being adversely affected – with the possibility of generating high levels of carbon monoxide (CO).

You will be familiar with the need to check gas valve settings by:

- measuring the burner pressure with a manometer
- comparing the result to the values supplied by the appliance manufacturer, either by referring to the appliance data badge or the installation instructions

Why you cannot use this procedure with many modern gas-fired condensing boilers:

It is because they use a different type of gas valve – even though they may be similar in appearance.

We now describe the air/gas ratio valves that are found on many condensing boilers and highlight the differences between these and 'standard' gas valves.

Recent legislation requires the use of high efficiency boilers both:

- for new installations; and
- for the majority of replacements in existing dwellings

To maximise performance over the full boiler operating range, you need to minimise the amount of excess air that passes through the boiler.

This has led to greater use of gas valves that control the ratio of gas to air. They are especially useful in modulating appliances where direct control of the fan has enabled a simple control over the gas pressure and combustion over a wide range of modulation, typically 3.5:1.

Differing names for these valves can be confusing:

They are most frequently called 'gas control valves', 'air/gas ratio valves' or simply 'gas valves'.

This different terminology can make it difficult for you to identify which boilers have this type of valve.

Typically, a SEDBUK (Seasonal Efficiency of Domestic Boilers in the UK) band 'A' boiler (see Note) will have these valves fitted but do not assume that a boiler which is not a SEDBUK band 'A' rating will not have an air/gas ratio valve.

Note: The latest 2009 SEDBUK uses a percentage – 90% – in place of Band A.

When you work on appliances with this type of valve, you must always refer to the appliance manufacturer's instructions.

Attention: Do not use the following information in place of the appliance manufacturer's instructions. Always refer to these for settings/adjustment requirements.

You must not attempt to adjust air/gas ratio valves unless:

- the appliance manufacturer's instructions clearly indicate they can be adjusted and show the procedure to follow
- you have the necessary equipment (regularly calibrated as required) to carry out the adjustments according to the manufacturer's instructions, e.g. an electronic combustion gas analyser (ECGA)
- you are competent to use that equipment and interpret any results you obtain

Figure 12.8 Air/gas ratio valve configuration

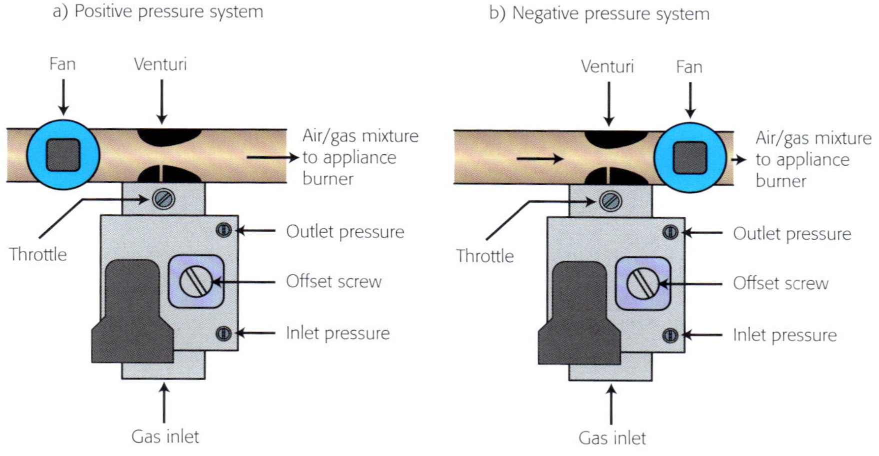

Air/gas ratio valves – can operate differently in new and old appliances

- most air/gas ratio valves used on new appliances operate on a similar principle
- however, it is important to note that some older appliances may have air/gas ratio valves that operate differently

As in all modern boilers, a fan is used to provide the airflow to the burner. The speed of the fan is accurately controlled by the electronic boiler control. This airflow generates a pressure drop across an air restrictor or venturi and this pressure is communicated to the gas control via a 'servo' regulator.

Most of the air/gas ratio valves used today are termed '1:1 valves' so that if the fan gives 5mbar of air pressure then the gas valve will deliver 5mbar of gas pressure. The gas is delivered through a controlled restriction, which typically has a relatively large diameter of up to 10mm.

This controlled restriction may be an orifice, traditional injector, a brass fitting or even an open calibrated pipe.

Two methods are used in current appliances to develop the driving pressure for the gas valve servo. They are usually referred to as positive or negative systems (see Figure 12.8).

1. In the positive system, the gas valve and pressure tapping are downstream of the fan and are therefore under positive pressure. The gas is injected into or just after the venturi to aid mixing.
2. In the negative system, these components are upstream of the fan and under negative pressure: the gas may be injected directly into the fan.

Figure 12.9 Air/gas ratio valve configuration – use of pressure connecting tube

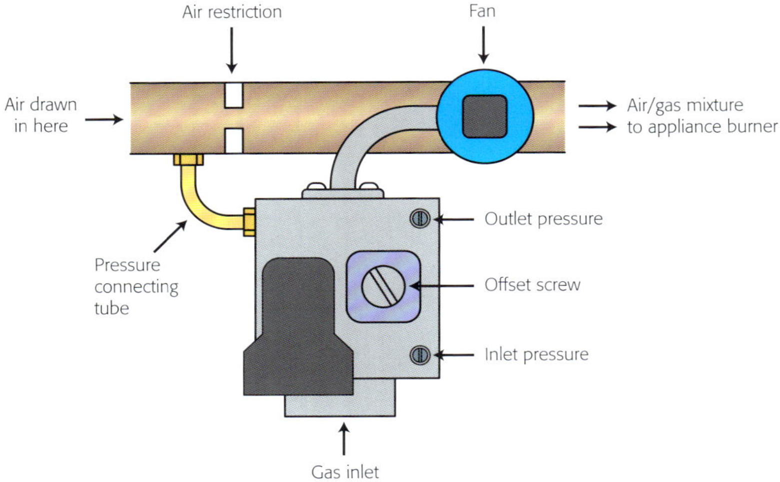

- there may be an air pressure tube connecting the valve to the venturi (see Figure 12.9). If this tube becomes disconnected then, with either system, the air/gas ratio will be incorrect
- alternatively, the gas valve may be positioned within a sealed compartment without a connecting tube on the servo connection. It is important to note that the pressures may be affected when you remove the cover door to gain access to the valve

Valve adjustments: two are possible – air/gas ratio and offset

The appliance manufacturer's instructions for specific boilers may indicate that these are factory set and must not be altered.

- before you make any adjustment to the valve setting, check to ensure the gas pressure at the valve inlet pressure test point is within the appliance manufacturer's specified range
- it is important that if you change/adjust either the air/gas ratio or the offset settings, you then check both settings as being correct
- you normally adjust the air/gas ratio before the offset

Air/gas ratio adjustment – to increase or reduce gas for airflow

The screw adjuster for the air/gas ratio may be referred to as a throttle or ratio adjuster. It may be on the valve, the venturi or a separate component. It can also be a fixed orifice or a combination of both orifice and throttle.

- strictly follow the appliance manufacturer's instructions
- some manufacturers specify and detail this adjustment, but some do not – as the adjustment may either be preset and sealed by the valve manufacturer, or at the factory by the boiler manufacturer

As explained, the gas control delivers a set amount of gas for a set airflow.

- the purpose of the throttle adjustment is to increase or reduce the amount of gas for a given airflow
- changing this setting alters the carbon dioxide (CO_2) in the combustion products

Why the correct CO_2 value is important:

Set the correct gas rate by changing the valve gas rate adjustment to give the correct combination of fan speed and CO_2. This is important so that the appliance functions correctly, otherwise:

- ignition, flame stability and resonance problems may occur; and
- the heat input may be incorrect

The manufacturer may quote a CO_2 value that corresponds to the maximum rate and the appliance manufacturer's instructions will indicate the position of this adjuster.

Carefully carry out the adjustment, with the boiler normally at maximum rate and with the aid of an ECGA in order to accurately set the CO_2 value to the manufacturer's recommendations. **Do not attempt this without one.**

Note: The appliance manufacturer's instructions will normally describe how to set the boiler to operate at maximum rate.

When you make adjustments, turn the throttle screw slowly, 45° turn at a time, checking the CO_2 value, after allowing sufficient time for the readings to stabilise (this time will vary depending on the appliance).

If the valve is in a negative pressure compartment, then replace the combustion box door before taking a CO_2 reading.

Offset adjustment – affects CO_2 and gas flow rate

The purpose of the offset adjustment is to set the CO_2 and therefore gas flow rate at the minimum of the appliance modulation range to allow for a lower or higher CO_2 than that measured at maximum.

Again strictly follow the appliance manufacturer's instructions. Some manufacturers specify and detail this adjustment but some do not, as the adjustment may be preset and sealed – by the valve manufacturer – or at the factory by the boiler manufacturer.

Some manufacturers specify that the valve is 'Factory sealed, no adjustment necessary'. If in any doubt, seek further advice from the appliance manufacturer.

As stated before, a 1:1 air/gas ratio valve modulates the gas pressure to follow the air pressure. In this way, by correctly sizing orifice diameters and resistances, the air/gas ratio can be kept constant for any value of the air pressure signal. When the gas pressure is set equal to the air pressure there is said to be zero offset.

Be aware that correct gas pressure is different in each installation

No two installations can be alike and every appliance will be slightly different. In the case of appliances with long flue pipes/chimneys, air throughput will differ greatly, so the correct gas pressure will be slightly different in every case.

This slight difference is important to enable these systems to effectively modulate, either deliberately or otherwise because of appliance build, or a blocked flue, for example. The air and gas pressures are very closely coupled to each other.

To ensure that combustion remains good over the full range of potential installations and throughout the modulating range, the gas outlet pressure may be 'offset' in relation to the air pressure.

Most appliances operate with a 'negative offset' whereby the gas outlet pressure is set slightly lower than the air signal pressure. This difference is usually set between 0 and -20 Pascals (PA) (0 and -0.2mbar) relative to the air, i.e. a greater air than gas pressure.

This value may be given on the appliance data plate instead of the usual gas outlet pressure. When set correctly this 'offset' usually ensures that at maximum rate the CO_2 value is about 9 – 9.5% and at the minimum about 8.5 – 9.5%. The values for LPG are normally 1% higher than these NG readings.

Confusion may arise over this adjustment

Many people are not fully conversant with pre-mix systems and their associated controls – and can get confused.

If the offset is wound in to try to increase the burner pressure with the appliance at maximum rate, then the increase in pressure measured at the outlet test point will be very small. However, at minimum rate this increase in pressure has more effect on the CO_2 and heat input – so you need to take care to fully understand the function of the adjustment.

If the manufacturer specifies the adjustment method and value for offset then follow their instructions.

To set the offset:

- you need to run the boiler at the minimum rate (the manufacturer's instructions should describe how to put the boiler into a test mode, so that it operates at the minimum rate)
- you can then adjust the offset setting to give the correct CO_2 readings
- turn the offset adjustment screw slowly 45° turn at a time:
  - clockwise to increase (more positive)
  - anti-clockwise to decrease (more negative)
- allow time for the readings to stabilise (this time will vary depending on the appliance)
- if the valve is in a negative pressure compartment then replace the combustion box door before you take a CO_2 reading

Where manufacturers require setting based on pressure differential:

Although setting the offset by means of a CO_2 reading is normally sufficient, individual appliance manufacturers may require the above setting and will detail this in their instructions.

If you adjust the valve by this means, you must have a micro-manometer available, that is calibrated and working correctly.

Chimney standards – 13

13 – Chimney standards

Chimney standards – 13

13 – Chimney standards

13 – Chimney standards

Figures

13 – Chimney standards

Tables

Introduction

This Part deals with the requirements for the evacuation of the Products of Combustion (POC) via a chimney system for domestic gas burning appliances, with rated heat inputs not exceeding 70kW (net) for 2nd and 3rd family gases in Permanent Dwellings (PD) and Residential Park Homes (RPH).

It reflects the general requirements of BS 5440–1: 2008 'Flueing and ventilation for gas appliances of rated input not exceeding 70kW net (1st, 2nd and 3rd family gases) – Part 1: Specification for installation of gas appliances to chimneys and for maintenance of chimneys'.

The safety factors to consider – and the appliances covered

The correct, efficient discharge of an appliance's POC is an essential safety feature, required under the GSIUR.

The type and construction of a chimney system depends on the type and size of the appliance. It is critical that its design allows for the discharge of POC into the atmosphere under all weather conditions.

Modern gas appliances are available with a great variety of chimney/flue pipe options. Although this offers you and gas users alike a great deal of choice whenever practical, you should choose a room-sealed (Type C) appliance. Also:

- changes in appliance design/requirements mean that, even if a chimney/flue pipe has worked satisfactorily in the past, do not assume that it will be adequate and comply with current standards for a replacement appliance
- always consult the appliance manufacturer's installation instructions
- you must always consider building regulations applicable to the geographical region where you are installing, as well as other relevant statutory documents

Note: Do not install new or previously used gas appliances unless you have the manufacturer's installation instructions.

Changes to industry standards – note new definitions

BS EN 1443 provide some new definitions (found in BS 5440-1), which alter common terms we have been used to within the gas industry, when we talk about flues and chimneys (see Figure 13.1).

In relation to this Part, the following definitions are used:

Chimney: This is a structure consisting of a wall or walls enclosing the flue or flues.

Chimney component: This is any part of a chimney.

Connecting flue pipe: Component or components connecting the heating appliance outlet and the chimney.

Flue liner: This is the wall of a chimney consisting of components, the surface of which is in contact with the POC. This is not just a flexible flue liner, but also any suitable material to convey these products.

Flue: This is the passage or space for conveying POC to the outside atmosphere.

Figure 13.1 Chimney components

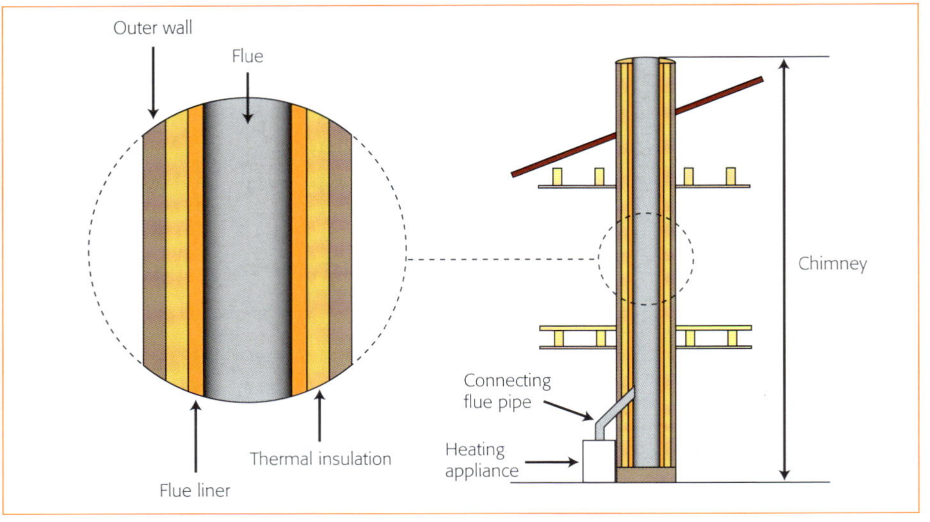

Appliance Classification – three types

Appliances and their method of evacuation of the POC are classified by Published Document – PD CEN/TR 1749 'European scheme for the classification of gas appliances according to the method of evacuation of the combustion products (types)' – to harmonise classification across Europe of appliances burning combustible gases.

PD CEN/TR 1749 separates gas appliances into three types:

1. Type 'A' Flueless – an appliance not intended for connection to a flue or a device for evacuating POC to the outside of the room in which the appliance is installed.
2. Type 'B' Open-flue – an appliance intended to be connected to a flue that evacuates POC to the outside of the room containing the appliance. The air for combustion is taken from the room.
3. Type 'C' Room-sealed – an appliance in which the combustion circuit (air supply, combustion chamber, heat exchanger and evacuation of the POC) is sealed with respect to the room in which the appliance is installed.

Table 13.1 and Figures 13.2 to 13.5 provide more details on the methods of evacuation of their POC/supply of combustion air available.

Table 13.1 Classification of gas appliances according to flue types

Appliance type	Primary definition	Natural draught or fanned draught		
		Natural draught	Fan downstream of heat exchanger	Fan upstream of heat exchanger
A – Flueless		A_1*	A_2	A_3
B – Open-flued	B_1 – appliance with a draught diverter	B_{11}*	B_{12}* B_{14}**	B_{13}*
	B_2 – appliance without a draught diverter (previously known as a "closed flue")	B_{21}	B_{22}*	B_{23}
C – Room-sealed	C_1 – appliance with a horizontal balanced flue/inlet air ducts to outside atmosphere	C_{11}*	C_{12}*	C_{13}*
	C_2 – appliance flue system connects to a common duct system for multi-appliance installations	C_{21}*	C_{22}	C_{23}
	C_3 – appliance with vertical balanced flue/inlet air ducts to outside atmosphere	C_{31}	C_{32}*	C_{33}*
	C_4 – appliance with flue system that connects to a 'U' duct flue system	C_{41}	C_{42}	C_{43}
	C_5 – appliance with a non-balanced flue/inlet air ducted system	C_{51}	C_{52}	C_{53}
	C_6 – appliance sold without a flue system	C_{61}	C_{62}	C_{63}
	C_7 – appliance connected to a vertical flue to outside atmosphere with the air ducts in the loft (draught diverter in loft above air inlets)	C_{71}	C_{72}* (commonly known as a 'Vertex' flue)	C_{73}* (commonly known as a 'Vertex' flue)
	C8 – appliance with a non-balanced flue system with an air supply from outside atmosphere and flued into a common duct system	C_{81}	C_{82}	C_{83}

* common types of appliances found in the UK

** appliance fan downstream of draught diverter

Figure 13.2 Open-flued appliance types

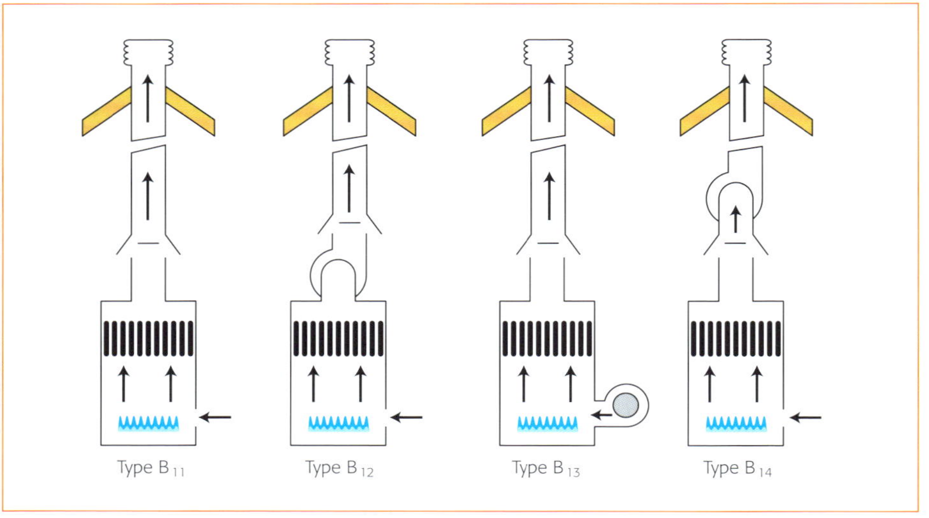

Figure 13.3 Room-sealed appliance types

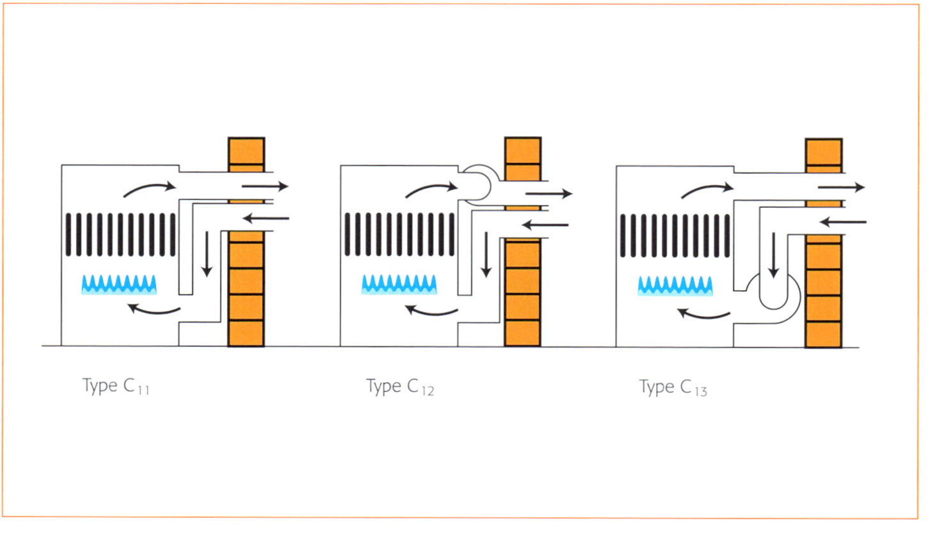

Figure 13.4 Room-sealed vertical termination

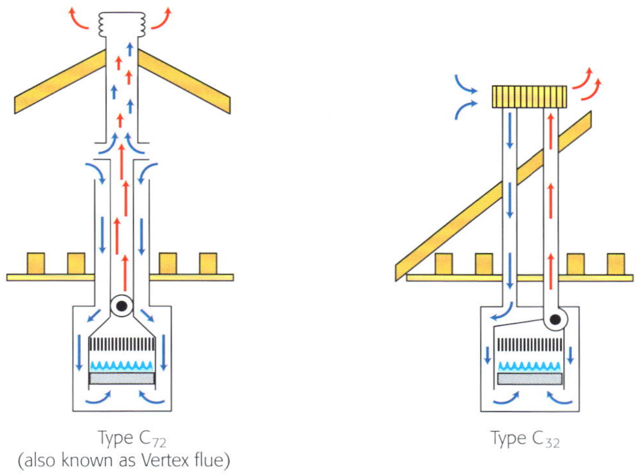

Figure 13.5 Se-duct/U-duct

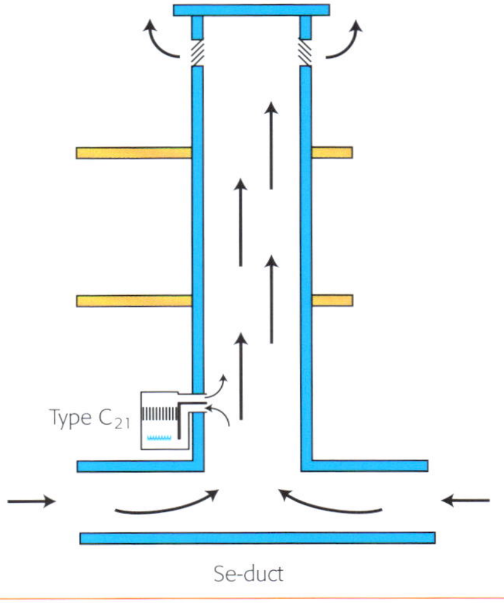

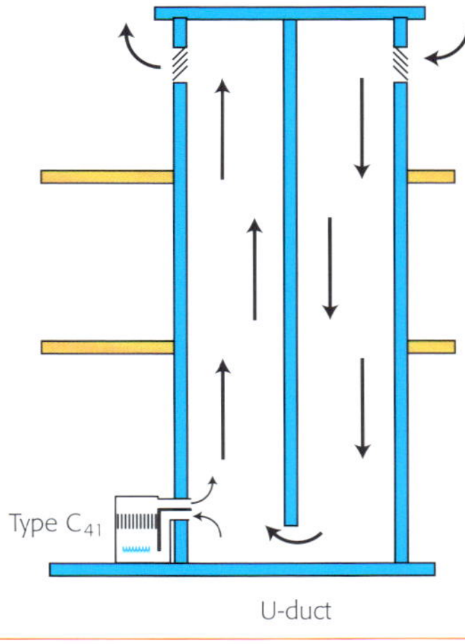

Exchanging information with the gas user – and planning

When installing the gas appliance, agree the important details of the installation with the gas user and document this.

Where the chimney system is part of the appliance, e.g. in the case of a room-sealed appliance, check the gas user agrees that the chimney system is suitable for the purpose and document this.

When you erect a new chimney system or modify an existing one, take into account the important details, to include:

- the type, size and route of the chimney system
- the type and size/heat input rating of the appliance

This is important when different trades are involved in the installation of the chimney system and the appliance.

Where you plan to reuse an existing chimney system, confirm that it's suitable for the installation of a new or replacement appliance.

Building regulations/standards – current requirements

On 1st October 2010 a suite of revised Approved Documents to the Building Regulations (England and Wales) were released. Among them is a revised Approved Document J (ADJ), which replaces the previous version published May 2006 (see **Part 1 – Gas safety legislation – Building Regulations/Standards – requirements you must follow for any building work**).

Similar requirements will be found in the Scottish Building Standards and other regions of the UK.

In particular, ADJ provides for:

- New provision for inspection hatches to be installed where chimney systems are installed within voids (see **Room-sealed fanned-draught chimney systems concealed within voids** in this Part)
- adequate supply of combustion air – see **Part 4 – Ventilation**
- the siting of flue terminals in relation to boundaries
- correct operation of chimney systems
- complete/correct combustion of the fuel used
- appliances being used safely without creating combustion problems i.e. spillage of the POC
- appliances being used safely without causing a danger to the premises into which they are sited
- appliances being suitable for the purpose for which they are intended
- chimneys being appropriately labelled to identify their capability

Requirements for the flueing of domestic gas appliances is covered by BS 5440-1.

Note: For further guidance on the siting of flue terminals in relation to boundaries, see 'Siting of flue terminals in relation to boundaries' in this Part.

Building regulations – when you must complete a compliance report

You are responsible for meeting the requirements of ADJ when you carry out the work. When you have completed the building work e.g. installation of a flue liner or chimney, you demonstrate this by filling in a compliance report (checklist) to indicate that you:

- have completed the work correctly and it complies with the regulations
- have detailed the materials and components used
- have detailed the tests that have been performed
- have provided this information to your client/customer (see **Part 10 – Emergency notices, warning labels and forms**)

When not to complete a compliance report

For appliances where the chimney system is supplied as part of the appliance, such as room-sealed boilers or space heaters.

If the work involves the like-for-like replacement of a component e.g. replacing a flue terminal, replacing a connecting flue pipe between a floor standing open-flued boiler and the chimney, there should be no need (see also **Identifying a hearth and chimney by means of a notice plate** in this Part).

Additional guidance/supplementary information on Approved Document J (2010)

This is aimed at specifiers, builders, manufacturers and installers of chimneys and associated products and appliances, amongst others.

It provides guidance during the transition away from the current system of using British Standards for specifying/installing chimneys – towards one based on European Standards (under the European Construction Products Directive (CPD)). It is about:

- the general issues of European standardisation for chimneys
- the overall framework of legislation under which this process is taking shape
- the new way of designating chimney products, according to essential performance characteristics

Identifying a hearth and chimney by means of a notice plate

One of the requirements of ADJ is the need to provide a notice plate giving details:

- of the characteristics of any new hearth, fireplace, chimney; or
- when a chimney is extended or lined

to ensure that any gas operative who visits the property in the future, knows what appliances are compatible/suitable to connect to it.

When do you complete a notice plate?

1. First establish if there is an existing notice plate. You are likely to find it next to the chimney structure, electricity consumer unit, or water supply stopcock as these are recommended positions within ADJ.
2. If there is an existing notice plate and changes to the specification restrict the use of the chimney or hearth further, you need to amend (if possible) or replace the notice plate accordingly. This includes where the chimney is part of refurbishment work and alterations to the specification further restrict the types of appliances that can safely be connected to that chimney.

There is no need to complete a notice plate

For chimney systems of room-sealed appliances with integral flue/air duct assemblies, which are dedicated to the appliance they serve.

Providing a notice plate – what you are responsible for

Where you carry out work which needs a notice plate, you are responsible for providing, completing and fixing the notice plate in position.

The minimum information you must provide on the plate is:

- the location of the hearth, fireplace (or flue box), or the location of the beginning of the chimney;
- the category of the chimney and type of appliance that can be connected to the chimney structure;
- the type and size of the chimney or chimney liner and the manufacturer's name;
- the date of installation

You can add more information (although you are not required to). This may include, for example:

- your name and address
- product trade names
- any warnings relating to (for example) specific limitations on the installation and use of a hearth

How you fix the notice plate

It needs to be durable and permanently marked. Fix it securely in a position that is obvious but not obtrusive. Suggested locations are:

- next to the electricity consumer unit
- next to the chimney or hearth to which the plate relates
- next to the water supply stopcock

If you are to install an appliance onto an existing chimney, check whether a notice plate already exists for it. Initially, this will apply to new buildings, but over time, as you come to renovate chimneys in older buildings, you will find they have notice plates.

When you find one, check that the chimney and/or chimney liner is suitable for the appliance you are installing. If any properties do not meet the requirements outlined in the appliance manufacturer's instructions, then you need to change them.

If you introduce a new chimney liner or replace an existing chimney liner, then it is important that you replace the notice plate to reflect these new conditions.

Note: Should you need up-to-date information for areas other than England and Wales (apart from Scotland) on all the requirements for building legislation raised in the Building regulations/standards and discussed in this Part, then contact the local BCB appropriate to the geographical region in which you are working.

Technical Handbook–Domestic 2015 – guidelines about need to obtain a warrant

All building work to which the The Building (Scotland) Regulations apply, requires you to obtain a building warrant prior to work commencing (except where specifically exempted – see Note).

Note: Work must conform to the building regulations, whether or not a warrant is needed. Check with the Technical Handbook – Domestic 2015 and/or the building standards department of the relevant local authority, if you have any doubt about this need for a warrant.

The following list is not exhaustive as to whether a warrant is needed. Technical Handbook – Domestic 2015 and the local authority will provide a more comprehensive and detailed list.

Further information on the requirements of building standards for Scotland, contact the Scottish Building Standards Agency at www.sbsa.gov.uk.

You do not normally require a warrant where

- the proposed gas appliance is less than 70kw (net)
- the house is no more than two storeys
- installation of a chimney system (room-sealed) or work associated with a flue outlet fitting/terminal
- you are installing a flue liner

You do normally require a warrant where

- the proposed gas appliance is greater than 70kW (net)
- the house is greater than two storeys
- any work associated with a chimney, flue pipe or constructional hearth

A chimney system operates thermally or by fan, to remove POC

The motive power for the flue of a chimney system is produced either:

- thermally, by the heat of the combustion products; or
- by means of a fan

An open-flued (Type B) appliance takes its combustion air from the space containing the appliance.

A room-sealed (Type C) appliance takes its combustion air from outside the space containing the appliance. An open-flued (Type B_5) appliance without a draught diverter (previously known as a 'closed flue') takes its combustion air from the space containing the appliance.

Note: See 'Incinerators' in this Part.

POC produced by gas burning appliances are non-toxic, provided:

- the appliance and burners are clean and in good working order; and
- there is an adequate supply of fresh air for combustion

If POC are not removed from the combustion space and the air used for combustion is not replaced with fresh air, the atmosphere surrounding the appliance will become contaminated (vitiated). This contaminated air will cause incomplete combustion.

Why it is necessary to remove POC

The POC contain traces of carbon monoxide (CO), which could be discharged into the appliance space or room. So it is necessary:

- to remove the POC as soon as they are produced through a chimney system to outside atmosphere; and
- to replace the used air with an equal quantity of fresh clean air

Figure 13.6 Parts of an open-flue chimney system

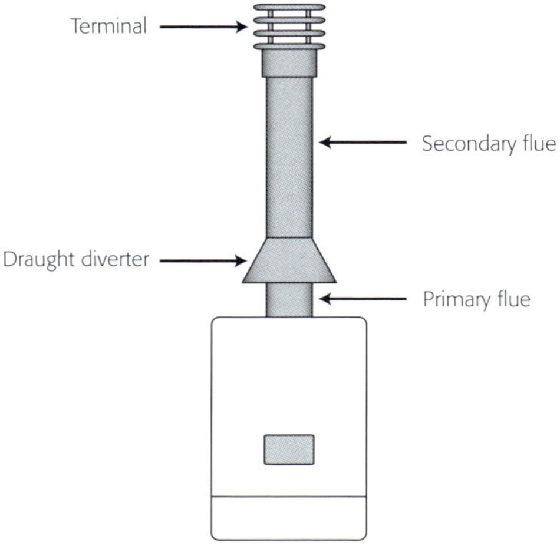

Individual open-flues – natural-draught (Type B) appliance

A typical individual open-flued system consists of four parts (see Figure 13.6):

- primary flue
- draught diverter
- secondary flue
- terminal

Primary flue – creates initial flue draught

Its length determines the overall efficiency of the appliance. It is generally incorporated into the appliance or comes as part of the draught diverter assembly, to provide motive force to overcome the resistance of the draught diverter.

Draught diverter – threefold function

This is normally an integral part of the appliance, but can be a separate component. It:

- prevents (diverts) any down draught in the secondary flue from entering the combustion chamber and interfering with the combustion process
- allows POC within the flue to be diluted with air
- prevents excess flue pull in the secondary flue from interfering with the combustion process

Where a separate draught diverter is fitted, always install it in the same space as the appliance.

Secondary flue – transfers POC/dilutant air to atmosphere

It is important to keep the resistances of the components in the secondary flue to a minimum.

Flue terminal – assists removal of POC and prevents blockage

Birds, leaves, snow, etc. can all block/restrict the flue, but as well as providing protection for the flue the terminal plays an important role in its correct operation.

- fit terminals to flues with a cross-sectional dimension of 170mm or less
- ensure that the terminal is suitable for the appliance type

Note: Terracotta chimney inserts (rain caps) are not suitable for use with gas appliances.

Figures 13.7 and 13.8 illustrate some suitable and unsuitable gas appliance flue terminals.

Individual open-flues – fanned-draught (Type B) appliance: allow flexibility

This system allows greater flexibility when siting the appliance and flue termination.

There are two types of fanned draught chimney system:

1. Where a fan is an integral part of the appliance (factory fitted).
2. Where a fan is fitted in the chimney system, which has been supplied or specified by the appliance manufacturer.

Note: Where the fan is not factory fitted:

- **connect it to the appliance in accordance with the manufacturer's instructions**
- **do not modify the appliance without obtaining prior written agreement from the appliance manufacturer**
- **if you use an independent fan, make sure it is acceptable to the appliance manufacturer**

Proprietary fan kits include fail safe features

These kits, often available from appliance manufacturers:

- prevent appliance operation if the fan fails to operate
- allow you to adjust the fan speed during installation and commissioning, to ensure adequate removal of POC

When the fan is separate from the appliance, take the fan's performance (output) into account when calculating the chimney system design. Specialist advice may be necessary regarding the sizing and specification of the fan.

Choose/design the fan to evacuate POC, even under adverse weather conditions (wind pressure) and take into account the chimney system route and termination.

Note: For information on flue terminal locations see 'Open-flued terminal positions – Flue termination open-flue – fanned draught' in this Part.

Due to the increased pressure within the chimney system, carefully check joints and the condition of the chimney system for leakage, see note.

Figure 13.7 Suitable flue terminals

Figure 13.8 Unsuitable flue terminals

Note: The joints of metallic chimney systems/flue pipes might show small signs of leakage. This is permitted as part of the certified product specification; metal to metal twist joints will not be hermetically sealed (airtight). If you are in any doubt about the suitability of the chimney system, contact a chimney expert or the manufacturer of the chimney system for guidance.

Safety devices – prevent appliance operating if fan fails or flue flow reduces

- chimney systems incorporating a fan must be fitted with a flow proving device, to prevent the appliance operating in the event of fan failure, or reduced flue flow condition
- the flue flow sensor in the safety control needs to fall to the 'no flow' position before the fan can operate
- following operation of the safety control, a manual reset is normally required, to re-establish the gas supply to the main burner

Chimney materials – must conform to relevant British Standard(s)

When constructing a chimney system, only use materials appropriate to its use and location.

Rigid metallic – factors to take into account

- rigid metallic chimney components should comply with BS EN 1856: 'Chimneys – Requirements for metal chimneys. Part 2 Metal liners and connecting flue pipes'
- do not use internal grade metallic chimney components externally – except in the uppermost section, which penetrates the roof line
- do not mix different makes of metallic chimney components in a chimney system construction, unless you use approved manufacturer's adaptors

Rigid non-metallic (asbestos) – requires specialist work/strict precautions

Warning: Asbestos removal is specialist work – only to be carried out by approved businesses defined in the Control of Asbestos Regulations i.e. licenced operators.

As required by the Health and Safety at Work etc. Act and The Control of Asbestos Regulations 2012, carry out a full risk assessment and apply special safety precautions when you work on an existing asbestos cement chimney, including and not forgetting surrounding structures (insulation boards and lagging, Artex finishes, fire proofing, etc.) which may also be manufacturered from Asbestos Containing Materials (ACMs) – see 'Asbestos – how/where you may find it' in this Part.

For further advice, consult your organisations operational procedures and/or contact the HSE information line on 0845 345 055, and/or visit www.hse.gov.uk/asbestos.

- you can re-use existing asbestos cement chimney systems complying with BS 567 and BS 835 if they meet current requirements for correct termination position, routing etc. see Note
- do not use asbestos cement chimney components when you construct new systems, or alter existing systems
- take stringent precautions when you work with asbestos – and be aware that non-asbestos content rigid chimney components are available
- in all cases consult the manufacturer's instructions

Note: An existing asbestos chimney/chimney component may only be re-used if it is in good condition and does not require cutting or machining.

Asbestos – how/where you may find it

Asbestos is a fibrous mineral, which in its various forms (Chrysotile, Amosite and Crocidolite) was used in insulating, fireproofing, pipes and sealing materials.

No form of asbestos is now used in products in everyday use, but different forms may occur in old installations and appliances. In the gas industry, the forms you may find are:

- Chrysotile (white asbestos) in the form of bonded cement chimney components, asbestos rope, seals and gaskets (the most common form you will encounter) – also in textured decorative finishes and 'Artex' type materials
- Amosite (brown asbestos) and Crocidolite (blue asbestos) as a constituent of some older type insulations, particularly asbestos insulating board (AIS) and 'soft' insulating material

Asbestos – a major health risk, so record of location/condition needed

The Control of Asbestos Regulations place a legal duty on owners, occupiers, managers and others having responsibility for non-domestic buildings and those 'non-private' parts of domestic buildings i.e. common hall ways, etc. of multi-occupancy premises (flats):

- to manage the risk from asbestos containing materials (ACMs); and
- to co-operate with whoever manages that risk

The risk is managed by:

- proper surveying of the building (workplace), which includes adequately documenting the presence or suspected presence of asbestos as well as providing that information to others who may come in to contact with ACMs (work within the premises i.e. tradesmen); and
- tradesmen being aware of and working correctly with and around ACMs, which will involve a degree of training to be undertaken

To help in this regard, the HSE has published various guidance on:

- how to survey workplace premises for ACMs; and
- how to record the results
- how to work with or around ACMs

One such guide – MDHS 100 'Surveying, sampling and assessment of asbestos-containing materials' – has been produced as a standard for people carrying out or commissioning asbestos surveys in buildings.

The guidance is part of the Methods for the Determination of Hazardous Substances (MDHS) series, produced by Health and Safety Laboratory covering all manner of dangerous substances including asbestos.

It is an important part of the HSE's strategy to prevent people dying from asbestos related diseases and adds to existing guidance on asbestos.

The key elements are:

1. Locating the extent and nature of any asbestos.
2. Assessing the condition of the material to determine whether it presents a risk.
3. Recording the information in a clear and easily usable form.

It describes three levels of survey on a building (of varying degrees of intensity) and advises which one is most appropriate for particular situations, for example:

- some surveys involve collecting representative samples of suspect materials for laboratory identification
- others rely on studies of building plans or other information sources

Other HSE guides such as 'Introduction to asbestos essentials...' (Ref: HSG213) and 'Asbestos essentials...' (Ref: HSG210) are aimed squarely at tradesmen, providing practical guidance and procedures for dealing with ACMs.

Pre-requisites needed before you can undertake asbestos surveys/take samples etc.

- you must have appropriate training, competence and experience
- you must be able to demonstrate independence, impartiality and integrity
- adequate quality assurance procedures must be in place for the individual or organisation carrying out the survey, in order to achieve consistently high standards

The guidance contains a wealth of information about where asbestos can be found in buildings and includes photographs to assist surveyors.

Pass information about asbestos to those at risk

- asbestos is the primary cause of occupational ill health in the second half of the twentieth century
- asbestos related diseases account for more than 3,000 deaths a year in the UK and an estimated 50,000 have died from mesothelioma and other conditions, since 1968
- around one quarter of current deaths occur in the building and maintenance trades – where people can be exposed to asbestos inadvertently when working in building

Note: The HSE has a wealth of information and guidance to assist all who manage, work or come in to contact with asbestos.

Visit – www.hse.gov.uk/asbestos – for further guidance.

Flexible metallic liners – installed within the fabric of building

Flexible metallic flue liners must comply with BS EN 1856-2 and are designed to be installed within the fabric of the building i.e. inside a brick/masonry chimney.

- do not join them to achieve the desired length
- ensure that flue liners do not form an angle greater than 45° from the vertical

Vitreous enamel – only to be used internally

Take great care when cutting vitreous enamel flue pipe, as it chips easily.

Plastic – mainly used with condensing appliances

Only use plastic flue pipes when the appliance manufacturer's specification allows. These pipes are mainly used with condensing appliances. Carry out jointing of the flue strictly in accordance with the manufacturer's installation instructions. See also **Room-sealed fanned-draught chimney systems concealed within voids** in this Part.

Brick/masonry chimneys – must be mechanically sound

Always check to ensure that the lining is mechanically sound before installing any appliance. Sweep brick/masonry chimneys with this type of lining, where they have been used with another fuel and carefully examine them before use with a gas appliance.

Where necessary, line them with one of the materials as follows:

- clay/ceramic flue linings complying with BS EN 1457
- concrete flue linings complying with BS EN 1857

- rebated or socketed lining made from kiln-burnt aggregate and high alumina cement
- metallic chimney liner complying with BS EN 1856-2
- factory made insulated chimney complying with BS EN 1856-1
- poured/pumped concrete linings installed by a method certificated by an accredited test house

Note: From 1987, poured concrete linings installed in existing brick/masonry chimneys, should meet the British Board of Agrément (BBA) Approval Scheme for installers.

Where no certificate of compliance with the BBA scheme exists, or no other documented evidence is available for the liner, install an additional liner e.g. flexible stainless steel liner inside the existing poured concrete liner.

Chimney flue blocks – ensure compliance

These should comply with BS EN 1806 (clay/ceramic) or BS EN 1858 (concrete). The cross-sectional area should not be less than 16500mm^2, with a minimum internal dimension of 90mm and never less than the area of the appliance flue outlet.

Be aware that before 1986, the cross-sectional area specification for chimney (pre-cast) flue blocks was 13000mm^2 with the minimum dimension 63mm.

A vertical height of 600mm is required above appliance connection or the recess created by the starter and cover chimney flue blocks before any raking blocks are fitted.

A choice of jointing material such as fire cement or silicone compound in cartridge form applied using a caulking gun may be recommended by the manufacturer of the blocks.

However, other possible recommendations for jointing could be ready mix fire cement, refractory hydraulically setting mortar or standard mortar subject to manufacturers advice.

The flue pipe section interconnecting the chimney flue blocks in the roof space to the ridge terminal should be a minimum size of 125mm double walled system to BS EN 1856-1 and 2.

Where chimney flue block system is constructed as part of external wall cavity, at that point the cavity should be insulated to a distance of 200mm either side of the vertical path of the flue.

A minimum of 50mm clearance between blocks and any adjacent structural timber e.g. floor joists needs to be provided.

Check to ensure that the gas appliance is suitable for fitting to the chimney flue block system.

Gas flue boxes – why and where to install

Suitable sheet metal flue boxes (see Figures 13.9, 13.10 and 13.11) enable you to safely install radiant convector gas fires, ILFEs, DFEs, combined gas fire/back boilers (BBUs) and gas heating stoves:

- within a builder's opening; or
- at a property built without a brick/masonry chimney

You may fit a chimney system that is either rigid twin wall or a flexible flue liner to the proprietary gas flue box, to allow the POC to be dispersed safely to the outside air.

Note: Install a flexible flue liner in one continuous length, with no joints or breaks and only use it inside the confines of a brick/masonry chimney.

Figure 13.9 Typical flue gas collector box

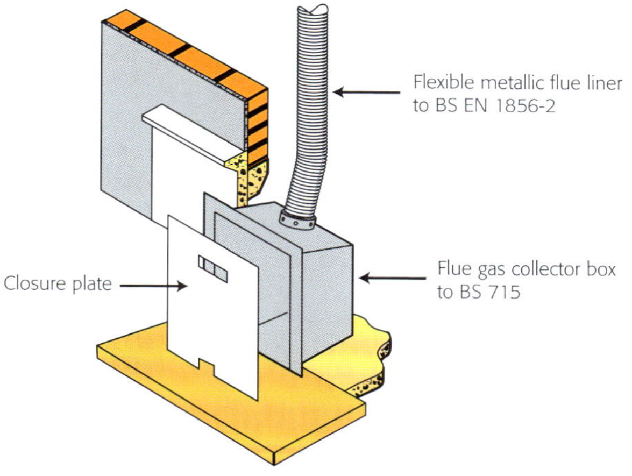

Figure 13.10 Typical gas flue box for a back boiler unit

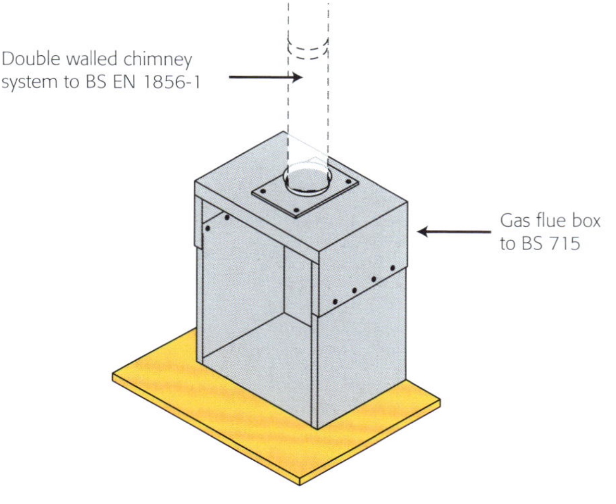

Figure 13.11 Method of installing a gas fire/decorative fuel effect (DFE) appliance in conjunction with a flue gas collector or a gas flue box

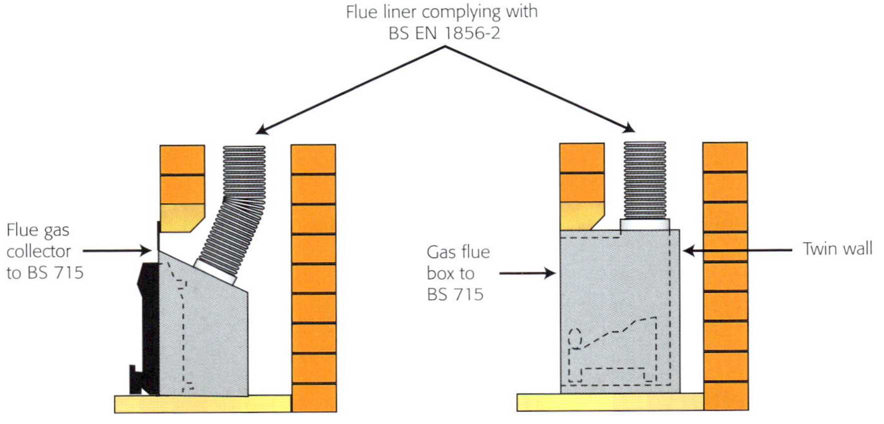

Gas flue boxes:

- are manufactured using corrosion resistant materials
- are either single walled or double walled, with or without insulation
- are normally rectangular in shape and constructed so they are sealed and mechanically strong
- should have a warning label affixed (displayed inside) stating 'fire risk' and indicating that they are only for use with gas appliances conforming to the relevant British Standard

Note: This is necessary to indicate that the gas flue box is not suitable for use with a solid fuel burning appliance.

In addition to appliances that incorporate a closure plate, you may fit specific appliances without a closure plate to this type of system.

Where you install an Inset Live Fuel Effect (ILFE) gas fire, you must suitably secure the appliance and seal it to the gas flue box. You should find detailed guidance in the individual appliance manufacturer's installation instructions.

Note: Unless otherwise specified in the manufacturer's instructions, you would normally install a gas fire, ILFE gas fire, BBU and gas heating stove to a minimum 125mm diameter flue.

DFE gas appliances – must be suitable for use

Before you connect a Decorative Fuel Effect (DFE) gas appliance to a gas flue box, you must ensure that the appliance manufacturer and/or gas flue box manufacturer approve it as suitable for that particular use.

Certain types of gas flue boxes are referred to as 'convector boxes' or 'hot boxes'. There is normally a 'chairbrick' shaped arrangement located within the gas flue box.

The convector box is designed to provide a level of convected heat output.

Note: Unless otherwise specified in the manufacturer's installation instructions, you would normally install a DFE gas appliance to a minimum 175mm diameter flue.

Dwellings built without a brick/masonry chimney

If there is no brick/masonry chimney installed within the property, gas flue boxes will normally accommodate the various types of gas fires, BBUs and gas heating stoves available.

Connect a rigid double walled chimney system manufactured and tested to BS EN 1856-1 to the proprietary gas flue box and pass it safely through the property, terminating through the roof, or at a ridge terminal.

You can install the chimney system internally or externally. However, to minimise condensation, consideration should be given to the length of external chimney system, in order to maintain the temperature of the POC above the dew point.

In these circumstances, use an insulated double walled chimney system. This is a particular requirement for external chimneys in excess of 3m in length. Normally, a false chimney-breast is constructed around a rigid chimney system and flue box.

Flue gas collector boxes – when and how to use

These are not true gas flue boxes, but flue gas collectors, built using corrosion resistant materials.

- where gas fires and other appliances incorporate a closure plate as a method of controlling the rate at which air is drawn from the room into the flue, then you may install the closure plate to the flue gas collector box
- provide a secure and long lasting seal between the closure plate and the flue gas collector itself (normally use a proprietory adhesive tape (PRS10) or other suitable sealing method: self-tapping screws and flue jointing compound, for example)
- install the flue gas collector box within a standard fireplace opening
- connect a flexible flue liner conforming to BS EN 1856-2 to the box within the confines of the brick/masonry chimney and suitably terminate it, to ensure that the POC will be evacuated safely, see Figures 13.9 and 13.11

Note: For further guidance on the installation of flue gas collectors, see 'Gas Fires and Space Heaters' from the current Gas Installer Manual Series – Domestic.

Chimney system design (domestic open-flued) – factors for effective working

- design the chimney system to ensure full clearance of the POC
- also consider flue resistance, temperature, condensation and aesthetic appearance

Factors which may have an adverse effect

1. A chimney system of inadequate size, height and route may cause spillage of POC at the appliance draught diverter.
2. If the terminal is sited incorrectly this can cause the same problem.
3. A flue that is oversized is more likely to be affected by condensation than a correctly sized flue.

   In general, a flue of the same size as the appliance flue spigot outlet will normally be of adequate cross-sectional area, provided that the flue height is also sufficient.

   The required cross-sectional area of an individual open-flue depends on a number of factors including appliance rated heat input, length of flue and the resistance of the chimney system components.
4. Passive stack ventilation systems, extract fans, tumble dryers, re-circulatory ceiling fans and warm air heaters (see also **Part 14 – Flue testing** for further guidance).

Calculating the equivalent flue height

This is an important part of chimney system design and is to allow comparison of the planned flue height against a known mathematical formula.

This formula also allows you to check existing chimneys systems to see if they will work.

Note: For the correct flue cross-sectional areas for gas fires, including ILFE gas fires and DFE gas appliances, see the manual entitled 'Gas Fires and Space Heaters' from the current Gas Installer Manual Series – Domestic.

General factors affecting flue performance are

- size (diameter or cross-sectional area) of chimney components
- flue height (actual and equivalent)
- terminal type and position
- number and angle of bends
- lengths of flue that run at an angle of up to 45° from the vertical
- frictional resistance created by the flue wall
- resistance of flue inlet and outlet

All these factors affect the equivalent or effective height of the flue.

Keep to the following points:

- size the flue in accordance with the appliance manufacturer's installation instructions
- ensure the cross-sectional area of the flue is not less than that of the appliance flue connection
- bends should not normally exceed 45° to the vertical

Note: You may use a 90° chimney bend where the flue pipe from a gas appliance enters an unlined brick/masonry chimney, to prevent debris from entering the chimney system/appliance. For further guidance, see 'Chimney connection – vary according to type/age' in this Part.

A newly constructed chimney system for a gas fire should not be less than:

- $12000mm^2$ in cross-sectional area, if the shape of the flue is round; or
- $16500mm^2$ in cross-sectional area, if the flue is rectangular, with a minimum dimension of not less than 90mm

For any other appliance, the flue should have a cross-sectional area equal to that of the flue spigot of the appliance, unless the appliance manufacturer specifies otherwise.

Do not connect the following appliances to chimney flue blocks with a cross-sectional area of between $12000mm^2$ and $13000mm^2$, or to flues with a minor dimension of 63mm or less:

- appliances with a flue outlet greater than $13000mm^2$
- BBUs unless you use a purpose designed adaptor/starter block, supplied and tested by the appliance manufacturer, and the installation instructions specifically state this is acceptable
- appliances which manufacturers state are not suitable for connection to chimney flue blocks
- drying cabinets

Note: Chimneys constructed of concrete blocks have a higher resistance to flow than traditional flue pipes of the same cross-sectional area. Not all appliances are acceptable for use on chimney flue block systems.

Consider the following points, as they may help the chimney system to perform correctly

- it should rise vertically for at least 600mm above the draught diverter
- it should take the most direct route practicable
- avoid horizontal and slightly inclined chimney/flue pipe routes
- keep the number of bends to a minimum
- avoid long external runs, particularly from appliances installed in basements and cellars (see Note)

Note: Where you cannot locate an open-flue internally within the building, the flue needs to have a minimum resistivity of R22 ($0.22m^2/kW$) where condensing of the POC are likely with a condensing/semi-condensing gas appliance.

An essentially vertical route from the draught diverter is recommended.

Acceptable exceptions may be:

- where chimney systems are constructed in buildings other than single storey and/or where equivalent flue height calculations indicate likely clearance of POC and where the flue passes the relevant safety checks (see also **Part 14 – Flue testing**)
- gas heating stoves where the manufacturer's instructions require a section of the chimney to be horizontal
- purpose designed modular boiler installations
- to avoid problems associated with down-draught on steeply pitched roofs, terminate flues at the highest point of the roof (e.g. at or near the ridge) rather than lower down it

Equivalent flue height – how to check/calculate to ensure full clearance

Use the following procedure (see Note) to determine if a given chimney system design, or existing chimney system, is likely to ensure full clearance of the POC:

1. Calculate the equivalent height of a straight, vertical, circular flue of specified size that will produce the same flue flow rate as the flue you are considering.
2. Check this height against Table 13.4 to see if it exceeds the minimum equivalent height requirement.

A flue is likely to be satisfactory if its equivalent height exceeds the minimum equivalent height given in Table 13.4.

Note: Do not use it for chimney systems designed for incinerators, see 'Incinerators have special flueing requirements' in this Part.

The calculation required to determine the equivalent height of the flue under consideration is detailed in BS 5440-1 as follows.

$$H_e = H_a \times \frac{(K_i + K_o)_e}{(K_i + K_o)_a - (K_e \times H_a) + \Sigma K}$$

The mathematical symbols represent:

H_e = height of the equivalent flue

H_a = vertical height of the actual or proposed flue, measured from flue spigot

K_i = inlet resistance of the flue

K_o = outlet resistance from the flue

$_e$ = equivalent flue diameter

$_a$ = actual or proposed flue diameter

K_e = resistance per unit length of the equivalent flue

ΣK = the sum of any other resistance factors: flue bends, flue terminal, flue type per metre length (circular or rectangular) of the actual or proposed flue

Two worked examples are given in this Part.

Note: Table 13.2 and Table 13.3 contain the resistance factors required for the calculations. Table 13.4 contains the minimum equivalent flue heights.

Remember this calculation cannot absolutely confirm correct flue operation. You must correctly test all flues as required by GSIUR. These mandatory tests are detailed in **Part 14 – Flue testing**.

Example 1 (using formula)

A typical floor standing boiler installation with a 100mm diameter straight chimney system, fitted to a typical approved terminal above the pitch of the roof is shown in Figure 13.12.

The calculation is as follows:

H_a = 2.2m (measured vertically from the top of the draught diverter to bottom outlet of terminal)

K_i = 2.5 (inlet resistance of actual flue as per Table 13.3)

K_o = 2.5 (outlet resistance of actual flue as per Table 13.3)

K_e = 0.78 (resistance per unit length, equivalent flue size as per Table 13.2)

ΣK = 2.316 (terminal = 0.60 from Table 13.2 + 2.2m of 100mm pipe = 1.716 as per Table 13.2 (2.2 x 0.78 = 1.716))

These figures will transpose into this formula:

$$H_e = H_a \times \frac{(K_i + K_o)_e}{(K_i + K_o)_a - (K_e \times H_a) + \Sigma K}$$

$$H_e = 2.2 \times \frac{(2.5 + 2.5)}{(2.5 + 2.5) - (0.78 \times 2.2) + 2.316}$$

Figure 13.12 Diagram for equivalent flue height calculation

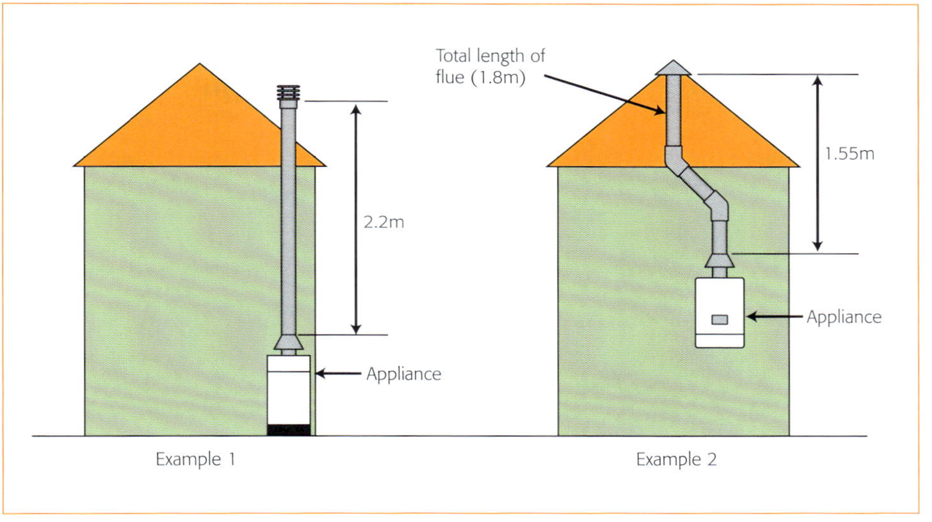

$$H_e = 2.2 \times \frac{5}{(5 - 1.716) + 2.316}$$

$$H_e = 2.2 \times \frac{5}{3.284 + 2.316}$$

$$H_e = 2.2 \times \frac{5}{5.6}$$

$H_e = 2.2 \times 0.892 = 1.96$

$H_e = 1.96$ (equivalent flue height)

From Table 13.4, we can see that 'other appliances' should have a minimum equivalent flue height of 1.0m.

The flue under consideration has an equivalent height of 1.96m and should therefore be satisfactory.

Example 2 (using formula)

In Figure 13.12, we can see a wall-mounted boiler with a 100mm double wall chimney system, which has two 135° bends, a 100mm ridge terminal and an actual height of 1.55m.

Taking values from Tables 13.2 and 13.3 and transposing them into the formula (in the same way as for Example 1) the resulting calculation looks like this:

$$H_e = 1.55 \times \frac{(2.5 + 2.5)}{(2.5 + 2.5) - (0.78 \times 1.55) + (0.61 + 0.61 + 1.404 + 2.5)}$$

$H_e = 0.86$ (equivalent flue height)

As the equivalent flue height is less than the 1m required in Table 13.4, the flue system is not likely to perform adequately.

Therefore consider re-designing the installation to satisfy the requirements for the minimum equivalent flue height.

Table 13.2 Typical resistance factors for flue components

Component	Internal size	Resistance factor (K_e in formula)	
Gas flue blocks	197mm x 67mm 231mm x 65mm 317mm x 63mm 140mm x 102mm 200mm x 75mm 183mm x 90mm	0.85 0.65 0.35 0.60 0.60 0.45	per metre run
Pipe	100mm 125mm 150mm	0.78 0.25 0.12	per metre run
Masonry chimney	213mm x 213mm	0.02	per metre run
90° bend	100mm pipe 125mm pipe 150mm pipe	1.22 0.50 0.24	per fitting
135° bend	100mm pipe 125mm pipe 150mm pipe 197mm x 67mm 231mm x 65mm 317mm x 63mm	0.61 0.25 0.12 0.30 0.22 0.13	per fitting
Raking block Adaptor block	any any	0.30 0.50	per block
Terminal	100mm ridge 125mm ridge 150mm ridge 100mm GC1* 125mm GC1* 150mm GC1*	2.5 1.0 0.48 0.60 0.25 0.12	per fitting

* GC1 is an industry term for a typical circular flue terminal.

Note: Where available, use manufacturer's stated resistance factors

Table 13.3 Inlet and outlet resistance factors

Appliance	Inlet resistance (K_i in the formula)
Gas fire with 12,000mm² equivalent flue size	3.0
Gas fire/back boiler unit	2.0
Other appliances with:	
• 100mm flue spigot	2.5
• 125mm flue spigot	1.0
• 150mm flue spigot	0.48
Flue	**Outlet resistance (K_o in the formula)**
100mm flue	2.5
125mm flue	1.0
150mm flue	0.48
Note: These factors do not apply to decorative fuel effect gas appliances (DFEs), or to inset live fuel effect fires (ILFEs).	

Table 13.4 Minimum equivalent flue heights required of planned or existing flue system

Appliance types	Minimum equivalent height
Gas fire * (suitable for pre-cast flue system)	2.0m of 125mm flue pipe
Other gas fires *	2.4m of 125mm flue pipe
Gas fire/back boiler unit *	2.4m of 125mm flue pipe
Other appliances **	1.0m of flue pipe of the same diameter as the appliance flue spigot
* For gas fires, effective height = equivalent height + 0.65m. ** Note: These factors do not apply to DFEs, or to ILFEs.	

Figure 13.13 Open-flued system with an air gap and fire-stop plate

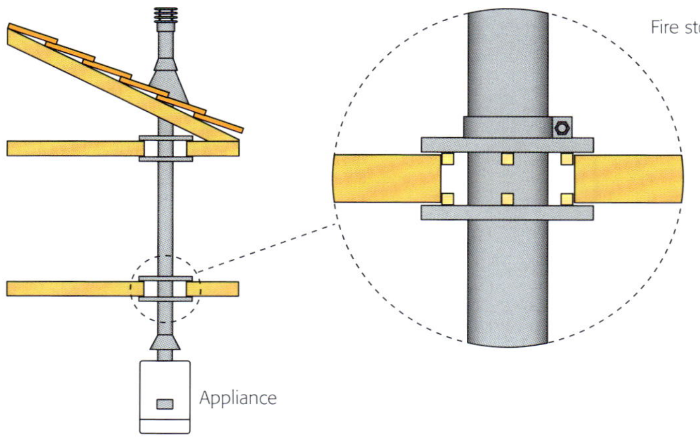

Protection from temperature effects

Any chimney system must be constructed of materials of a type and thickness to prevent damage to, or ignition of the building.

Chimney products – how to ensure separation

If the chimney system passes through any dwelling (existing - see Note) other than that in which the appliance is fitted:

- install it in a fire resistant enclosure to separate it from that dwelling
- the enclosure must comply with Building Regulations. Seal it to prevent any leakage of the POC from entering that dwelling
- surrounding combustible material must not exceed a temperature of 80°C.

Achieve this:

- by ensuring an air gap around the chimney system of at least 25mm
- if a sleeve is used to create the required gap, by filling its annular space with non-combustible material and covering the ends with wall or ceiling plates (fire stop plates), see Figure 13.13.

Note: ADJ requires chimney systems not to be routed through other dwellings as access for periodic inspection cannot be guaranteed. This however, doesn't apply to communal areas having purpose designed ducts where inspection access is provided.

If a chimney system passes through a cupboard in which combustible material is stored (e.g. an airing cupboard):

- first separate the appliance from the airing compartment with a non-combustible partition
- also protect the first metre of the chimney system by an air gap of at least 25mm. This air gap may be provided by a non-combustible guard, which forms an annular space around the chimney system of not less than 25mm
- ensure any clearance between the guard of the chimney system and the compartment partition does not exceed 13mm
- wherever possible, access should be provided to enable the chimney system to be inspected throughout its whole length

Expanded metal or rigid wire mesh is suitable materials for the partition and guard. These may be perforated, if required, by apertures, with major dimensions no greater than 13mm.

Note: For double wall chimney systems complying with BS EN 1856-1, the external skin and air gap may provide insulation to a standard where no extra precautions are necessary, other than normal installation tolerances.

Chimney flue blocks – points to note

- do not directly plaster over chimney flue blocks: excessive heat may cause the facing plaster to crack
- you can face chimney flue blocks with concrete blocks or plasterboard, as long as they are not in direct contact with the chimney flue blocks (i.e. fixed to the chimney flue blocks using the 'dot and dab' method)
- you need to seal any air gaps created between the chimney flue blocks and the plasterboard at the chimney opening

Condensation – when it occurs and when it is a problem

Condensation occurs when the temperature of the POC (which contain water vapour) falls below the dew point, i.e. the point at which the water vapour content condenses. Condensation is not a problem in small quantities and is often unavoidable.

Condensation becomes a problem when:

- the flue length is excessive
- excessive lengths of the chimney/flue pipe route are exposed externally, or occur in unheated areas
- excessively oversized flue (cross sectional area)
- the chimney system cannot retain sufficient heat

Keep condensation to a minimum to prevent

- corrosion of both the chimney system and the gas appliance
- condensation mixing with soot deposits left in flues, causing staining to inner walls and outer brick work

Make provision for collecting and disposing if condensation occurs.

When designing a chimney system, consider the known maximum condensate-free lengths as provided in Tables 13.5 and 13.6.

These Tables show the limits of the various chimney system options and indicate occasions when upgrading the chimney system construction are likely to solve the problem.

Note: An internal chimney system should have no more than one quarter or 1.5m of its total length exposed above the roof. Be guided by whichever of these measurements is the shorter.

Table 13.5 Other appliances up to 88% efficiency (Maximum condensate-free lengths for unlined masonry and other chimney systems)

Heat input rating	Masonry chimney (225mm x 225mm) and standard pre-cast concrete flue block systems. Individual flues.		125mm Metal chimney – double wall. Individual flues.	
kW (gross)	External	Internal	External	Internal
6	2m	4.5m	7m	18m
8	3m	5m	10m	20m
10	3.5m	6m	14m	22m
12	4m	7m	16m	23m
14	5m	7.5m	17.5m	25m
16	5.5m	8m	18m	26m
18	6m	10m	20m	27m
20	7m	13m	22m	28m
30	10m	17m	27m	33m
40	16m	18m	33m	38m
50	17m	19m	37m	42m
60	18m	21m	42m	45m

Note: All figures are approximate.

Table 13.6 Gas fires up to 77% efficiency (Maximum condensate-free lengths for unlined masonry and metal chimney systems)

Chimney system type	External	Internal*
Masonry chimney (225mm x 225mm) and standard pre-cast concrete flue block systems** Individual flues	up to 10m	up to 12m
125mm metal chimney – single wall (not insulated) Individual flues	not allowed	up to 20m
125mm metal chimney – double wall Individual flues	up to 28m	up to 33m

* An internal chimney is one where the only surface exposed to external temperatures are the length of chimney above the roof which should not exceed 1.5m or one quarter of the flue height, whichever is the shorter.

** Applies to pre-cast concrete flue blocks with a cross sectional area of 13000mm² and an aspect ratio (see **Part 19 – Definitions**) of up to 4:1 or 20000mm² with an aspect ratio of up to 5:1.

Insulated double walled chimney components are designed to prevent condensation problems. On other systems the limitations for condensate-free lengths are more critical.

Existing exposed chimney system can be wrapped with insulating material, e.g. in a roof space. For a brick/masonry chimney, the solution could be to install a flue liner.

If you take action to prevent condensation, you may not need to provide for the collection and disposal of condensate.

Note: Double walled chimney components that only have an air gap between the inner and outer duct should not be used externally for lengths in excess of 3m. In this situation use insulated double walled chimney components.

Installation of open-flued systems – factors to consider

Check compatibility with the existing or planned chimney system: it is the major consideration when you install any open-flued (Type B) gas appliance.

Always take into account appliance maximum heat input and periods of continuous use.

Plan new chimney systems to ensure the route allows for mostly vertical runs, correct angle bends, adaptors for changes in materials and correct termination (termination should not cause a nuisance to adjacent properties).

The need to line an existing brick/masonry chimney depends upon the appliance type, the condition of the chimney, its age and height.

Comply with the following when relevant:

- before installing the gas appliance, thoroughly sweep any chimney system previously used by an appliance burning any fuel other than gas
- the chimney system must serve only one fire place opening/appliance
- you need to visually check the chimney system, to ensure that it complies with the requirements of this Part
- carry out a flue flow check (see **Part 14 – Flue testing**)
- remove any dampers or restrictor plates. Where removal is not possible, permanently fix the damper/restrictor in the fully open position, to ensure a clear and unobstructed flue way
- where the appliance manufacturer's installation instructions require the chimney system to be lined (see **Condensation – when it occurs and when it is a problem** in this Part), sweep it before starting work

Brick/masonry chimney systems – check when constructed, before you work

- brick/masonry chimney systems constructed after 1965 are generally lined with clay, salt-glazed clay or high alumina cement pipes. These will normally be suitable for use with most open-flued gas appliances
- those constructed prior to 1965 are also suitable, but may require lining, depending on condition and appliance type

Before use:

- check the route of any existing chimney system along its whole length for continuity and obstructions. Remove register plates, restrictors or dampers, or permanently secure them in the open position. Seal any under-grate air ducts
- sweep an existing brick/masonry chimney system, unless previously used for a gas appliance
- connect a gas appliance directly into the chimney system and not through a redundant fuel burning apparatus (which you must remove – see Note).

Table 13.7 Masonry chimney minimum: catchment space (voids) – any appliance

Situation	Lined		Unlined
	new or previously used for a gas appliance	used previously for an oil or solid fuel appliance	
Minimum depth **to be measured from the bottom of the flue spigot**	75mm	250mm	250mm
Minimum volume **to be measured from the bottom of the flue spigot**	$0.002m^3$ (2 litres)	$0.012m^3$ (12 litres)	$0.012m^3$ (12 litres)
Note: This table does not apply to decorative fuel effect (DFE) gas appliances.			

- where a solid fuel back boiler is left in situ, do not leave it in a condition where pressure can build up (this could make the installation unsafe)
- where possible, leave the open vent pipe from the back boiler complete to its termination above the cold feed cistern, or to the point at which it connects to the hot water storage vessel
- If you do need to disconnect, or remove, the open vent pipe, then:
  - remove the waterway access covers on the back boiler
  - alternatively, drain the system as far as practical and drill a 6mm (minimum) hole into the back boiler to prevent a build up of pressure

Note: Gas Safe Register produced a Technical Bulletin (TB), which was drafted with the HSE on the safety risks from redundant solid fuel back boilers.

TB 101 'HSE Safety Alert Risks from redundant solid fuel back boilers' can be freely downloaded by registered operatives – visit https://engineers.gassaferegister.co.uk

Catchment space (void) – provide at base of chimney system

Provide a catchment space (void) at the bottom of any brick/masonry chimney system serving a gas appliance.

1. Ensure provision for clearance of debris from this space.
2. Seal all openings other than the flue outlet and front opening.

Note: Table 13.7 and 13.8 detail catchment space requirements for gas appliances.

Chimney system connection – vary according to type/age

Unlined

1. When an appliance is not connected to the base of an unlined brick/masonry chimney system, but is connected at a higher level (see Figure 13.14), seal the brick/masonry chimney system not less than 250mm and not more than one metre below the entry of the flue pipe. Provide a means to clear debris from the catchment area.

Table 13.8 Flue block chimney/metal chimney system minimum: catchment space (voids) – any appliance

	Minimum void dimensions below appliance connection	
	Depth	Volume
Any appliance (previously used with solid fuel or oil)	250mm	$0.012m^3$ (12 litres)
Any appliance (new or unused, or previously used with gas)	75mm	$0.002m^3$ (2 litres)
Note: This table does not apply to decorative fuel effect (DFE) gas appliances.		

Figure 13.14 Connection to an unlined chimney system

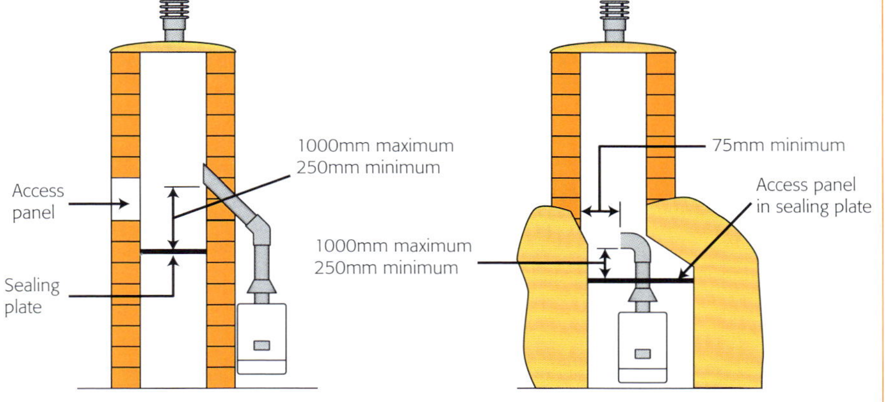

a) Typical side view with appliance external to chimney

b) Typical front view with appliance installed in a recess

2. When an appliance is connected with a flue pipe to the base (recess) of an unlined brick/masonry chimney system (see Figure 13.14), it should project into the brick/masonry chimney system by a minimum of 250mm and it should be designed to stop the entry of debris into the flue. There should be a minimum clearance of 75mm between the outlet of the flue pipe and any brick/masonry chimney wall.

Lined (post 1965)

- connect appliances other than gas fires, to a brick/masonry chimney system of this type, using a short length of rigid flue pipe
- only in the case of a BBU, use a flexible flue liner to connect the appliance to the brick/masonry chimney system
- seal any annular space around the connection and fit a suitable terminal at the outlet of the flue

Single/multi-wall system chimney products – what you need to know

Single wall

Note: Do not use single wall chimney components externally – except for the protrusion above a roof line.

- joints on spigot and socket (male and female) chimney components should have the sockets facing uppermost
- you need to caulk (fill in) all the joints, using non-combustible fibre rope and seal with a suitable fire cement

Note: Use suitable adaptors when changing flue size or material.

- ensure the chimney system is supported appropriately – at a maximum of 1.8m intervals. This is particularly important in roof spaces. Do not use the terminal, including ridge terminals or ridge terminal adaptors, as a support for the chimney system

With improvements to heat insulation requirements in buildings, roof spaces for roofs built to the 1985 (or subsequent) Building Regulations can effectively be considered to be outside air. Therefore new chimney systems run within roof spaces should be of double walled construction as a minimum.

Where a new or replacement appliance is connected to an existing single walled chimney system, it is recommended the section of chimney sited within a roof space should be:

- of double wall construction; or
- insulated separately

Therefore consider wrapping the chimney system within the roof space with mineral wool insulation, held in place with a wire netting material.

With regard to the chimney system, ensure

- it is weather-proofed when it passes through a roof
- it includes a purpose made plate incorporating a minimum upstand of 150mm above the highest junction with the roof
- other than for tiled or slated roofs, the chimney system passes through a sleeve, which is secured to the roofing material
- there is an air gap of 25mm between the chimney system and any combustible material
- that chimney systems in areas used to store combustible materials (e.g. airing cupboards) have a protective guard with a 25mm space around the chimney system

Double wall

Note: Do not use internal grade chimney components in external locations. Chimney system support and weather proofing should be the same as that for single wall chimney systems.

- you must follow the manufacturer's instructions when jointing double wall chimney components
- check each section and do not use if there is damage to the joints or internal wall
- do not strain adjustable bends, to avoid damage to the internal wall
- do not cut double walled flue pipe, unless permitted by the chimney system manufacturer
- you must use the flue pipe manufacturer's adaptor to connect appliances, or different makes/types of flue pipe

- double walled chimney systems manufactured to BS EN 1856-1 may satisfy the requirement for a 25mm air gap from combustible material. Any clearance required around a double walled chimney system can normally be measured from the outer edge of the inner flue wall
- adequately support double walled chimney systems and seal according to the manufacturer's instructions

Existing chimney systems and new or replacement installations

Take care when you install a new or replacement appliance to an existing chimney system. Due to improvements in appliance design, a chimney system that has operated safely with an existing appliance may not operate satisfactorily with a replacement appliance of the same type.

Before you connect a new or replacement appliance to an existing chimney system

Ensure that the existing chimney system meets the requirements of the current British Standard (BS 5440-1).

When you install a replacement appliance onto an existing chimney system

- take care not to disturb the chimney system generally
- secure the chimney system in place, before you attempt to remove an existing appliance connected to an independent chimney system
- take care (including handling with care) because with asbestos cement chimney systems, existing joints/connections can become loose or, even worse, they can come apart further up the chimney system
- metallic chimney systems also can become loose, or come apart, if not handled carefully. Manufacturers use special connections for these, particular to their own brand of fittings
- when disconnecting chimney components, take care not to disturb any joints further up the chimney system

Following the installation of the replacement appliance

- inspect the chimney system throughout its length, including upper floors and roof space, to ensure that it is complete and in a sound condition
- where independent chimney systems are run within buildings and are 'boxed in', remove the duct covering and inspect any joints inside
- where chimney systems run through other premises (such as flats or maisonettes) you must gain access to inspect the entire chimney system – to ensure that it is intact and in a sound condition – this installation practice is discouraged under ADJ
- where appropriate consult the landlord or managing agent

Stainless steel flue liners – note restricted use

Only use a flexible stainless steel flue liner to line an existing brick/masonry chimney system, or to connect a BBU to an existing brick/masonry chimney system. Install as follows:

- fit the flexible stainless steel flue liner in one complete length from the gas appliance to the terminal
- ensure the metallic flue liner is adequately supported at the top and bottom with suitable clamp/register plates
- seal any annular spaces between the liner and brick/masonry chimney system (see Figure 13.15)

Figure 13.15 Method of securing a flue liner in position serving a gas appliance

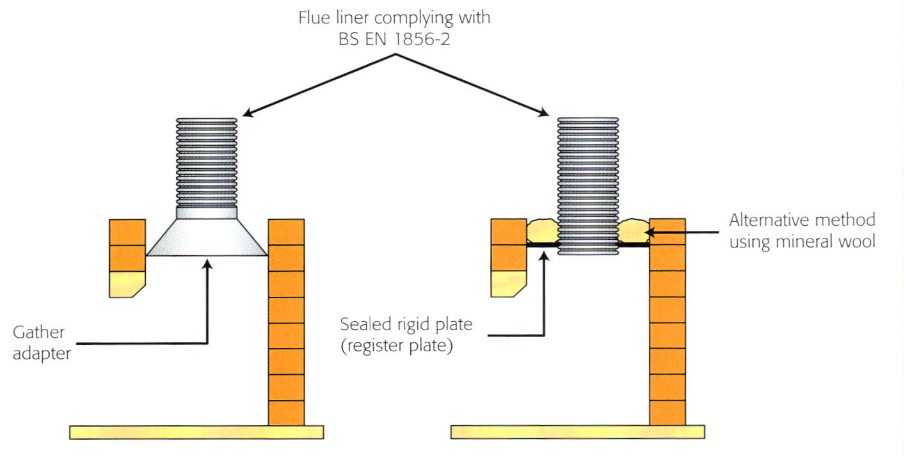

- follow the flue liner manufacturer's advice particularly with regard to the location of the liner where it passes through and around bends – relating to necessary spacing and support
- fit a suitable terminal to the flue liner's outlet and ensure it is weather-proofed

Note: When you install a new appliance, preferably replace an existing stainless steel flexible flue liner. If this liner is in a good condition and you are confident it will continue to operate safely throughout the lifetime of any new appliance (normally 10-15 years) – then you may think about re-using it.

Unlined brick/masonry chimney systems – where they cannot be used

All brick/masonry chimney systems serving BBUs need to be lined. Although not specified in British Standards, most gas boiler manufacturers also require brick/masonry chimney systems to be lined.

A flue liner is required for brick/masonry chimney systems serving gas appliances if they exceed the flue lengths given in Table 13.9.

Table 13.9 Appliances and chimney combinations which require the chimney to be lined

Appliance type	chimney Length
Gas fire	> 10m (external wall) > 12m (internal wall)
Gas fire/back boiler unit	any length
Gas fire with circulator	> 10m (external wall) > 12m (internal wall)
Circulator	> 6m (external wall) > 1.5m external length and total length in excess of 9m
Other appliances	flue lengths greater than those given in Tables 13.5 and 13.6

Poured/pumped concrete chimney system liners – limitations on use

The use of a poured/pumped concrete liner is a suitable alternative to a flexible stainless steel flue liner and must be installed by a competent installer approved by the British Board of Agrément (BBA).

You are advised to check the lining is mechanically sound and suitable for the gas type and appliance.

Poured concrete linings are not acceptable for lining a new brick/masonry chimney system. The brick/masonry chimney system must be constructed to conform to current relevant Building Regulations.

Poured concrete linings installed into existing brick/masonry chimney systems after 1987, need to meet the BBA 'approved scheme for installers'. If no certificate of compliance with this scheme (or other documented evidence) is available, install a further liner i.e. a flexible stainless steel flue liner.

Note: Before use with a gas burning appliance, sweep brick/masonry chimney systems lined this way if they have been used with solid fuel or oil.

Chimney flue block (concrete or clay/ceramic) systems – more resistance to flue flow

Only install gas appliances to a chimney flue block system if the appliance manufacturer's installation instructions allow.

Ensure, as far as possible, that:

1. The chimney flue blocks have been correctly installed and they are free from damage or obstruction.
2. You carefully remove any excessive jointing extrusions.
3. You make any change of direction in the chimney flue block system by raking blocks and do not exceed 30° from the vertical.
4. When you connect a flue pipe to the chimney flue blocks, use a purpose-made adaptor block (transition piece). The angle of the transition should not be greater than 45° to the vertical. The flue pipe should not project into the chimney flue block system, as this could restrict the cross-sectional area of the flue.
5. You provide a catchment space (void) below a gas fire spigot (see Table 13.10) which is accessible for clearing debris.

Table 13.10 Chimney flue block systems: voids for gas appliances

	New or previously used for a gas appliance	Used previously for an oil or solid fuel appliance
Minimum depth from bottom of flue spigot	75mm	250mm
Minimum volume from bottom of flue spigot	2dm^3 (0.002m^3 or 2 litres)	12dm^3 (0.012m^3 or 12 litres)

In the case of a gas fire, the cross-sectional area of the chimney flue blocks should be:

- no less than 13000mm^2 with a minimum dimension of 63mm if the chimney system was installed prior to 1986
- no less than 16500mm^2 with a minimum dimension of 90mm if the chimney system was installed after 1986

Before you install a gas fire onto a chimney flue block system, consult the manufacturer's installation instructions, to ensure it is suitable for the dimensions of that particular chimney system.

For BBU installations, ensure the correct recess/starter blocks have been installed.

Note: Chimney flue blocks have more resistance to flue flow than circular flue pipes. You therefore need to ensure correct operation of the chimney system by carrying out the necessary tests (see Part 14 – Flue testing).

Flue terminal design – guidelines to comply with

Flue terminals design:

- should assist the passage of POC to the atmosphere
- should minimise down draught
- should prevent entry of leaves, snow, rain, etc
- where a proprietary terminal has not been used, the free area of the outlet openings should be a minimum of twice the cross-sectional area of the flue size required by the appliance
- outlet openings should allow entry of a 6mm diameter ball but not a 16mm diameter ball
- outlet openings should be uniformly located around the terminal, or arranged on two opposite sides

Proprietary flue terminals should conform to one of these standards:

- BS EN 1859: 'Chimneys. Metal chimneys. Test methods'
- BS EN 13502: 'Chimneys. Requirements and test methods for clay/ceramic flue terminals'

Where you use a non-proprietary terminal:

- the free area of the outlet openings should be a minimum of twice the cross-sectional area of the flue size required for the appliance
- locate outlet openings uniformly around the terminal, or arrange them on two opposite sides

Note: Always size the cross-sectional free area of the flue and terminal to at least the minimum specified by the appliance manufacturer.

In some instances, the cross-sectional free area of the terminal may be smaller than that of the cross-sectional area of the flue, e.g. a 125mm diameter BS EN 1859 terminal may be fitted to a 225mm x 225mm solid fuel designed brick/masonry chimney system, serving a radiant convector gas fire.

Sizing the openings for a non-propriety flue terminal

As described, the total free area of the outlet openings of a non-propriety flue terminal should normally be at least equal to twice the cross-sectional area of the flue that it is serving, or 12000mm^2 (per flue) multiplied by two located on two opposite sides, whichever is the greater.

For example, a 125mm diameter flue has a cross-sectional area of 12273mm^2. This is based on the following formula (πr^2):

3.142 (π) x 62.5mm x 62.5mm = 12273mm^2

The total free area of the openings for a non-propriety terminal to terminate a 125mm diameter flue should therefore be at least 12273mm^2 x 2, i.e. 24546mm^2.

Flue terminal positions for open-flued (Type B) appliances affect performance/safety

A terminals location has a significant effect on a chimney systems performance. Wind movement/air flow currents around a building may create pressure zones.

These will create negative or positive conditions so it is vital that you locate a flue terminal outside of the influence of any pressure zone that could adversely affect:

1. The flue performance.
2. Its ability to clear POC fully.
3. Any other factors affecting flue performance (see Figure 13.16 and **Part 14 – Flue testing** for further guidance).

To help you correctly locate the flue terminal, consider:

- the location of the terminal in relation to the roof
- the effect of any adjacent structures such as dormer windows, brick/masonry chimney stacks, etc and/or topographical features (hills and valleys, for example)
- the type and use of the chimney system (natural, fanned draught etc.)

Note: Only use the terminal positions in Tables 13.11 and 13.12 when the appliance manufacturer's instructions do not state a specific location.

Figure 13.16 Building pressure zones

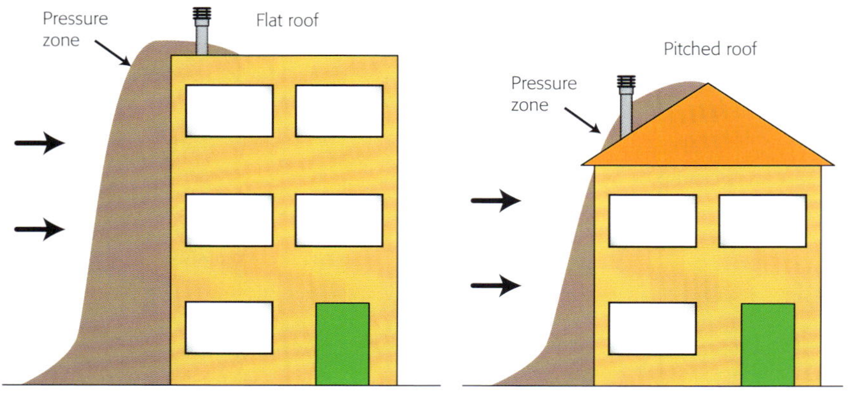

Table 13.11 Siting of roof mounted terminals for both individual natural draught and fanned draught open-flued appliances

Type of construction		Siting not within 1.5m from a vertical surface of a structure on the roof *			Siting within 1.5m from a vertical surface of a structure on the roof *
		Internal route		External route (height of base of terminal above the level of adjacent roof edge)	Internal and external route (height of base of terminal above the top of the structure
		On ridge	Not on ridge		
Pitched roof		At or above ridge level (see Figure 13.18)	No part of the flue outlet should be closer than 1.5m measured horizontally to the roof surface, or 600mm above the ridge (see Figure 13.17)	See Figure 13.17	600mm (see Figure 13.17)
Flat** roof	With parapet	Not applicable	600mm above flue/roof intersection (see Figure 13.19)	600mm (see Figure 13.19)	600mm above structure (see Figure 13.19)
	Without parapet		250mm above flue/roof intersection (see Figure 13.19)		

* For example, a chimney stack, dormer window, parapet, etc.

** A flat roof is defined as being within 20° of horizontal.

Note: The heights given above should be increased by 500mm when there is a sloping hill or embankment within a distance of 4 times the height of the building and having a height between 0.5 and 1 times the height of the building. Further precautions are required when there is a combination of external chimney, steep pitch, high efficiency appliance, complex roof geometry, nearby hills, adjacent buildings, etc. The height should be 2m and an extractive terminal fitted.

Table 13.12 Minimum clearances for open-flued fanned-draught terminal locations, appliances up to 70kW (net)

Terminal location	Minimum distance
Directly below, above or horizontally to an opening, air brick, opening window, etc. (see Note below)	300mm
Below temperature sensitive (plastic) gutters, soil pipes or drain pipes	75mm
Below eaves	200mm
Below balconies or car port roof	200mm
From a vertical drain pipe or soil pipe	150mm
From an internal or external corner*	200mm
Above ground, roof or balcony level	300mm
From a surface facing a terminal	600mm
From a terminal facing the terminal	1200mm
From an opening in a carport into the dwelling e.g. door or window	1200mm
Vertically from a terminal on the same wall	1500mm
Horizontally from a terminal on the same wall	300mm
From the wall on which a terminal is mounted	50mm
Above intersection with roof	150mm
From a vertical structure on the roof	300mm

Note: In addition to the above requirements, the terminal should be no closer than 150mm to any opening in the building fabric for the purpose of accommodating a built-in element such as a window frame.

* This dimension does not apply to protrusions in buildings of 450mm and below. For example, external chimneys or walls.

Attention: It is recommended that terminals of fanned draught appliances are not sited within 2m from any opening into a building, which is immediately opposite, or where POC can drift across a boundary into neighbouring properties.

Figure 13.17 Open-flue terminal positions

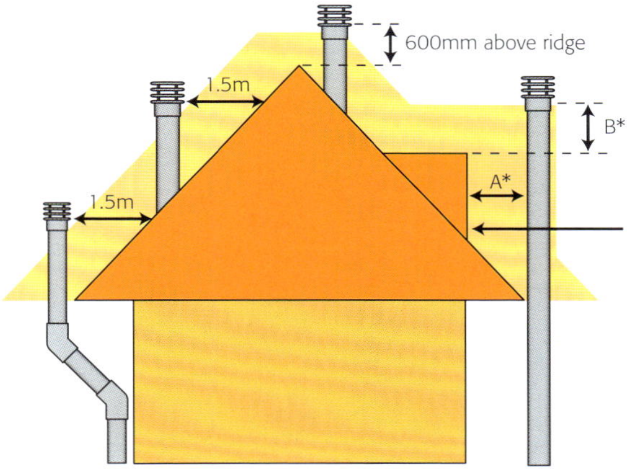

Open-flued terminal positions

Natural draught

The termination position of the chimney system based on horizontal/vertical measurements.

A flue should not terminate within 1.5m measured horizontally from the roof to the nearest outlet of the terminal (see Figure 13.17 and Table 13.11).

If the flue termination is sited within 1500mm (measured horizontally from the ridge), the base of the terminal should be a minimum of 600mm above a line drawn through the ridge (see Figure 13.17 and Table 13.11).

Where the flue terminal is sited within 1.5m measured horizontally from a structure such as a dormer window, brick/masonry chimney stack or parapet, the base of the terminal needs to be a minimum of 600mm above the highest point of that structure (see Figure 13.17 and Table 13.11).

Ridge terminal locations

Do not site ridge terminals that are open to atmosphere on all four sides closer than 300mm to each other or closer than 1.5m to an adjacent structure (see Figure 13.18 and Table 13.11).

Terminal locations – flat roofs

A flat roof is one that is horizontal, or which rises less than 20° from the horizontal.

For terminal locations above flat roofs, see Figure 13.19.

Figure 13.18 Siting for ridge terminals

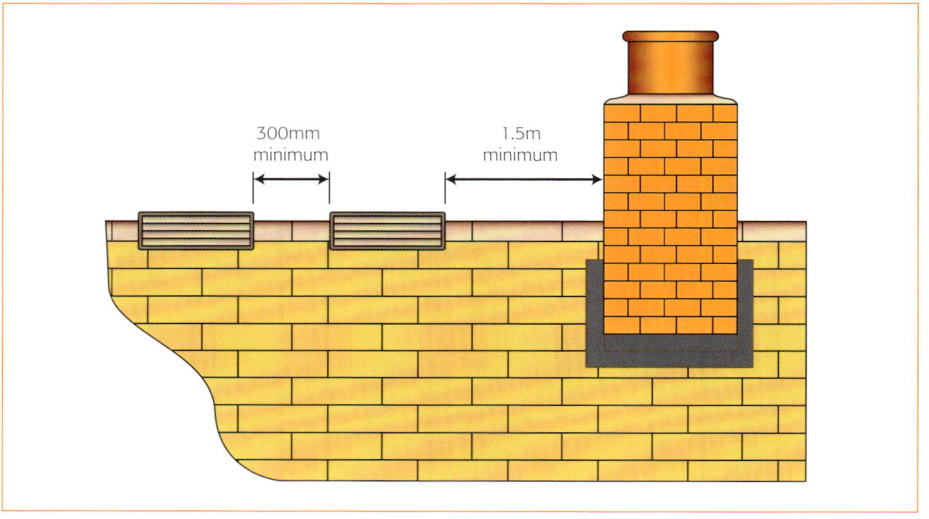

Figure 13.19 Flat roof terminations

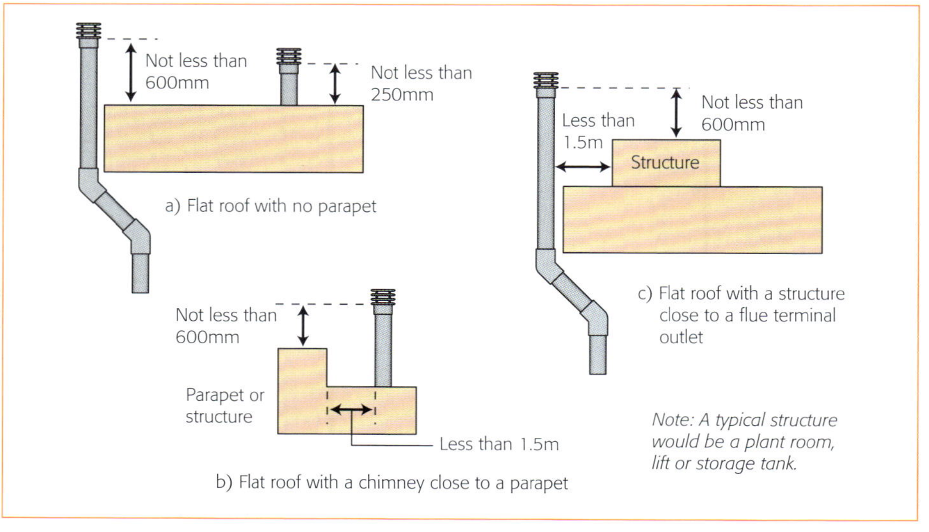

Figure 13.20 Siting a flue terminal near to a tall structure

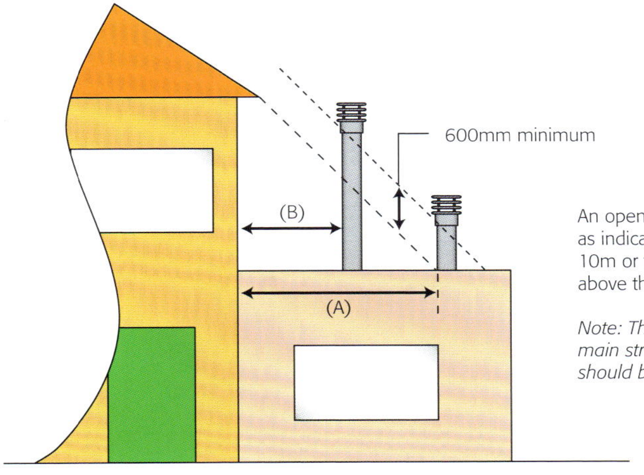

Terminal locations near to tall structures

If the chimney is located on a small structure alongside a tall structure, for example in the case of a single storey extension attached to a two storey pitched roof house, follow this procedure:

- where there is a higher adjacent structure that may affect the performance of the chimney system, do not site the height of the base of the terminal less than 600mm above an imaginary line drawn from the highest point of the tall structure to the edge of the adjoining low structure
- if the adjoining low structure is greater than 10m in length, draw the imaginary line to a point measured to a maximum of 10m from the adjoining tall structure
- the chimney system should not terminate within 2.3m of the adjoining tall structure (see Figure 13.20)

Flue terminations in close proximity to each other

Where three or more chimney systems terminate less than 300mm from each other, they should terminate at the same height.

Where numerous flue outlets terminate next to each other, for example in the case of a brick/masonry chimney stack with three or more outlets, the minimum distance measured between the outlets should not be less than 50mm.

Flue terminations adjacent to openings into roofs of buildings

If the chimney system terminates adjacent to a window such as a 'Velux window' or other opening on a pitched roof, do not site the terminal in the area 2m immediately below the opening and 600mm beside or above the opening (see Figure 13.21).

Figure 13.21 Siting a flue terminal near to openings on a roof

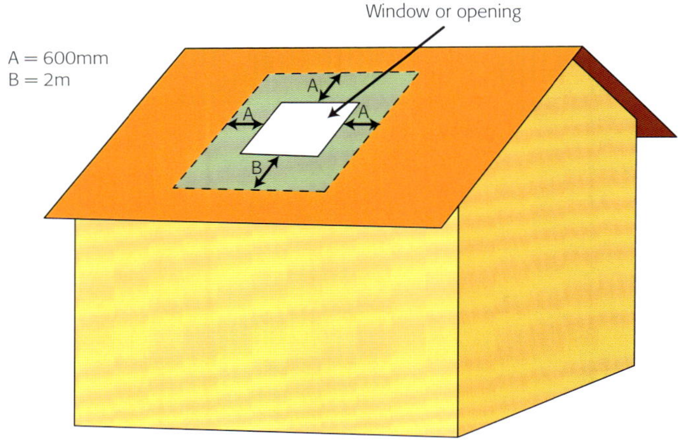

No chimney should penetrate the shaded area around the window or opening. Chimneys should also conform to the requirements outlined in Figure 13.17.

Flue termination open-flued fanned-draught

Design and position terminals so that POC can safely disperse at all times.

If you locate the terminal to comply with Table 13.11, the fan needs to ensure total clearance with adverse wind conditions of 0.15mbar.

If the terminal location does not comply with Table 13.11, but complies with the requirements of Table 13.12, the fan needs to ensure total clearance with adverse wind conditions of 0.75mbar.

Individual open-flued fanned-draught systems do not normally require a terminal of the types specified for open-flued natural-draught. However, a terminal may be recommended, for example to prevent water ingress (rain).

Any opening in the termination should be such that it will admit a 6mm diameter ball, but not a 16mm diameter ball.

Position the terminal so that the POC do not cause a nuisance to neighbours, or cause damage to the fabric of the building.

It is recommended that you position a fanned flue terminal as follows:

- at least 2m from any opening in a building directly opposite; and
- so that the POC are not directed to discharge across a boundary

Basements, light wells and retaining walls – points to note

Do not site flue terminals within the confines of a basement area, light well or external space formed by a retaining wall, unless you take steps to ensure that the POC can disperse safely at all times.

- It is possible to locate open-flued fanned-draught (Type B) appliances in such a location, provided that they are not sited lower than one metre from the top level of that area to allow the POC to disperse safely; and
- the basement area or lightwell has a single retaining wall creating an uncovered passageway at least 1.5m wide, allowing the free passage of air beyond the ends of the structure and retaining wall

Light well – an area that may inhibit dispersal of POC

A lightwell can be described to cover areas where for example, there are walls opposite and/or around the wall on which a flue terminal is located, that may inhibit the free dispersal of POC.

An example is the common victorian urban terrace property, where the front door opens onto the street footpath across a narrow well, which provides light and possibly access to a cellar. It may also be the small central area in a square building designed to give light to the rooms on the inside of the building.

Be aware of the dangers of locating flue terminals where the POC cannot freely disperse. This could be accentuated by the possible accumulation of these products caused by several terminals all discharging into a single confined space.

Bird guards/terminals – important in areas with a known problem

Where you know that a brick/masonry chimney system has been used by nesting birds and other animals, or when there is a known problem in the area, you can fit a brick/masonry chimney system with a suitable bird guard, or terminal, to prevent them getting in.

This is particularly important in areas where birds such as jackdaws and seagulls are known to roost.

Take care before fitting a terminal or bird guard to the brick/masonry chimney system to ensure that it will support this. You may need to additionally reinforce the structure.

Once you have installed a terminal or bird guard, you need to carry out the relevant safety checks to ensure that the POC are being cleared (see also **Part 14 – Flue testing**).

Wall-faced/wall-adjacent terminations

Wall-faced/wall-adjacent flue terminations, see Figure 13.22, are not acceptable for any new or existing open-flued natural-draught gas appliance installations.

Wall-faced/wall-adjacent terminations should be treated as 'AR' in accordance with the current GIUSP.

See also **Part 8 – Gas Industry Unsafe Situations Procedure** and **Part 10 – Emergency notices, warning labels and reporting forms**.

Figure 13.22 Wall faced and wall adjacent open-flued natural-draught termination

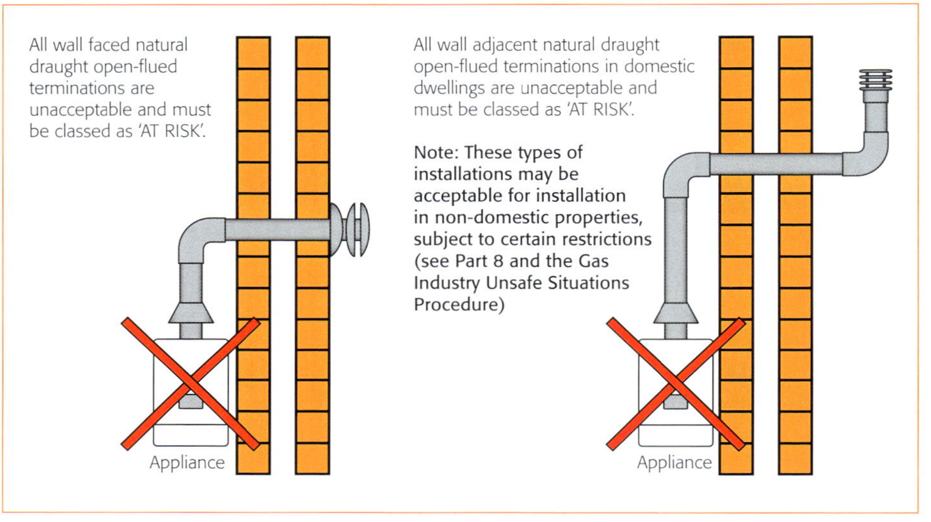

Room-sealed (Type C) appliances – flexible but ensure they have clean supply of fresh air

General

Room-sealed (Type C) appliances are available in a number of flue options:

1. Room-sealed – balanced flue.
2. Room-sealed – fanned-draught.
3. Room-sealed – non-balanced flue.

They allow for greater flue flexibility and you should specify them whenever possible.

Many appliance manufacturers provide for vertical fan flue termination, in addition to the more traditional horizontal termination, with specially designed chimney systems.

When you install a room-sealed appliance in a cupboard or compartment, additional ventilation may not be required (see **Part 4 – Ventilation**).

How room-sealed (Type C) appliances differ from open-flued (Type B) appliances:

They take the air for combustion from outside the room in which they are installed. In most cases, the air for combustion will be sourced from the area into which the POC are discharged.

How to avoid problems

Generally, this arrangement is satisfactory. However, for an appliance to burn gas correctly, it needs to have a clean supply of fresh air.

Problems may arise when the room-sealed flue termination is positioned too close to an obstruction, e.g. soffit board, soil pipe, brick pillar, etc.

If the terminal of the room-sealed appliance is located too close to an obstruction, POC may not disperse adequately and may re-enter the chimney system via the air duct. They may contaminate (vitiate) the incoming air.

If this occurs, there will be insufficient oxygen to fully support the combustion process. (see **Part 3 – Characteristics of combustion**). As a result, both the pilot and burner flames may have a tendency to 'lift-off'.

In such circumstances it is not unusual for the appliance to 'fail-safe' and cause nuisance shut down, leading to customer dissatisfaction.

It is important therefore that you strictly adhere to the installation requirements detailed in the appliance manufacturer's instructions. Where appropriate, also see Table 13.13.

Room-sealed fanned-draught chimney systems concealed within voids

Modern gas boiler design has seen the introduction of new chimney system materials (PVC for example), different chimney system configurations (twin pipe as well as concentric) and consequently, longer flue runs to the point of termination.

With the flexibility offered by modern chimney system design, the choice of boiler position is no longer constrained to 'anywhere as long as it's an outside wall' and as such, boilers can be located almost anywhere in a property.

This fact is good news for developers/architects, particularly those of multi-occupancy dwellings (flats) where the boiler can be 'hidden' in a compartment/enclosure or other space not so frequently visited/viewed by the home owner.

The chimney system of these boilers is then routed within voids – ceilings, floor, wall, etc. – to the point of termination, either on an outside wall or, if vertical termination, the roof.

However, many of these chimney systems cannot be periodically inspected once in position and enclosed by finished décor, causing a real issue for gas operatives/businesses who have a duty of care to discharge under the GSIUR.

Worse still, cases have been recorded whereby occupants of affected properties have been exposed to varying levels of CO as a result of defective/broken chimney systems within these voids.

The incidents of CO poisoning has mobilised the industry – construction, gas, HSE, etc – to review current installation practices and to address the issues raised.

This review culminated in new guidance being produced, both in the form of up-dated information included in to ADJ 2010 (1.47 Concealed flues) and manufacturers' installation instructions for new installations, and via a Technical Bulletin (TB) issued by Gas Safe Register and a Safety Notice issued by the HSE for existing installations.

TB 008 (Edition 3) 'Existing concealed room-sealed fanned-draught boiler chimney/flue systems in domestic premises' provides important guidance which needs to be understood by gas operatives/businesses – visit https://engineers.gassaferegister.co.uk to download a copy of the TB.

Additionally, Gas Safe Register have a series of Fact Sheets that can aid registered engineers in explaining the problem to customers; one such fact sheet deals with 'Flues concealed in voids – gas safety concern' and is freely available from Gas Safe's website.

New installations

ADJ requires that where chimney systems are routed within voids that appropriate means of access be provided at strategic locations along the route of the system in order that periodic visual checks can be undertaken – servicing/maintenance visits for example.

The checks required are to confirm that:

- the chimney system (flue pipe and air duct, either concentric or twin pipe design) is continuous throughout its length
- all joints appear to be correctly assembled and sealed (refer to manufacturers instructions for guidance)
- the chimney system is adequately supported throughout its length
- the chimney system has the required fall back to the appliance (condensing appliances) and/or additional drain points, where required, are installed

The route of the chimney within the void needs to be confined to the dwelling concerned and not pass through other properties, given that access for inspection purposes cannot be guaranteed in these locations – communal areas are acceptable where inspection hatches are provided.

The means of access needs to be sufficiently sized to allow inspection, but they are not intended to be of such a size that would allow full physical access.

An access hatch size of 300mm x 300mm is suggested, located at strategic points along the chimney route – changes of direction and within 1.5m of each joint (see Figure 13.23).

It's important that any access hatches do not impair the fire, thermal or acoustic characteristics required under Building Regulations.

This may mean using access hatches with resilient seals and exhibiting the same characteristics for fire, thermal and sound as the structure surrounding them.

Existing installations

TB 008 (Edition 3) is of real importance when dealing with existing gas installations, as it is highly likely that access provisions for inspection purposes – as required by ADJ – haven't been provided for.

The TB continues to evolve since its introduction in 2010 with the latest version – Edition 3 – issued April 2013, providing further guidance for registered engineers on existing installations.

The previous editions – Edition 2 & 2.1 – introduced a cut off point whereby any installation encountered after the 31st December 2012 that do not have suitable inspection hatches installed are to be classified as 'AR' in accordance with the current GIUSP.

The TB (Edition 2 & 2.1) also introduced a risk assessment process (regardless of whether hatches had been installed or not), which gas operatives needed to follow in order to discharge their 'duty of care' as defined within the GSIUR; this risk assessment remains a requirement in the latest, 3rd Edition of the TB as does the 'AR' classification.

Preferred option

The preferred option for an existing installation is the same if it were new – suitably sized and correctly located inspection hatches (see Figure 13.23) to be installed along the route of the chimney system, along with the installation of CO detector alarms complying with BS EN 50291 in every room along the chimney system route – including any neighbouring/adjacent property, as necessary (see **Adjacent properties** in this Part).

This is in addition to regular servicing/maintenance checks as stipulated by the appliance manufacturer and Regulation 26(9) of the GSIUR – including, where appropriate, combustion gas analysis.

Figure 13.23 Example locations of access panels for concealed horizontal chimneys in voids

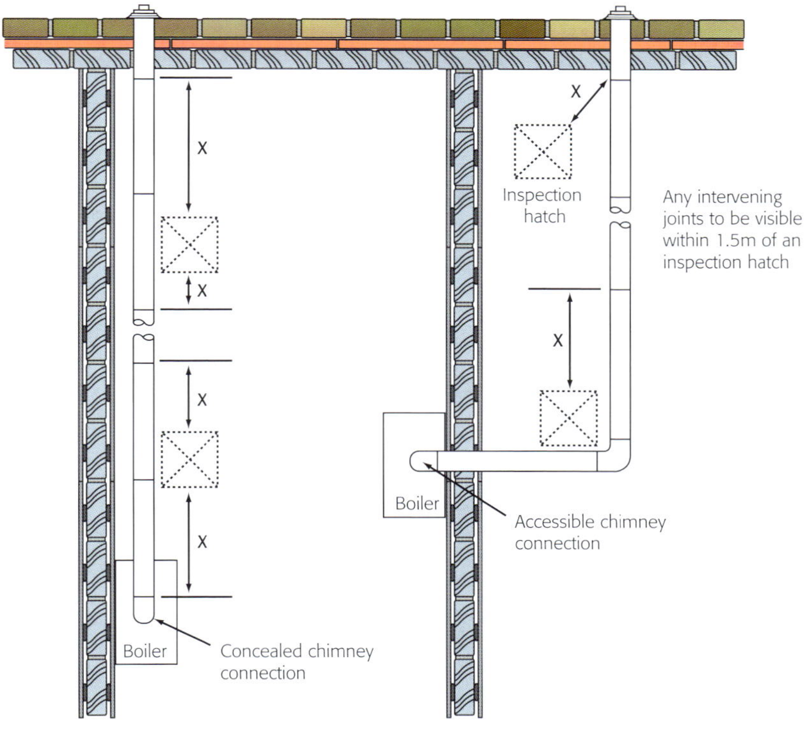

All voids containing concealed chimney systems should have at least one inspection hatch measuring at least 300mm square.

No chimney joint within the void should be more than 1.5m distant from the edge of the nearest inspection hatch, i.e. dimension ' X ' in the diagram should be less than 1.5m

Locate inspection hatch to optimise the ability to inspect the chimney system, e.g. inspection hatches located directly below the route of the chimney system may restrict inspection.

Where possible, inspection hatches should be located at changes of direction.
Where this is not possible then bends should be viewable from both directions.

Therefore, where the chimney system can be inspected and no defects (see Note) of that system are found (including a satisfactory combustion gas analysis, where appropriate), combined with the installation of the requisite number of CO detector alarms, the installation would be considered safe for continued use.

Note: Types of defects, which would cause the installation to be regarded as unsafe include:

- **incorrect chimney components (joints, materials, etc.) not specified by the manufacturer**
- **signs of water leakage (condensate) at joints**
- **incorrect/inadequate chimney system support**
- **inadequate gradient/fall back to the appliance**

If no inspection hatches are provided however, it doesn't automatically mean that the installation is unsafe, i.e. 'AR' or 'ID' - a second option exists, which although are exceptional, may be encountered.

The option is the installation of what's termed 'CO Safety Shut-Off Void Monitoring' system (COSSVM).

COSSVM

It is recognised within the TB that in some exceptional circumstances, the installation of inspection hatches for existing installations may not be practicable, for instance where the enclosure is to small or where the buildings fire precautions would be adversely affected.

In such cases, a system that automatically and continually monitors the void and which automatically isolates the gas appliance in the event of COSSVM being activated, combined with those checks/test required by Regulation 26(9) on the appliance, would be acceptable.

Note: It must be stressed that COSSVM are not acceptable for new or replacement installations, which must comply fully with the requirements of ADJ.

Classification of non-compliant installations

For existing installations, which do not have the appropriate inspection hatches and CO detector alarms installed, or COSSVM systems installed, then the installation remains as an 'AR' classification* in accordance with the current GIUSP.

*Where there are signs of distress to the ceiling/enclosure along the route of the chimney system, the installation needs to be classified as 'ID' and not 'AR' in accordance with the current GIUSP. A RIDDOR report – see **Part 1 – Gas safety legislation – Reporting of Injuries, Diseases and Dangerous Occurrences Regulations (RIDDOR)** – also needs to be completed.

Gas operatives should seek to gain approval from the responsible person to make the installation safe until such time as suitable inspection hatches and CO detector alarms are installed.

However, if the responsible person refuses to allow the installation to be made safe or for the remedial work to be undertaken, gas operatives should continue to service/maintain the appliance, recommending that CO detector alarms be installed as a minimum to ensure continued safety whilst continuing to apply an 'AR' classification to the installation.

Exception

There may be some boiler installations, which employ a short, vertical section of flue pipe that passes through a void/enclosure before terminating externally without inspection hatches being provided.

In such cases and providing the following is met, these installations would be deemed as being acceptable:

- no changes in the flue pipe direction
- no signs of distress to the void/enclosure; and
- the flue pipe length doesn't exceed the maximum single chimney component length supplied by the manufacturer; or
- documented confirmation is available from the installing business responsible for the installation that states no joints are within the void/enclosure

Adjacent properties

Some existing installations may be encountered which have chimney systems that run through neighbouring/adjoining properties before terminating externally (this practice isn't allowed for new installations).

Such installations are problematic for registered engineers, as gaining access to these properties isn't guaranteed.

Under these circumstances, gas operatives need to take 'reasonable steps' to discharge their duty of care.

The appliance and its chimney system needs to be inspected/tested within the property containing the appliance as stated previously.

The reasonable steps are defined within the TB as:

- making enquires with the occupants of adjoining properties in order to gain access
- requesting to see evidence, where it exists, of reports/examinations undertaken previously on behalf of occupants, building owners, managing agents, etc
- leaving documentation with the occupants of adjoining properties, explaining the need to gain access for inspection (see Note); and
- providing information to the 'Flues in Voids' database, hosted by Gas Safe Register at https://engineers.gassaferegister.co.uk/FluesInVoids.aspx

Note: Gas Safe Register have a series of Fact Sheets that can aid registered engineers in explaining the problem to customers; one such fact sheet deals with 'Flues concealed in voids – gas safety concern' and is freely available to downloaded from Gas Safe's website - www.gassaferegister.co.uk/learn/consumer_factsheets.aspx

Openings into buildings

Situate room-sealed flue terminals away from any opening into a building, including doors, opening windows and air vents.

The location of a room-sealed flue terminal also depends on the rated heat input of the appliance. The requirements are shown in Figure 13.24 and Table 13.13 overleaf.

Note: You cannot reduce the minimum distance from an opening window, door, air vent or opening in the building structure, below those dimensions given in ADJ and BS 5440-1.

For other distances, appliance manufacturers may specify alternative minimum dimensions that you should follow.

Figure 13.24 Room-sealed terminal positions

Room-sealed flue terminal position (see Table 13.13).

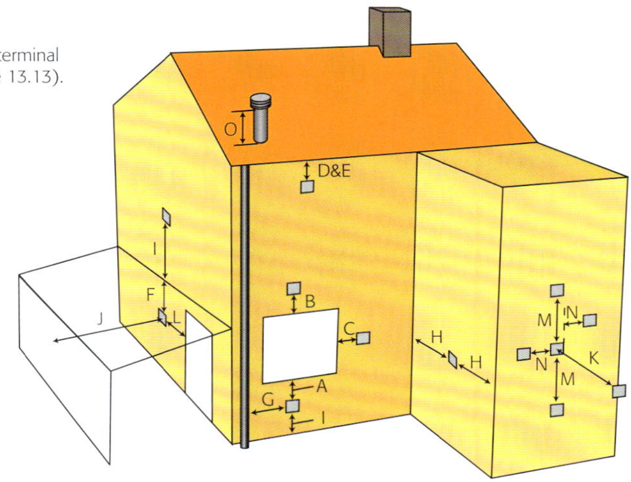

Figure 13.25 Minimum clearances for structural/temperature considerations

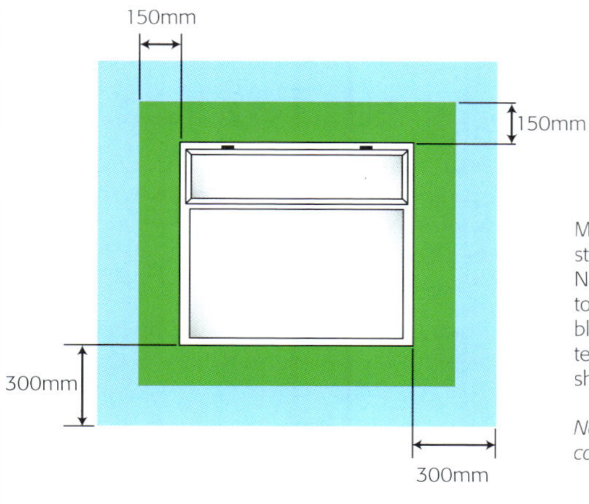

Minimum clearances for flue terminations for structural and/or temperature considerations. Natural draught room-sealed terminals need to be outside of the shaded area indicated in blue and fanned draught open-flued/room-sealed terminations need to be outside of the shaded area indicated in green.

Note: All wall terminations also need to comply with the requirements of Table 13.13.

Table 13.13 Room-sealed appliances: minimum dimensions for flue terminal positions

Dimension as defined in Figure 13.24	Terminal position	Heat input rating expressed in kW (net)	Natural draught	Fanned draught
A *	Directly below an opening, air brick, opening window, door, etc.	0-7kW >7-14kW >14-32kW >32-70kW	300mm 600mm 1500mm 2000mm	300mm 300mm 300mm 300mm
B *	Above an opening, air brick, opening window, door, etc.	0-7kW >7-14kW >14-32kW >32-70kW	300mm 300mm 300mm 600mm	300mm 300mm 300mm 300mm
C *	Horizontally to an opening, air brick, opening window, door, etc.	0-7kW >7-14kW >14-32kW >32-70kW	300mm 400mm 600mm 600mm	300mm 300mm 300mm 300mm
D	Below temperature sensitive building components e.g. plastic gutters (see Figure 13.28)	0-70kW	300mm	75mm
E	Below eaves	0-70kW	300mm	200mm
F	Below balconies or car port roof	0-70kW	600mm	200mm
G	From a vertical drain pipe or soil pipe	0-70kW	300mm	150mm†
H #	From an internal or external corner	0-70kW	600mm	300mm
I	Above ground, roof or balcony level	0-70kW	300mm	300mm
J	From a surface facing a terminal ‡)	0-70kW	600mm	600mm
K	From a terminal facing a terminal	0-70kW	600mm	1200mm
L	From an opening in the car port into the dwelling	0-70kW	1200mm	1200mm
M	Vertically from a terminal on the same wall	0-70kW	1500mm	1500mm
N	Horizontally from a terminal on the same wall	0-70kW	300mm	300mm
O	Above intersection with roof	0-70kW	N/A	300mm

* In addition, (for temperature and structural reasons) the terminal should not be closer than 150mm (fanned draught) or 300mm (natural draught) from an opening in the building fabric for the purpose of accommodating a built in element such as a window frame, (see Figure 13.25). Separation distances are linked to rated heat inputs as shown.

† This dimension may be reduced to 75mm for appliances up to 5kW (net) rated heat input.

‡ Fanned flue terminal should be at least 2m from any opening in a building which is directly opposite and should not discharge POC across adjoining boundaries.

This does not apply to building protrusions less than 450mm e.g. a chimney or an external wall, for the following appliance types: fanned draught appliances, natural draught appliances up to 7kW (net), or if detailed in the manufacturer's installation instructions.

> = Greater than.

Installation

Room-sealed (Type C) natural-draught appliances are always located on or adjacent to an external wall.

Fanned-draught types with extended chimney systems may be located some distance from the termination position on the external wall. Install as follows:

- make good the wall face around the chimney to stop the ingress of rain, draughts or POC
- if the inner wall face is constructed of a combustible material, a fireproof plate may be required between the appliance body and wall surface
- also refer to particular appliance manufacturer's installation instructions
- where the chimney passes through a wall of combustible construction, the assembly should pass through a non-combustible sleeve with a minimum 25mm air gap.

Siting of flue terminals in relation to boundaries – care needed

The boundary of a property is defined in ADJ as 'the boundary of the land or buildings belonging to and under the control of the building owner'.

For the purpose of determining suitable flue terminal positions for gas appliances, the boundary can be considered to extend to the centre line of any adjacent routes or waterways e.g. paths, streets, rights of way, canals, rivers or railways.

To ensure that the installation of chimneys does not affect neighbouring properties, BS 5440-1 recommends that care be taken when siting terminals in relation to boundaries.

The distances from the boundary are derived so that in the event that a neighbouring property is extended up to the boundary line, the performance of the flue will not be adversely affected.

The principle you need to follow when you select a flue terminal position is this:

Imagine that the neighbouring property had already been extended up to the boundary. If this means that the selected flue terminal position does not meet the minimum distance requirements shown in Figure 13.24 and Table 13.13, that would then apply and do not proceed with the installation, see Figure 13.26.

Room-sealed terminal positions after 1st January 2001 – factors to note

- site flue terminals to ensure total clearance of the POC in accordance with Table 13.13 and Figure 13.24
- ensure that when POC are discharged, they do not cause a nuisance to adjoining or adjacent properties
- position terminals so that damage cannot occur to other parts of the building
- if the outer wall surface is constructed of combustible material, fit a non-combustible plate behind the terminal, projecting 25mm beyond its external edges
- do not site flue terminals in a covered passageway between properties
- it is recommended that you site fanned flued terminals at least 2m from any opening into a building, which is directly opposite and should not discharge POC across adjoining boundaries (see **Pluming – two main problems arise** in this Part)

Figure 13.26 Siting of flue terminals in relation to boundaries

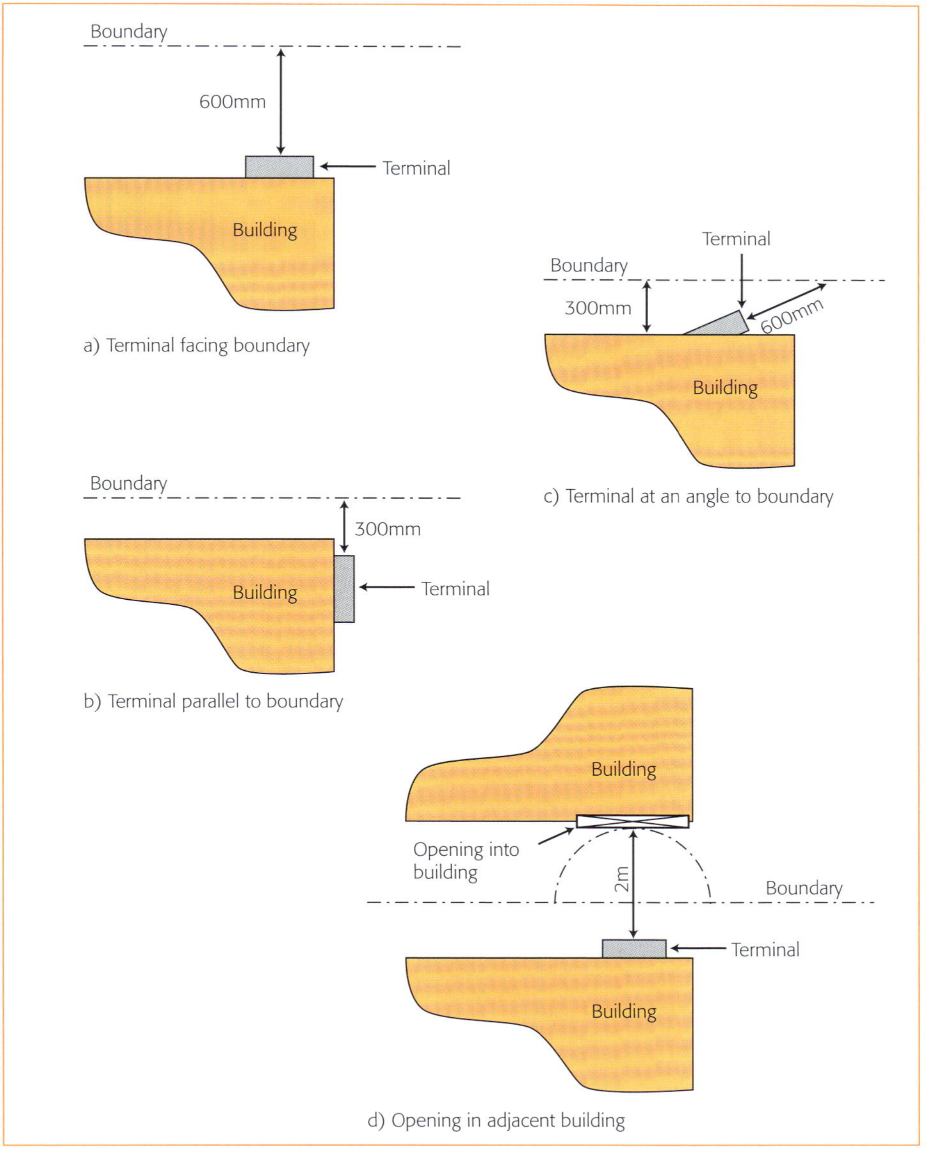

Note: Figure 13.24 shows terminal positions that comply with British Standards and appliance manufacturers' installation instructions.

These instructions however, are based on a detached property. In reality, room-sealed appliances are often installed in link or terraced properties, which may have an extension (actual or planned) to the rear.

To maximise use of available ground, the extension wall is often built directly on the boundary or close to it.

Neighbourhood considerations

If a room-sealed gas appliance is fitted on that boundary wall, the flue terminal may extend into the neighbouring property. This may lead to a customer complaint, with allegations of trespassing and complaints of POC entering the adjacent premises.

The situation is compounded if the gas appliance is a room-sealed, fanned-draught or condensing boiler where POC are blown onto neighbouring doors and windows. To add to the misery, in colder weather, the POC condense in the cold air to form a plume, which can be particularly irritating for the neighbour.

Be aware that installing an appliance in such a position may contravene the GSIUR.

In addition, the neighbour may also have plans to develop or build an extension on their property, which may ultimately obstruct the flue outlet.

When you are requested to install a gas appliance in these positions, advise the customer of the consequences. Such installations often offend neighbours who may contact local Planning or Environmental Health Officers (EHOs) who in turn may have powers under the Environmental Protection Act 1990 to shut down an offending appliance.

See also **Exchanging information with the gas user and planning** – and **Siting of flue terminals in relation to boundaries – care needed** in this Part.

Carport or building extension

Where you site a flue terminal within a carport or building extension, it should have at least two completely open and unobstructed sides.

The distance between the lowest part of the roof and the top of the terminal should be at least 600mm for natural-draught appliances or 200mm for fanned-draught appliances (see Figure 13.24 and Table 13.13).

Note: Do not treat a covered passageway (e.g. between terraced houses) as a carport and do not install new appliances in this location.

If you find an existing gas appliance flue terminal within a covered, passageway, the engineer will need to carry out a risk assessment of the installation in accordance with the procedure outlined in the current Gas Safe Register Technical Bulletin - TB 007 Room-sealed (natural / fanned) and open-flued (fanned draught) appliance flues terminating within covered passageways (ginnels) - which will help to classify the installation.

In older pre-war housing stock, many covered passageways were used as makeshift air raid shelters during the Second World War and often still have corrugated iron sheeting fixed to the ceiling. The integrity of the ceiling above the corrugated iron sheeting is often unknown and may need to be considered when determining if there are possible entry points into the property.

Figure 13.27 Heat-shield for fire protection clearances room-sealed appliance flue terminal (natural-draught)

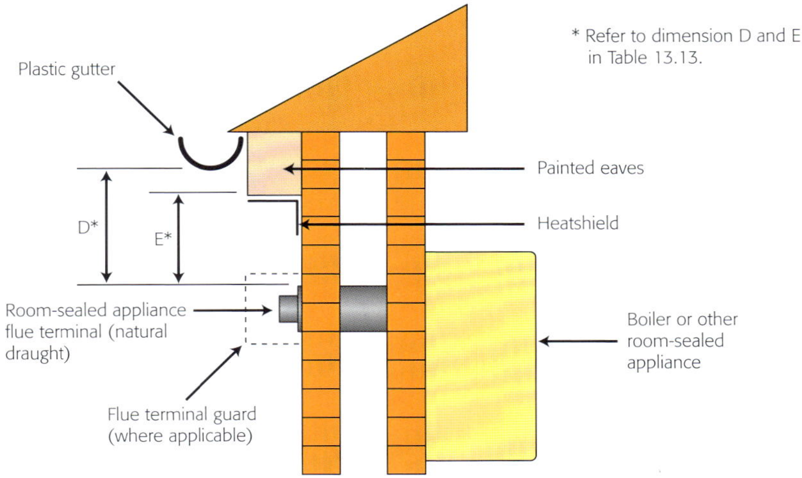

Basements, light wells and retaining walls – points to note

Do not site flue terminals within the confines of a basement area, light well or external space formed by a retaining wall, unless you take steps to ensure that the POC can disperse safely at all times.

- It is possible to locate room-sealed (Type C) appliances in such a location, provided that they are not sited lower than one metre (1m) from the top level of that area to allow the POC to disperse safely; and
- the basement area or lightwell has a single retaining wall creating an uncovered passageway at least 1.5m wide, allowing the free passage of air beyond the ends of the structure and retaining wall

Light well – an area that may inhibit dispersal of POC

A lightwell can be best described to cover areas where for example, there are walls opposite and/or around the wall on which a flue terminal is located, that may be inhibiting the free dispersal of POC.

An illustration of this might be the common victorian urban terrace property, where the front door opens onto the street footpath across a narrow well, which provides light and possibly access to a cellar.

It may also be the small central area in a square building designed to give light to the rooms on the inside of the building.

Be aware of the dangers of locating flue terminals where the POC cannot freely disperse. This could be accentuated by the possible accumulation of POC caused by several terminals all discharging into a single confined space.

Heat shield

When a room-sealed appliance flue terminal (natural-draught) is sited less than 1m below a plastic gutter, or less than 500mm below a painted surface, fit a suitable 1m long heat shield (see Figure 13.27 and Table 13.13).

Terminal guards – when to use

Fit a flue terminal guard over a terminal:

- if anybody could come into contact with that terminal - injury through touching excessively hot surfaces – where the terminal is less than 2m from the ground, balcony or flat roof
- if there is a risk of blockage, for example by birds nesting
- if it could be subject to damage

The appliance manufacturer will specify a suitable terminal guard, where required.

No part of the guard should be less than 50mm from any part of the terminal when fitted in accordance with the manufacturer's instructions. The guard should not have any sharp edges, which could cause injury and any opening should be no larger than 16mm. Proprietary guards are generally available.

Timber frame construction – ensure no adverse effect on weather proofing

Whilst it is possible to install both open-flued and room-sealed appliances in timber frame properties, take great care to ensure that the chimney does not interfere with the weather proofing qualities of any outer wall it may penetrate.

Before you attempt this work and for further installation details, refer to the manual entitled 'Gas Installations In Timber/Light Steel Frame Buildings' from the current Gas Installer Manual Series – Domestic.

Shared chimney systems – for shared or multi appliance installations

The installation of new shared chimney systems needs to comply with the manufacturer's instructions and the appropriate British Standards, BS 5440-1 for example.

For existing shared chimney systems, its use, maintenance and de-commissioning should follow the requirements of IGEM/UP/17 'Shared chimney and flue systems for domestic gas appliances'.

In general:

- open-flued or room-sealed appliances can be used on an appropriate shared chimney system
- never mix appliance chimney types on a shared chimney system

For open-flued appliances, a subsidiary flue provides the flue pull to clear the appliance's POC and discharges them into a main flue.

Shared chimney systems can be categorised as:

- common flues
- shunt duct (branched) flues
- communal flue systems
- Se-duct flues
- U-duct flues

Common flue systems in the same room

Common flue systems are used where two or more open-flued appliances are to share the same chimney system, e.g. for a warm air heater with circulator.

- each appliance should have a draught diverter and a flame supervision device
- each should also incorporate a safety control designed to shut down the appliance before there is a dangerous

quantity of POC in the room concerned, i.e. an Atmosphere Sensing Device (ASD), see **Part 19 – Definitions**.

- each flue should have at least 600mm of vertical secondary flue before it joins at the common flue. The common flue needs to be large enough to accept the combined appliance outputs.
- gas-fired appliances should not share the flue with appliances burning other fuels such as oil or solid fuel. The flue serving the most used appliance should be connected closest to the main flue.

Note: Seek advice from the appliance manufacturer as to suitability for installation on a common chimney system.

When two or more appliances are to be installed in the same room or space, you may need specialist advice to design the chimney system.

Shunt duct flue systems

You can use shunt duct flue systems when you need to install open-flued appliances on different floors, or in different rooms with similar levels of ventilation.

- the main flue of a masonry shunt duct flue system should not be part of an external wall
- similarly, do not enclose a shunt duct flue pipe system in such a way that the flue duct becomes part of an external wall or is located externally
- the nominal cross-sectional area of the flue for two or more appliances should not be less than 40,000mm^2
- an appliance should be connected to the main flue by a subsidiary flue that is at least 1.2m long measured from the flue outlet of the appliance

Note: For a gas fire, the connection of the subsidiary flue to the main flue should not be less than 3m above the appliance outlet.

All appliances connected to the main flue need to be of the same flue type. This is important to remember when you install replacement appliances.

The main flue should have no offsets and the flue route should be no more than 10° from the vertical.

Any ventilation openings in the rooms where the appliances are installed should be similar in design and aligned on the same walls.

Label all appliances indicating:

- that they are fitted to a shared flue system
- the name of the person responsible for the building, together with telephone and contact details

Terminal positions

Position any terminal so that it is a minimum of 6m above the highest appliance connected to the main flue. To achieve this, you may have to flue the top floor appliance separately.

- when the chimney system passes through a pitched roof, the base of the terminal outlet opening should be above the level of the ridge of the roof
- when the chimney system passes through a flat roof, site the base of the terminal outlet opening following the guidance given in Table 13.11

Communal flue systems (CFS)

These generally metallic systems are used for the connection of room-sealed appliances (condensing and non-condensing) in communal housing/flats such as multi-storey buildings and other multi-dwelling premises.

The CFS is designed with a condensate collection unit at its base, which is intended for collection and removal of condensate in the flue itself and not intended for the removal of condensate from the individual appliances connected to it.

Individual appliances will still need their own condensate drain connection not connected to the CFS installed in accordance with manufacturers' instructions.

CFS can be sub-divided in to three main categories:

- CFS/NV (naturally vented)
- CFS/FA (fan assisted); and
- CFS/EO (exhaust only), which can be NV or FA

Naturally vented (NV) CFSs provide combustion air and removal of exhaust products from multiple gas-fired heating appliances using a twin pipe or concentric system operating under naturally vented negative pressure and can be used generally for between 1 and 4 appliance connections at each floor level with connections of 80mm twin pipe, or 60/100mm or 80/125mm concentric.

Fan assisted (FA) CFSs provide combustion air and removal of exhaust products from multiple gas-fired heating appliances using a twin pipe or concentric system operating under fan assisted positive pressure and can also be used generally for between 1 and 4 appliance connections at each floor level with connections of 80mm twin pipe, or 60/100mm or 80/125mm concentric.

Exhaust only (EO) CFSs, which can be either NV or FA provide removal of exhaust products from multiple gas-fired heating appliances using a single exhaust duct.

Ventilation for the connected room-sealed appliances is provided via separate air duct installed within the dwelling but communicating with outside air for each appliance (although a second CFS/EO can be installed to provide combustion air for connected appliances).

Where appliances are connected to fan assisted systems (CFS/FA or CFS/EO/FA) they will require a non-return valve (NRV) fitted to prevent recirculation of exhaust gases to appliances not in operation.

When replacing a gas-fired appliance connected to the CFS it needs to be of the same type and input as the original although both condensing and non-condensing appliances can be fitted to the same CFS if its kW rating is not exceeded.

The CFS should not be altered, e.g. by cutting new or enlarging existing holes to enable the connection of the new appliance and where an appliance is removed the redundant connection points will need to be sealed with appropriate blanking cap.

Se-ducts and U-ducts – require specialist design as sizing of duct is critical

Many different types of flue systems are currently used with gas burning appliances. Two are the 'Se-duct' and 'U-duct' systems. You will find these at various locations nationwide, usually in multi-storey buildings (particularly blocks of flats) and you must think carefully when you install or maintain gas appliances connected to them.

The Se-duct derives its name because it was developed by the then South Eastern Gas Board. The U-duct is a variation of it.

Only use NG burning appliances that the manufacturer confirms are suitable for fitting to these flue systems (see Figure 13.5). These include both natural-draught and fanned-draught room-sealed appliances.

However, it is important – to ensure their safe operation – that you ensure that fanned-draught appliances are not fitted back to back with any other appliance in the same duct. You must follow the relevant manufacturer's installation instructions when you install these flue systems.

Most natural-draught room-sealed appliances require the following from shared flue systems:

- the air inlet connection needs to be flush with the main duct inner wall
- the flue outlet duct of the appliance should extend into the flue, to the dimension required by the appliance manufacturer
- both the air inlet and the flue outlet spigots need to be properly sealed where they enter the duct wall

Label all appliances indicating:

- that they are fitted to a shared flue system
- the name of the person responsible for the building together with telephone and contact details

Note: Always refer to the appliance manufacturer's installation instructions.

Se-duct systems – how they work

Se-ducts are designed to serve a number of room-sealed appliances in multi-storey dwellings and consist of a vertical duct rising up through the building. Se-ducts need to be vertical and should discharge the POC above roof level.

The vertical duct is usually constructed from pre-cast concrete sections in standard sizes. The duct can also be constructed using fire resistant materials, as long as they are of adequate strength and thickness.

Combustion air is taken from the base of the system in one of the following ways:

1. By a horizontal duct from one side of the building to another.
2. By a single duct from a neutral pressure zone.

All combustion air ducts should be open to the atmosphere at each end.

Where you use a low level duct (at the base of the building), fit a removable grille to prevent debris from falling into the area below. The grille openings should allow the admittance of a 10mm diameter ball but not a 16mm diameter ball.

Label all grilles and the duct base:

- detailing the purpose of the duct
- giving contact details of the responsible person for the building

The responsible person for the building should:

- provide labels for the ducts and grilles, which indicate that they supply air to the building's gas appliances and their flueing system and ensure they should not be blocked or restricted in any way

U-duct systems – how they work

These are similar to Se-ducts and are used where it is not possible to obtain combustion air from the base of the building. So the air is brought down from the top of the building by a vertical duct adjacent to the flue discharge point.

U-duct systems consist of two vertical ducts connected at the base. Both ducts rise above roof level. One duct takes in air for combustion and the other duct discharges the POC.

How you install to Se-duct and U-duct

Connect appliances to the vertical flue on each floor, through pre-formed holes or holes you cut during installation of the appliance

To ensure the appliances are installed correctly, you must follow the manufacturer's instructions at all times, paying particular attention to the correct commissioning procedure.

Apply the following guidelines:

- seal all unwanted holes so that the inner wall of the duct has a smooth finish
- repair or replace cracked or damaged sections of duct
- prevent rubble from falling into the duct during installation
- make new holes in the duct with a core drill or by stitch drilling (do not use rotary percussion drills)

- to prevent combustion products entering the room from other parts of the building, keep the duct open only as long as you need for installation or servicing
- the base of the terminal should be at least 250mm above roof level
- wherever possible, position the terminal away from parapets, storage tanks or other structures
- condensing appliances must not be installed into existing Se-duct/U-ducts, unless replacing condensing appliances
- if the terminal is within 1.5m of any structure, raise the base of the terminal above the structure
- for new buildings, carry out commissioning in accordance with the appliance manufacturer's installation instructions, with all the space heaters working, plus 33% of the instantaneous water heaters – those in the top third of the building.

  The combustion air supplied to the highest appliance should contain no more than 1.5% CO_2 by volume
- for single appliance installations into existing buildings, commission the appliance as if it is connected to a single flue and with due regard to the appliance manufacturer's installation instructions

Flue/appliance identification – label clearly

Attach a label to all the gas appliances:

- stating that they are fitted to a Se-duct/ U-duct flue system
- stating that they should not be removed or replaced without reference to the person responsible for the building
- providing contact details

Replacement appliances

Only consider 'Like for like' and appliances with similar gas burning consumptions for the replacement of existing appliances to either Se-duct or U-duct flue systems.

Common faults

Look for:

- appliance air inlet blocked or restricted caused by debris at base of flue system
- replacement appliances fitted not compatible with original specification

Diagnosis/inspections – specialist knowledge and equipment needed

Following stringent changes to the UK gas safety legislation, landlords and property developers have a legal obligation to inspect all flues and ensure that these systems are maintained to a safe condition.

Visual external inspection of gas flue systems cannot always establish the cause of a problem. More specialist knowledge and equipment is needed to provide an in-depth internal examination of the flue system using CCTV equipment.

Note: Further detailed guidance on various shared chimney systems can be found in both BS 5440-1 and IGEM/UP/17 'Shared chimney and flue systems for domestic gas appliances'.

Special chimney systems – an outline for you

Fan dilution systems

The principle of this type of flue system is to mix the POC with fresh air drawn from the outside atmosphere to dilute the POC reducing CO_2 concentrations below 1% and disperse them at low level to atmosphere in what would otherwise be unacceptable or undesirable locations.

The individual appliances are connected by their vertical flue pipes to a common header, which is typically open to the external air at both ends.

One end of the flue acts as the intake for the dilutant air (a variation is to have the air intake located within the space the appliances are housed, with additional ventilation provided to the space to account for configuration) and the other end as the discharge point for diluted combustion products to the outside atmosphere.

The design of fan dilution systems is a specialist activity and therefore, needs to be undertaken by persons who are competent in their use and designed in accordance with the manufacturers' requirements, as well as those of the appropriate standards – e.g. BS 6644 or IGEM/UP/10 (Edition 4).

Fan dilution is more commonly used in non-domestic installations and as such, detailed guidance can be found in CORGI*directs* 'Essential Gas Safety – Non-domestic' manual (see **Part 22 – CORGI*direct* Publications**).

Vertical fanned draught room-sealed chimneys – Type C_{31}, C_{32} and C_{33} appliances

The majority of fanned flue appliances now offer the vertical option. Position the terminal in strict accordance with the appliance manufacturer's instructions.

There are two main types of vertical fanned room-sealed chimney systems:

1. Those taking air directly from outside the building, where air is drawn into the combustion chamber through the intake section of the chimney system and the POC are forced up the exhaust section of the chimney system to the outside.
2. Those taking air from a correctly ventilated roof space (Vertex flues). Air is drawn into the combustion chamber through the suitably protected air intake section of the chimney system in the roof space.

   POC are forced outside via a secondary chimney section separated from the primary flue by a draught break.

Always follow the appliance manufacturer's installation instructions, see also Figure 13.4.

Here is general guidance on the information you need:

- the horizontal clearance between the terminal and any adjacent vertical walls/structures or obstructions
- the clearance between the terminal and an openable window, air vent or any other opening into the building
- the horizontal distance between any flue terminals at the same height

In addition these requirements will also need to be met:

- the terminal should extend a minimum distance of 300mm above the surface of the roof. This is measured from the exterior surface of a flat roof, or the higher point of intersection with a pitched roof, to the lowest outlet of the terminal
- when the chimney system passes through a ceiling or wooden floor, there needs to be an air gap of 25mm between any part of the chimney system and any combustible material. Using a proprietary ceiling plate will normally provide for this
- when the chimney system passes from one room to another, you must fit a fire stop to prevent the passage of smoke or fire, irrespective of the structural material through which the chimney system passes
- the chimney system must be adequately supported to prevent the weight of the chimney system being transferred to the appliance flue connection. You can achieve this by fixing brackets, installed at a maximum distance of 1.8m apart
- where the chimney system passes through the roof, use the correct type of flashing kit to ensure a waterproof joint

Vertex flues (Type C_7) – take air for combustion from roof space

With this system, a concentric air/flue duct (flue inside the air duct) is run vertically from the boiler through the roof space, where the air duct ends and the flue duct continues to a suitable terminal above the roof.

As air for combustion is taken from the roof space:

- ensure this space is adequately ventilated to give the manufacturer's free air requirements
- ensure it conforms to current Building Regulations and does not communicate with other premises
- locate the draught break at least 300mm above the finished level of roof insulation

Note: These systems allow you to position the appliance (usually a boiler) centrally inside a property, (e.g. airing cupboard) – and you need to position the appliance close to a ceiling adjoining a roof space, or in such a position within the appliance manufacturer's recommended maximum flue length requirements.

Ensure that the overall flue length does not exceed the maximum length indicated by the appliance manufacturer's installation instructions, see also Figure 13.4.

Two pipe unbalanced flue systems (Type C_5) – where to locate

With this system, you use two separate pipes, one for the flue and one to supply air for combustion.

- the pipes can usually be connected at the top, base or sides of the appliance and can be taken in any direction
- typical locations may be with a vertical terminal through the roof and the air inlet on a side wall. It is also acceptable with horizontal terminals to locate the terminal on a wall with the air inlet separate from it.

  If you fit the flue terminal to a different wall from the air inlet, these should not be opposite walls

Generally, the appliance manufacturer's installation instructions specify the individual requirements for their flueing systems.

In many instances, the appliance specific flueing installation instructions are supplied within the box containing the flue components, rather than within the main installation instructions.

- as with all fanned-draught flue systems, the length of pipe depends upon the number of bends required. Always follow the appliance manufacturer's installation instructions.
- where the manufacturer's instructions do not provide any specific information, then position the terminal in accordance with the requirements for fanned-draught flues generally

Chimney systems for condensing boilers – lower flue gas temperatures

Condensing boilers and other high efficiency appliances produce lower flue gas temperatures. The POC are also slightly acidic.

For these reasons, the use of plastic chimney systems is a common sight for these appliances. The materials, jointing method, flue length, number of fittings and type of terminal used should be as supplied, or as specified by the appliance manufacturer.

Condensate drain – chimney must provide for collection/disposal

With appliances of this type, you must make provision for the collection and disposal of condensate in the flue.

The condensate drainpipe should:

- be made of corrosion resistant, 22mm outside diameter pipe – generally plastic installed with a minimum down slope of 2.5° from the horizontal (50mm per metre of pipe length)
- preferably drain internally into a stack via a trap with a 75mm seal and connect downstream of any sink waste trap, or although not preferred drain externally directly or indirectly into the soil or rain water pipe via a hopper, soak away, gully or stack pipe.

Note 1: Further detailed information on condensate discharge can be found in CORGI*directs* 'Central Heating Wet and Dry manual – Part 5 – Condensing boilers – Condensate discharge'.

Note 2: All trap arrangements should comply with Building Regulations/Standards.

Pluming – two main problems arise

Carefully consider where you locate appliances liable to pluming, particularly condensing boilers.

The POC from condensing boilers are considerably cooler than those from a conventional boiler and therefore can produce an effect, which is visually similar to that of a boiling kettle.

This effect is at its worst when the boiler initially fires up and most of the heat is extracted from the POC.

Note: The POC from a condensing boiler flue is slightly acidic.

There are two main problems associated with pluming:

1. If POC discharge onto an adjacent property, this may result in nuisance and disputes between neighbours. It is always desirable to site the flue terminal and any plume within the boundaries of the property in which the appliance is fitted.
2. If POC discharge adjacent to any woodwork, this may cause localised decay (rotting). Follow the manufacturer's clearances. Ensure that all combustion products are likely to clear the building and disperse without dew forming on woodwork.

See also **Exchanging information with the gas user – and planning** and **Building regulations/standards – current requirements** in this Part.

Note: Plume kits have been produced by most boiler manufacturer's (see 'Plume management kits – additional information' below), which if used may resolve nuisance issues.

However and it's important to note, these kits are not to be used as a substitute for correct/compliant flue terminal locations.

Plume management kits - additional information

Plume management kits are produced by appliance manufacturers to facilitate a different flue terminal location and avoid nuisance pluming over neighbouring properties, boundaries and public spaces etc.

The kits must be installed strictly in accordance with the manufacturer's instructions and normally consists of suitably sized plastic pipe with separate 87° or 45° elbows and a bird protection guard for the terminal outlet.

The kit should be installed with a continuous fall back to the adapter to allow and condensate to drain away.

Flue termination and air intakes will usually be located on the same wall or building face to balance wind effects although the flue terminal may well be redirected above roof level in some cases.

However and as the previous note explains, the standard restrictions on terminal location still apply.

Open-flue (Type B_2) without a draught diverter (formerly known as a closed flue)

This type of system is usually used with appliances such as incinerators, where open-flues could result in spillage of contaminated POC from the draught diverter (see **Incinerators – have special flueing requirements** in this Part).

Due to the specialised nature of these appliances:

- you must strictly follow the manufacturer's instructions on installation, ventilation and flueing. If you deviate, it could result in an unsafe installation
- you need to carry out a spillage test on all closed flue appliances, in accordance with the appliance manufacturer's installation instructions
- size and position flue terminals for appliances without a draught diverter as specified in the appliance manufacturer's installation instructions

Note: This type of chimney system is often used to flue gas fires installed in caravan holiday homes. You can find further guidance in CORGI*directs* 'LPG – Including Permanent Dwellings, Leisure Accommodation Vehicles, Residential Park Homes and Boats' from the current Gas Installer Manual Series – Domestic.

Figure 13.28 Typical balanced compartment

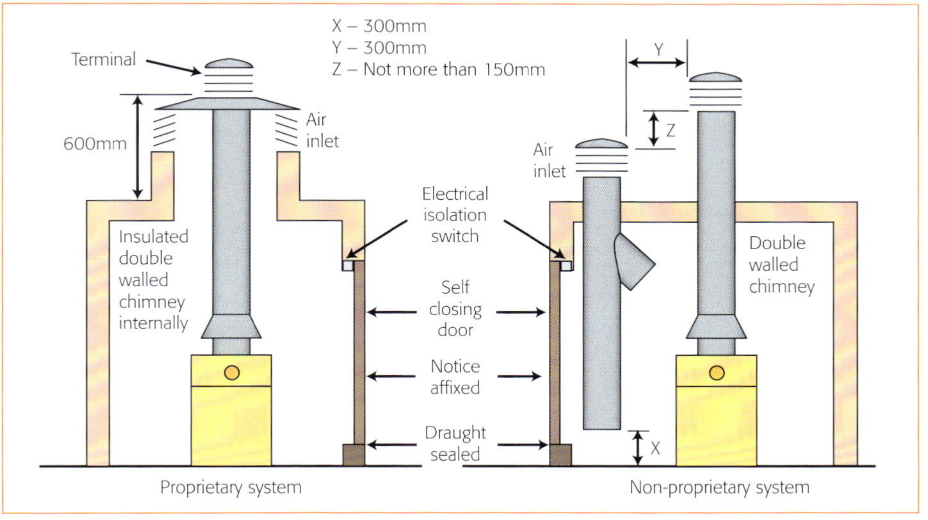

Balanced compartments up to 70kW (net) rated heat input

A balanced compartment provides a means of installing open-flued appliance(s) in a room-sealed situation and it:

- is particularly suited where room-sealed appliances are not generally available
- may also provide an alternative solution to long external flue runs
- may also be used when a boiler house/compartment is sited adjacent to a taller structure

Flueing and ventilation are arranged within a sealed compartment to produce a 'balanced flue' effect, with the flue termination and the provision for high level ventilation positioned closely together. They are therefore subject to the same atmospheric conditions.

There are two main types of installation:

1. Proprietary systems designed and installed by specialist companies.
2. Individually designed systems installed utilising standard materials.

The basic design of both systems is similar, with an open-flue from the appliance(s) and a ventilation air duct provided at high level from the outside air (see Figure 13.28).

Where the air inlet to the boiler compartment is in accordance with the proprietary system the size of the air intake duct needs to be at least 12.5cm^2 per kW net of the appliances maximum rated input.

Alternatively where the chimney system is of the non-proprietary type (i.e. air is ducted to low level) the size of the air intake duct needs to be at least 7.5cm^2 per kW net of the appliances maximum rated input.

Non-proprietary installations – note differing requirements

For these installations, flue sizing is the same as for standard open-flued installations, but ventilation requirements differ.

The arrangements for supplying air and discharging POC to and from the balanced compartment should ideally follow the basic flueing principles for individual open-flues.

The free area of the ventilation duct or pipe should be 1.5 times the normal requirement when ducted to low level and 2.5 times the normal requirement when ducted to high level only (see **Part 4 – Ventilation**).

The air inlet pipe should terminate not more than 150mm below the flue terminal (see Figure 13.28).

Within the compartment, if the ceiling is more than 300mm above the level of the base of the draught diverter, the ventilation duct or pipe should have a 'T' or 'Y' piece at high level.

The high level opening should have the same cross-sectional area as the duct or pipe. It should finish not more than 300mm above floor level in close relation to the appliance.

- fit both terminations with approved flue terminals
- however, if you use two different types, then use the greater extractive type for the flue
- if it is impracticable to terminate in a conventional location, you must seek the advice of the appliance manufacturer or chimney system manufacturer
- ensure the compartment itself has a self-closing, flush fitting door, which is draught-proofed and incorporates a switch/device, which shuts down the appliance when the door is opened
- ensure there is a notice on the door advising that the door remains closed at all times, except for re-setting the appliance controls
- ensure the door does not open onto a bath/shower room
- if the balanced compartment is to open into an area used, or intended to be used, as sleeping accommodation, the appliance must have a heat input rating below 12.7kW (net) or 14kW (gross) and incorporate an ASD designed to shut down the appliance before there is a build up of a dangerous quantity of POC.
- insulate any exposed lengths of flue pipe (not double walled) or hot water supply pipes within the compartment, to minimise temperature rise in the boiler compartment

Incinerators – have special flueing requirements

The chimney system (not covered by BS 5440-1) from the appliance to the terminal should have no breaks (i.e. no draft diverter).

This ensures there is no contamination of the appliance location by smoke or incinerated material and it will provide the maximum flue pull. Whenever possible, the chimney system should comply with the following guidelines:

- it should take the shortest, most vertical route
- chimney components should comply with the appliance manufacturer's specification
- the terminal should have openings to permit entry of a 25mm ball
- it should have a minimum 1.5m vertical height above the appliance
- it should have a removable panel to allow for cleaning
- the manufacturer's installation instructions should always be followed

Maintenance of chimney systems – must be regular to ensure safe operation

Chimney systems must be regularly maintained to ensure they continue to operate safely with the gas appliance installed.

Note: You may use a portable electronic combustion gas analyser (ECGA) to assess whether the combustion of the appliance is satisfactory. However, this is only an aid to the servicing of the appliance: you must carry out other checks as well.

See Part 3 – 'Characteristics of combustion – Portable electronic combustion analysers – Appliance testing – helping you assess safe operation/post-work check'.

Open-flues (Type B) – what to check

- the chimney system should only be used if it is in a safe condition
- carry out a visual check, flue flow check and spillage check – with the appliance connected, in accordance with the appliance manufacturer's instructions
- existing chimney flue block systems can be prone to the build-up of insect nests and spiders webs, particularly during the summer months when the gas appliance is not in use. You may notice this when the appliance is first operated after a period of non-use.

  It may be necessary to sweep the chimney flue block system to remove any build-up – and clear any extrusions from the flue joints that may have been left when the flue system was constructed.

Room-sealed (Type C) – what to check

The room-sealed chimney system should only be used if it is in a safe condition. Also check that:

- the room-sealed chimney system is installed according to the appliance manufacturer's installation instructions
- the securing screws or fastenings, supplied with the appliance for the combustion circuit seal, are undamaged
- the material/gasket used to seal the combustion circuit is in a satisfactory condition otherwise replace it with one specified by the appliance manufacturer
- the appliance case, back panel, etc. shows no sign of distortion or corrosion that may affect the combustion circuit seal or soundness

See also **Part 14 – Flue testing – Testing appliances with positive pressure cases – inadequate seals can be a danger.**

Shared flue systems – how to maintain/who is responsible

The maintenance of a shared flue system is the responsibility of the landlord or the responsible person for the building, who must arrange for annual checks to be carried out on the whole of the flue system.

Because inadequate installation or operation of any single gas appliance connected to a shared flue system may affect the safe operation of others, it is essential that you carry out routine inspections on all gas appliances.

- as far as is reasonably practicable, check the design and installation features, including the roof termination and air inlet ducts/grilles
- where practicable, inspect the appliance connections to ensure the chimney installations are installed correctly and the appliances are operating safely

- where an appliance is not operating correctly, for example persistent pilot outage or poor combustion, you may have to inspect the whole flue duct
- if a replacement appliance has been installed elsewhere on the chimney system, check that no problems are occurring if the flue connection and/or air inlet duct protrude excessively into the duct (this may have an adverse affect on the safe operation of existing appliances connected above the replacement appliance)
- to prevent combustion products entering the room from other parts of the building, ensure the duct is only kept open as long as needed for installation or servicing (See also **Part 14 – Flue testing – Flue testing in multiple dwellings including Landlords' responsibilities**)

Balanced compartments – checking their correct use

- only use the chimney system if it is in a safe condition
- there should be no spillage or leakage of combustion products when the chimney system and appliance is in operation
- carry out a visual check, flue flow check and spillage check with the appliance connected, in accordance with the appliance manufacturer's instructions

For further details of the visual check, flue flow check and spillage check see **Part 14 – Flue testing.**

Flue testing – 14

14 – Flue testing

Figures

Tables

Figures

Tables

Introduction

We explain here the principles of how a chimney system operates. It will help you understand how:

- to test and commission the appliance
- to identify and rectify intermittent or more permanent problems associated with troublesome chimney systems

A chimney system acts to assist the safe operation of an appliance by:

1. Removing the products of combustion (POC) to the atmosphere; and
2. Inducing an adequate supply of fresh air.

With an open-flued appliance, the chimney system is open to the room or space in which it is installed. So it is vital for the occupants' well-being that the chimney system discharges its POC to the atmosphere under all weather and operating conditions.

So the chimney system must be designed and installed correctly before you can finally carry out the necessary flue tests
(see **Part 13 – Chimney standards**)

Operating principles – pressure differentials cause movement of gas/air

To produce movement in a quantity of flue gas, a pressure differential must be applied to the gases in (flue) and around the chimney system.

This differential may be:

- mechanical, as in a fanned-draught flue; or
- thermal, as in a natural-draught flue

Simple physical laws govern how a flue of a chimney system operates. In the simplest form, it operates like our weather system, i.e. cold, heavy air rushes in from Polar Regions to replace hot, lighter air, heated by the sun at the equator. This heated air is displaced upwards by the cold, heavier air, before circulating around the earth.

In the case of a gas appliance, the heat created during the combustion process has a similar effect to that of the sun – and is responsible for the differential in densities between the hot flue gases and the cooler surrounding air. Thus the hot combustion products are displaced up the flue to the atmosphere.

As Figure 14.1 overleaf illustrates, when combustion takes place, POC are produced which are hotter and therefore lighter than an equal volume of the cooler surrounding air. A pressure differential is therefore created between (A) and (B) and is known as the flue draught.

It is equal to the difference in pressure between two columns, one consisting of cool, heavy air (C) and the other of hot, lighter combustion products (E). This causes air to flow from (A) to (D), into the base of the flue, displacing the hot gases up the flue to atmosphere, via the terminal.

Figure 14.1 Operating principles of an open-flue chimney

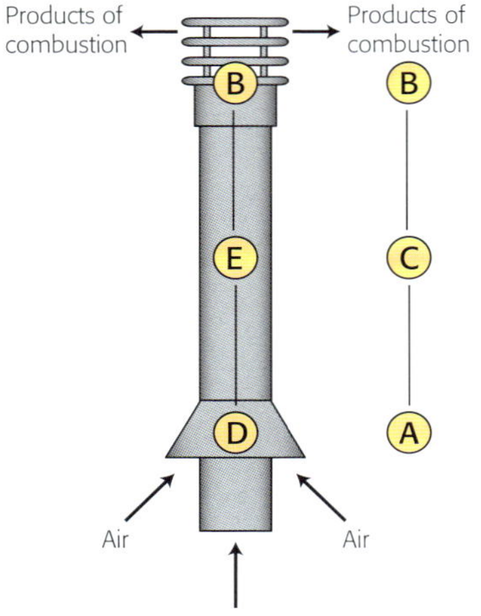

Pressure at (A) (a point just outside the base of the flue) = Atmospheric pressure at the top of the flue (B) = Pressure due to a column (C) of colder dense surrounding air

Pressure at (D) (a point just inside the base of the flue) = Atmospheric pressure at the top of the flue (B) = Pressure due to a column (E) of hot less dense gases

Many factors affect performance

Very little motive power is produced to create a flue draught. Factors that can affect performance include:

- the chimney system's height and the cross-sectional area of the flue
- thermal input, route, heat losses and wind effects
- whether there is adequate ventilation into the combustion area (see Note)
- the depressurising effects of air extract fans in the same or adjoining rooms including upstairs rooms (see **Flue checks – for open-flued appliances – Spillage test – to verify POC being safely discharged** in this Part and **Part 4 – Ventilation**).

Note: POC including excess air, cannot be evacuated to the atmosphere via the chimney system unless the air used in this process is replaced by adequate and permanent ventilation, preferably sited at low level in the room.

Down-draught – usually an intermittent problem

Down-draught is a flue condition where the POC are forced back down the flue into the room. It is usually an intermittent problem, caused by external wind conditions acting on a poorly sited flue outlet, ventilation openings or chimney system route/configuration.

Figure 14.2 Adverse pressure and suction zones

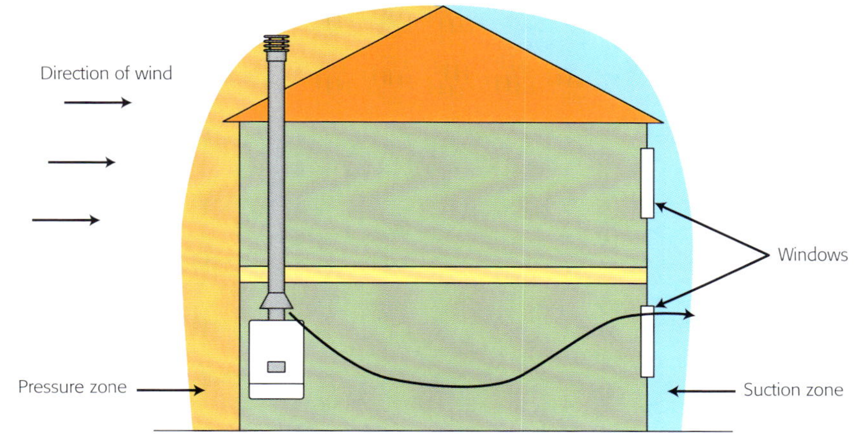

Wind effects – how to minimise

Wind blowing across a building is likely to produce a pressure differential between the bottom and top of the flue, at the terminal.

This differential depends on the wind speed and direction, the position of the flue outlet in relation to the building and the presence of neighbouring structures and geographical features.

All of these may assist or prevent a natural flue draught and cause one of the following:

1. Increased flow up the flue.
2. Reduced flow up the flue.
3. Intermittent down-draught.

Wind pressure can be many times greater than flue draught.

It is important you minimise the effect, by carefully positioning both the flue outlet and combustion ventilation. For recommended terminal positions see **Part 13 – Chimney standards.**

Figure 14.2 shows that the zone of greatest pressure is windward of a dwelling.

A chimney system with an outlet in this (high pressure) zone, where the room containing the appliance can communicate with the leeward (low pressure) side of the building via internal doors, windows, ventilation or other openings, is likely to be subject to down-draught.

This can result in spillage of POC from the appliance. Any chimney system should therefore terminate outside the high pressure zone.

To benefit from the wind effect, consider siting the combustion ventilation opening on the windward side (prevailing wind) so as to assist the chimney system – as opposed to positioning it on the leeward side, which could affect the safe operation of the chimney system.

Just as wind speed can significantly affect the performance of a chimney system, it can affect testing too.

An open-flued appliance tested on a windy day may very well pass a spillage test. The same appliance might fail the test if carried out on a calm/still day.

Thermal inversion – hinders desired flue draught

This inversion generally occurs when the temperature inside the dwelling is raised to a temperature exceeding that found outside the building – and there are 'openings' at high level in the dwelling for this heated air to be displaced upwards to atmosphere by cooler heavier air from outside. In effect, the dwelling becomes a 'flue'.

The symptoms of thermal inversion are very similar to down-draught, but it is caused by different conditions.

The following explanation may help resolve a 'down-draught' problem that occurs even in still air conditions and on a flue that was previously operating satisfactorily.

Following a fairly cool spell, a heating appliance or warm morning sunshine can very quickly warm the air inside a house. Yet, in the case of open-flued appliances, the chimney walls and the air both inside and outside the chimney system remain relatively cool.

This condition is the exact opposite of that which is required to create a flue draught as described in **Operating principles** in this Part.

An ill-fitting roof access door, a fireplace opening in an upstairs room or a fan light, etc. will allow this heated air to 'escape' from the dwelling.

This escaping air is as a direct result of cool heavier air entering the building. Normally, this cooler air will be adventitious air from cracks around doors and windows, etc.

It may also enter the dwelling via an appliance flue (especially where the dwelling has well fitted or draught-proofed doors and windows etc). If a chimney system is the source of the incoming air, this cycle will continue.

The chimney system will remain cold and will be detected at the appliance as an apparent prolonged down-draught.

This often causes nuisance shutdown if the appliance is fitted with a TTB heat sensor or an atmosphere sensing device (ASD). See also **Part 12 – Checking of appliance gas safety devices/controls** and **Part 19 – Definitions**.

When operated under these circumstances, an appliance may:

1. Emit all its POC into the room.
2. Consequently, the chimney system may fail to reach its operating temperature and may therefore not be subject to an 'up-draught'.

Other (most common) causes of thermal inversion are:

- a house heated by radiators or gas fire(s); or
- a house with passive stack ventilation (see **Passive stack ventilation (PSV) – extracts air from rooms in which installed** in this Part)

Draught-proofed doors, windows and other openings aggravate the condition.

Trends in newer housing stock also have an affect:

Houses in the UK were traditionally built with more than adequate ventilation, in the form of:

1. Purpose-designed outer wall grilles communicating directly with internal bedrooms.
2. Adventitious ventilation via ill-fitting windows, doors and floor boards etc.

The recent trend is to increase comfort in homes alongside reducing running costs and emissions of CO_2. Measures can include air grilles being removed, draught-proofing and double glazing installed, insulation of walls, floors and roofs, fitted carpets laid, etc. As a result, homes are far more airtight.

It does mean that, in today's houses, warm stale air can make upper rooms uncomfortable. To overcome the problem, the occupier may open a small window or fanlight – generally at high level – allowing the stale air to escape.

If however, there is an open-flued appliance present and the dwelling is heated, cooler heavier air may flow down the flue serving the appliance, displacing the warm stale air to the atmosphere.

This creates reverse flue conditions (thermal inversion). It is particularly likely where the open window is at a similar height to, or higher than, the flue terminal.

Passive stack ventilation (PSV) – extracts air from rooms in which installed

Modern homes may have PSV fitted: a ventilation system designed to extract air from the rooms in which it is installed (normally wet rooms such as kitchens, bathrooms, WC, etc).

It consists of a series of ducts, one in each room to be ventilated. These ducts pass vertically through the building and terminate above roof level.

They operate through a combination of the natural stack effect (i.e. the movement of air due to a difference in temperature between the inside and outside) and the effect of the wind passing over the roof of the dwelling.

When they operate, the duct(s) may create so much up-draught as to overcome the chimney system of an open-flued appliance.

This may be due to the tall height of the PSV duct in relation to the appliance chimney system, or it may be due to differential wind effects at the PSV termination point and flue terminal.

- so position the terminals of PSV systems and open-flued systems so the PSV system does not adversely affect the safe operation of the open-flued system
- where possible locate the PSV and open-flued system on the same face of the building – and the open-flued system terminal at the same height or higher than that of the PSV system termination point

Figure 14.3 Thermal inversion – house with roof conversion

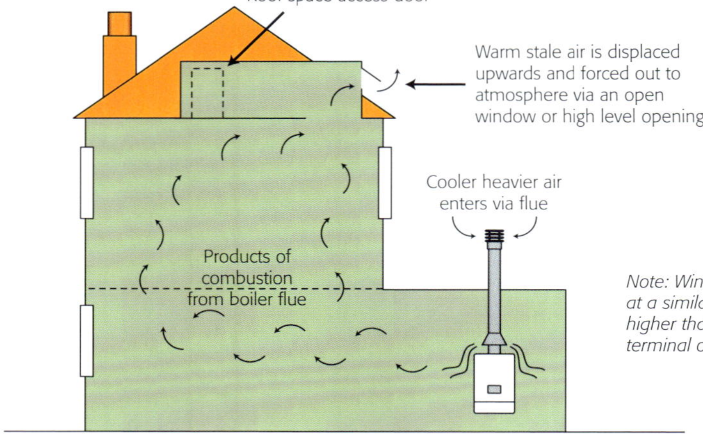

Complaints of periodic fumes – the signs to look for

If there are complaints about these fumes from an open-flued appliance, they may be caused by one of the following:

- modifications to the house – such as an open-flued appliance or open fireplace in an upper room or roof space converted into a room/bedroom.

  These often have a small openable window, fanlight, roof window or other opening, such as the new roof space access doors. These openings often terminate at a similar or higher level than the flue outlet (see Figure 14.3)
- the fitting of double glazed windows and outer doors, including draught-proofing
- an open-flued gas fire or other appliance fitted in a new or converted house that incorporates an open-plan staircase (see Figure 14.4)
- the installation of a PSV system

Can the problem be solved?

Thermal inversion is generally a direct result of warm, lighter air escaping from the building via a fanlight, open fireplace or other opening(s) and inadequate, low level ventilation to an open-flued appliance.

You may find a solution by:

- sealing all unnecessary high level openings in the rooms; and/or
- increasing the amount of low level ventilation to the house generally, or simply to the room containing the affected chimney system

Figure 14.4 Thermal inversion: open-plan link house with draught proofed doors and windows

Remember: thermal inversion can occur even when the chimney system has been installed and terminated in accordance with British Standards.

- in the case of intermittent down-draught, do not be tempted to try a selection of specialised terminals in an attempt to solve the problem. You may have limited success
- the right course of action is to fully diagnose and rectify the true cause of the problem

Note: Many of the specialist anti-down-draught terminals available are designed to reduce the pressure or create an up-draught in the flue. In most cases they depend on wind energy to achieve this.

Thermal inversion can and does happen in still air conditions.

Inspection – what is involved

1. Before you commence installation or following any work on an open-flued appliance under the GSIUR:
   - you must verify that the flue system is suitable, i.e. that it is in proper condition for the installation or continued safe operation of an existing gas appliance
   - where faults are identified, do not proceed with the appliance installation
2. In the case of an existing appliance, where the chimney system is failing to clear its POC safely, you must inform the responsible person.
3. If appropriate, make the appliance safe and follow the current GIUSP (see **Part 8 – Gas Industry Unsafe Situations Procedure** and **Part 10 – Emergency notices, warning labels and forms**).

14 – Flue testing

You must follow Building Regulations

The Building Regulations for England and Wales – ADJ – provides guidance on flue testing.

Note: Building Regulations/Standards applicable to other geographical areas of the UK provide similar guidance.

ADJ confirms that the commonly used tests – flue flow and spillage tests – described in BS 5440-1 and other relevant British Standards are acceptable for testing flues for gas appliances.

Appendix 'E' of ADJ, describes other methods of testing existing, relined or new, open, natural-draught flues for compliance with Regulation J2 of ADJ.

In particular, 'Smoke Test 1' (Paragraphs E13-14) is primarily written for chimney systems serving solid fuel or oil-fired appliances. However, you can also use it for chimney systems serving gas appliances, where you have any doubt over the condition of chimney system.

Appendix 'E' of ADJ (Paragraphs E9-10) also describes methods of proving the internal dimensions of ridged chimney systems – not applicable for flue pipes or metallic flue liners – using 'coring balls' (helpful when flue tests indicate a blockage in a chimney system).

The guidance recommends:

- you use a heavy coring ball (with dimensions about 25mm less than the flue you are checking) and lower this (on a rope) down the chimney system
- if an obstruction is encountered, remove it and repeat the coring ball test

Note: Circular shaped coring balls are not generally suitable for checking rectangular or square section chimney systems, as they may not make contact with obstructions in corners, such as mortar extrusions. So use purpose designed coring balls when you check chimney systems shaped this way.

Flue checks – for open-flued appliances

Do the following checks, following any work on the appliance:

1. Carry out a visual inspection of the whole length of the chimney system, where practicable.
2. Check there is provision of adequate combustion/compartment ventilation.
3. Conduct a flue flow test.
4. Conduct an appliance spillage test.

Use Figure 14.5 as a quick reference guide for checking open-flued systems.

Visual inspection

Carry out visual inspection checks to ensure that the chimney system is:

- constructed of suitable materials
- unobstructed, complete and continuous throughout its length (see Note 1 & 2)
- correctly sized
- where appropriate, suitably terminated and fitted with an approved gas flue terminal which is correctly sized and sited – and has a weather tight joint between the terminal and the chimney system (see **Part 13 – Chimney standards**)
- correctly jointed

Note 1: This check must ensure that the chimney system is continuous throughout its length (from appliance to terminal). This includes checking that areas such as roof spaces.

Figure 14.5 Open-flue checking procedure

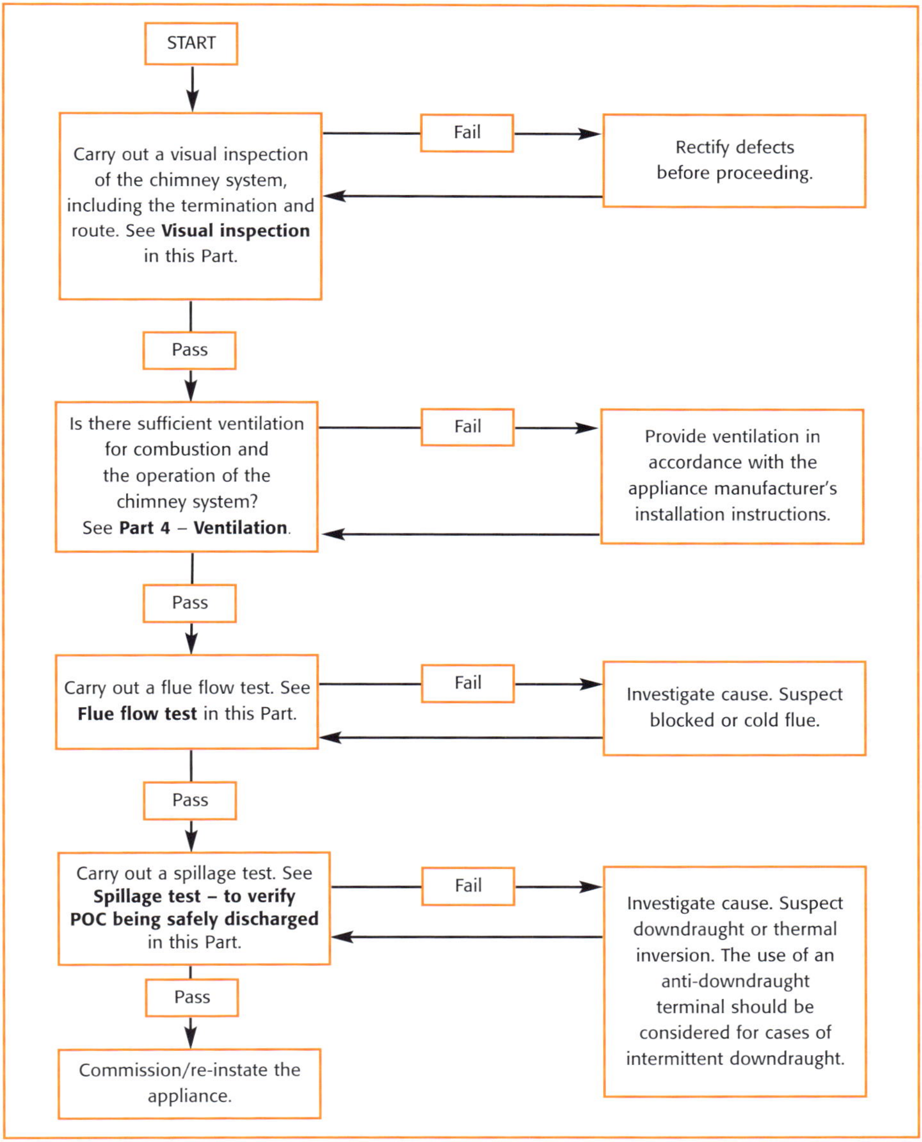

Note 2: Sometimes a brick/masonry chimney stack in upper storeys, particularly in roof spaces, is altered (reduced in length) bringing the chimney stack down below the roof line of the property.

Where this is done and assuming the chimney system is to remain in operation, a flue pipe is employed to connect the now reduced chimney stack to the point of termination (e.g. ridge tile adaptor or other suitable method of termination).

Where this is the case, ensure any seal between the flue pipe and brick/masonry chimney system is visible for periodic inspection.

Checks on the chimney system will include the following:

- it is in good condition
- it is adequately supported
- it serves only one room or appliance (unless designed for multiple appliance installation)
- it has been swept prior to installation (if previously used with other fuels)
- it is free of any dampers/restrictor plates. These need to be removed/fixed in a permanently open position
- where appropriate, any catchment space is of the correct size, free of debris. Gaps into that space are sealed
- it is designed to comply with the GSIUR (see Note)

Additionally, you must fully investigate any visual signs of spillage (i.e. fireplace/wall staining).

Note: For power-assisted chimney systems, ensure the operation of the fan is interlocked with the gas supply – to prevent the operation of the appliance if the draught fails.

Combustion/compartment ventilation – adequate air supply essential from outset

An adequate air supply is an intrinsic part of the chimney system (see **Operating principles – pressure differentials cause movement of gas/air** in this Part):

- at best, inadequate ventilation to the room will impair flue performance
- at worst, the chimney system may not function at all

So before you conduct any appliance or flue tests, it is vital to establish this adequate air supply (see **Part 4 – Ventilation**).

Flue flow test – follows after visual inspection

A flue flow test on the chimney system must follow a satisfactory visual inspection, to verify that:

- the flue is unobstructed and that the chimney structure is sound, i.e. that there are no cracks, missing bricks to brick/masonry chimney systems, or corrosion taken place to chimney components (flue pipes, bends) or flue liner that could admit air and impair flue performance
- there is adequate ventilation to the room

How to carry out the flue flow test

Use a suitable smoke pellet. For chimney systems covered by this part, a pellet should normally generate 5m^3 of smoke in approximately 30 seconds (burn time).

Place the smoke pellet at the base of the flue and close all doors and windows in the room.

When carrying out a flue flow test for a gas fire, which is designed to be installed with a closure plate, keep this plate in situ.

Note: A flue flow test is extremely subjective. It is to ascertain whether the chimney system is of a suitable condition to safely remove POC to atmosphere when the appliance is operating.

Weather conditions, the temperature of the flue and a combination of materials used to construct the chimney system can all influence the flue flow test.

You do not normally need to cap the chimney during the normal flue flow test.

Carry out the test as follows:

1. Check an adequate supply of air for combustion has been provided in accordance with the appliance manufacturer's installation instructions (see also **Part 4 – Ventilation**).
2. Close all windows and doors in the room or space in which the appliance is installed.

Note: You do not need to operate extract fans for a flue flow test.

3. Place an appropriate smoke pellet at the position (or intended position) of the appliance. Smoke should enter the flue and be discharged from the correct terminal only and there should be no discharge or leakage into the room (or any other room or space, including a roof space along the chimney system route).

   If the flue is reluctant to draw the smoke away from the appliance position, apply heat to it (minimum 10 minutes), using the flame of a blow torch or similar into the base of the flue. Then repeat the test.

Note: The pre-heating process may require as much as half an hour before the flue operates correctly, as a blow torch does not supply a sufficient volume of heat that is consistent with the normal operation of a gas appliance (see Attention).

If the flue flow test still fails after pre-heating and there is no obvious reason, you may need to place the appliance in position: but do not connect it to the gas supply. This should allow the flue flow test to be carried out with representative flue flow conditions.

Attention: ADJ suggests an alternative method of warming a flue by using an electric heater.

4. The chimney system will have passed the test if:
   a) There is no significant escape of smoke from the appliance position.
   b) Smoke is discharged from the correct flue terminal.
   c) There is no seepage of smoke over the length of the chimney system.

If any of these requirements are not achieved, the chimney system has failed the test.

Note: To avoid misleading results, the appliance should not normally be in use when you carry out the above test.

There is a risk that the flue will be unable to cope with the combined volumes of smoke from the pellet and POC from the appliance.

Signs to look for

- once ignited, the test smoke should be seen to pass in and up the flue to the atmosphere with little or no smoke entering the room
- examine the brick/masonry chimney stack or termination point. Smoke should be seen to flow only from the flue terminal or brick/masonry chimney pot to be used
- smoke emanating from any other source, e.g. from slates/tiles, masonry, roof space, rooms above the appliance or chimney system under test, will indicate that the flue is incomplete.

  It is therefore either not suitable for continued use or, in the case of a new installation, it is not acceptable. In this case, complete remedial work before the installation can proceed
- in cases where there is an existing appliance, follow the current GIUSP (see **Part 8 – Gas Industry Unsafe Situations Procedure** and **Part 10 – Emergency notices, warning labels and forms**)

Spillage test – to verify POC being safely discharged

Following a satisfactory:

1. Visual inspection.
2. An examination for combustion air; and
3. A flue flow test.

you can:

4. Install and/or re-instate the appliance.
5. Then carry out a spillage test.

Note: In the case of a new or replacement installation, only commission the appliance if you have the appliance manufacturer's installation instructions available.

How to carry out the spillage test

Part of the commissioning procedure is to verify that POC are being safely discharged to atmosphere and the spillage test shows this; You use a smoke producing device, e.g. smoke match, joss stick, etc. in accordance with the manufacturer's instructions.

Where the manufacturer's instructions do not give specific advice, you may check the appliance as follows:

1. Close all windows and doors in the room or space in which the appliance is located. Switch off any mechanical ventilation supply to the room other than any that provides combustion air to an appliance. Where appropriate, operate at maximum speed any fan(s) within the room (see Note 1) and ensure that any PSV system is fully open.

Note 1: Examples of fans that may affect a chimney systems performance are:

- **room-extract fans (at all speeds)**
- **fan assisted open-flued appliances**
- **tumble dryers**
- **cooker hoods (at all speeds)**
- **warm air heaters (gas and electric)**
- **Radon gas extraction systems**

  See 'Part 4 – Ventilation – Air supply from under-floor (Radon gas) – particular care needed regarding ventilation' for further guidance
- **landfill gas extraction systems**
- **ceiling fans (i.e. paddle fans), at all speeds and in both directions where appropriate**
- **heat recovery re-circulatory systems**

2. Check for spillage at the appliance draught diverter or gas fire canopy within 5 minutes of lighting the appliance. The majority of the smoke should be drawn into the flue (you may ignore the 'odd wisp' of smoke).
3. Close any PSV system and repeat the test.
4. If spillage occurs, leave the appliance in operation for an additional 10 minutes and then re-check.
5. If the appliance fails the test, isolate the appliance and rectify any fault.

If there is a fan(s) elsewhere in the building, this can adversely affect the performance of the chimney system, so

1. When you have completed a successful spillage test with all doors and windows shut, repeat the test with all internal doors open and the fan(s) operated at maximum speed.
2. Then carry out a second spillage test to ascertain if the fan(s) has an adverse affect on the correct operation of the chimney system.

Where the appliance fails the spillage test (with or without a fan(s) present)

1. Repeat the test with a window partially open.
2. If the appliance then clears its POC, measure the area of the window opening. This should be the amount of additional ventilation that needs to be provided.
3. Alternatively, as a general rule, provide an additional 50cm^2 free area of ventilation in the room or space where the appliance is installed and repeat the spillage test.

Note: Where the fault cannot be rectified, disconnect the appliance and follow the current GIUSP.

Modern gas fires generally have the spillage testing procedure displayed on a badge, which is either attached to the appliance or appliance case, or included in the manufacturer's installation instructions.

For Decorative Fuel Effect (DFE) gas appliances

Place the smoke producing device inside the top edge of the fireplace opening – unless individual manufacturer's installation instructions state otherwise (see Figure 14.6).

For warm air heaters:

You can find further advice in CORGI*directs* manual entitled 'Central Heating Wet and Dry' from the current Gas Installer Manual Series – Domestic.

Testing of open-flues (fanned-draught)

Fans integral with appliance: carry out the commissioning in accordance with the appliance manufacturer's installation instructions.

Fans installed on site: commissioning of open-flued fanned-draught systems installed on site, should generally follow the requirements for flue testing natural-draught open-flues (see **Flue checks – for open-flued appliances** in this Part).

However, you must make the following checks as well:

- ensure that the fan speed is set to the requirements of the fan manufacturer's commissioning instructions
- ensure all safety controls are checked for safe operation
- check that the safety control shuts off the gas supply to the main burner within 6 seconds of any spillage occurring from the appliance draught diverter/flue break, in accordance with the manufacturer's instructions
- ensure that with all external doors and windows closed, and where necessary with interconnecting doors open, clearance of POC from any other open-flued appliance in the room (or adjoining room or space) is not affected by the operation of the fan draught flue system

Figure 14.6 Spillage testing

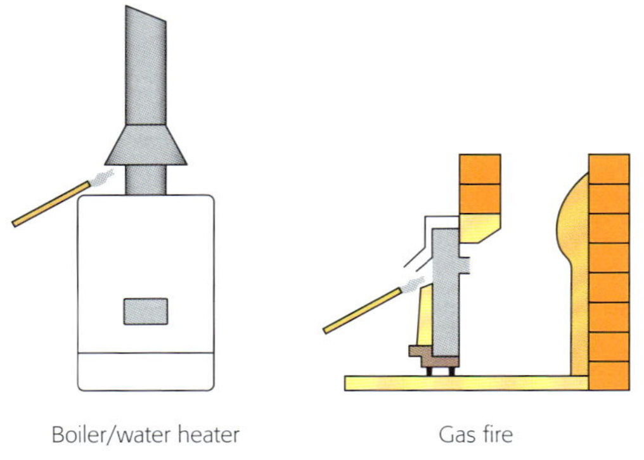

How to assess the effects of passive stack ventilation

For PSV systems, assess spillage with the PSV duct(s):

- in the open and closed positions; and
- where necessary, with the interconnecting doors open

Note: Air in the PSV system can move in either direction. That is why you need to conduct spillage tests with the PSV vents in the open and closed positions.

Flue testing with decorative re-circulatory ceiling fans present

When you carry out a spillage test on an open-flued appliance (particularly gas fires; radiant convector, ILFE's and DFE gas appliances), where there is a re-circulatory ceiling fan fitted in the same room, carry the test out:

- with the fan both on and off, at all speeds; and
- where appropriate, with the fan operating in each direction

Tests show that these fans can disturb air movement in the room – to the extent that they can cause spillage to occur where none was present with the fan in the 'off' position.

You are reminded that if you are in any doubt regarding the safe use of the appliance, follow the current GIUSP.

Open-flued appliances and mechanical ventilation systems

When either extraction or infiltration ventilation systems are installed into a building, there are other aspects to consider (problems that may arise) when such systems are used in conjunction with open-flued gas appliances.

Extract fans – problems you may find

Extraction systems are useful in removing odours from a room, but occasionally they can create problems with the safe removal of POC, if an open-flued appliance is fitted in the same or adjacent room as the extract fan.

If you find this problem, you may need to install additional ventilation to prevent the de-pressurising effect caused by the fan.

Examples of mechanical extract fans that may affect the performance of a chimney system are:

- room-extract fans (at all speeds)
- fan assisted open-flued appliances
- tumble dryers
- cooker hoods (at all speeds)
- warm air heaters (gas and electric)
- Radon gas extraction systems
- landfill gas extraction systems
- heat recovery re-circulatory systems

Radon gas – the tests needed

Where Radon gas extraction systems are fitted, you need to carry out flue spillage tests to ensure the safe operation of open-flued appliances.

Undertake these with the Radon gas extraction system in operation, to ensure that all POC are safely removed from the building.

Heat recovery systems – becoming more popular

Mechanical ventilation systems are becoming more popular. The two types normally used are:

1. Balanced supply and extract – Mechanical Ventilation with Heat Recovery (MVHR).
2. Mechanical Extraction Ventilation (MEV).

MVHR units – why these are used

- to provide control of condensation, indoor smells, such as tobacco odours
- improve the comfort conditions of the occupants of the building

In recent years, the use of MVHR has been combined with warm air heating systems.

This allows for the same ducting to be used for heating and ventilating, as well as providing the opportunity to recover useful energy from inside the building.

How MVHR systems work

1. They extract stale air from the kitchen, bathroom, toilet and any utility rooms and pass this 'damp air' through the heat recovery unit (usually fitted in the roof space).
2. Before it is discharged outside the building, the heat contained is given up to the cold fresh air coming into the building.
3. This fresh air is heated, without any mixing between the different air flows and is now heated to room temperature by the heating unit, before being passed to the various rooms inside the building.

MEV units – why these are used

- to allow recovery of heat by the use of more complex heat pump technologies

Spillage testing – both MVHR and MEV systems

Where open-flued appliances are situated in buildings fitted with MVHR and MEV systems, you will find specific advice relating to spillage testing of such appliances given by the British Research Establishment (BRE) Digest 398 document.

See also **Flue checks – for open-flued appliances – Spillage test – to verify POC being safely discharged** in this Part.

Flue testing in multiple dwellings – including landlords' responsibilities

Landlords have a duty under the GSIUR to ensure that any 'relevant gas fitting' and flue that serves that 'relevant fitting', is checked for safety:

- at intervals not exceeding 12 months, from when the last check was undertaken; or
- from when a new appliance is installed

A 'relevant fitting' is regarded as one that the tenant would not be entitled to remove from the property once the tenancy agreement has finished.

Note: Gas appliances left in the property by the out-going tenant (e.g. gas cookers, fires and cabinet heaters), would become 'relevant gas fittings' as defined by the GSIUR.

In addition, a landlord may have a 'duty of care' under the Health and Safety at Work etc Act (HSWA) to maintain any chimney system serving the tenants own appliance(s).

Types of chimney systems – problems can arise when shared

In a normal domestic dwelling e.g. house, bungalow or self-contained flat, this inspection would not normally present a problem, as long as all the appliances to be checked have their own dedicated chimney system and are accessible from within the property.

However, problems can arise when the chimney system:

1. Is shared with other tenants serving other appliances; or
2. The chimney system passes through another property.

Examples of shared chimney systems are:

- Se-ducts
- U-ducts
- shared open-flues (common flues)
- communal flue systems (CFS)

Se-ducts, U-ducts and CFS utilise room-sealed (Type C) appliances such as central heating appliances, whereas common flues serve open-flued (Type B) appliances such as central heating appliances, gas fires and water heaters (see **Part 13 – Chimney standards – Shared chimney systems – for shared or multi appliance installations**).

Landlords' responsibilities where flue system is shared

The responsibility of a shared flue system in rented accommodation covered by Regulation 36 of the GSIUR, lies firmly with the landlord/managing agent (duty holder).

The landlord must make the necessary arrangements with tenants to enable access for inspection, tests and maintenance works to be undertaken.

However, in a multi-occupancy building where there is mixed ownership, e.g. both rented and owner occupied properties, this is not as clear-cut and:

- in the case of rented accommodation, the landlord is responsible for the chimney system attributable to the rented accommodation

- in the case of owner occupied properties, arrangements need to be set in place by contractual means, to make sure that the chimney system passing through those properties is safely maintained – to ensure that the whole of the chimney system passing through the building continues to operate safely

Chimney system checks – shared open-flues

Include the normal flue flow and spillage tests and remember that, because of the special construction and operation of these chimney systems, not all appliances are suitable for use with them.

Therefore, it is vital to inspect and check the ventilation requirement and the operation/suitability of the appliance once you have gained access.

If you fail to carry out a satisfactory inspection of other appliances utilising the same chimney system in other properties, this could affect safe operation.

As far as is reasonably practicable, check the chimney system for design, termination and compliance with current requirements i.e. BS 5440-1 and IGEM/UP/17.

Shared open-flued systems – check ventilation openings/pressure zones

Each room that has an appliance installed should have similar sized ventilation openings, including any openable windows.

Failure to meet this requirement can result in different pressure zones being created throughout the building. Under certain conditions, this could result in POC from one property spilling into another.

Access to properties – what to do when this is barred

Sometimes occupiers of properties refuse access. The landlord/managing agent, or duty holder is then responsible for taking all reasonable steps to ensure the continued integrity of the chimney system.

This may involve:

- sending letters (by recorded delivery as appropriate) to the occupiers of all dwellings in the building to which the chimney system communicates; or
- by making personal contact

whatever is deemed necessary for the inspection to be undertaken

If the occupiers totally refuse inspection of the chimney system

- the landlord/managing agent, or duty holder may need to demonstrate they have taken all reasonable steps to gain access and therefore, should keep records of all correspondence that has taken place, before their duties can be discharged

Note: This does not mean that entry to the property can be gained by force. Proper action through the courts would be appropriate.

Poor flue performance – the appliance must not be used, till remedied

When there is good reason to suspect a problem with the chimney system e.g:

- poor flue flow
- spillage/leakage of POC
- pilot outage; or
- clear medical evidence of occupants suffering from the effects of CO poisoning

The appliance(s) must not be used until the appliance and the complete length of the chimney system is examined and any remedial work required has been undertaken.

Where necessary, implement the current GIUSP. See also **Part 8 – Gas Industry Unsafe Situations Procedure** and **Part 10 – Emergency notices, warning labels and forms for further guidance**.

Reports of Carbon Monoxide (CO) poisoning

You must report these to the Health and Safety Executive (HSE) through RIDDOR (Reporting of Injuries Diseases and Dangerous Occurrences Regulations). See also **Part 8 – Gas Industry Unsafe Situations Procedure** and **Part 10 – Emergency notices, warning labels and forms** for further guidance.

- do not carry out any remedial work deemed necessary to rectify faults, until after the supplier of gas to the property (or a third party appointed by them) carries out a full investigation
- then seek permission from the HSE to re-commission the installation following the investigation

Chimney inspection – specialist check can be needed

A visual external inspection of any chimney system, will not always establish the cause of a flueing problem.

Consider employing a specialist organisation to undertake an in-depth examination of the chimney system using specialist equipment and to provide the duty holder with a full report on the internal condition of that system.

One method could be to use CCTV technology to give a comprehensive recorded inspection, identifying faults on chimney systems: chimney flue blocks, brick/masonry chimney stacks, Se-ducts, U-ducts and shared open-flued systems.

Testing appliances with positive pressure cases – inadequate seals can be a danger

There are still a large number of gas appliances in use in customers' homes that utilise positive pressure case technology.

Some years ago the gas industry recognised the dangers presented by inadequately sealed positive pressure case appliances.

So an agreed procedure was produced to assist you to recognise the signs and assess the condition of this type of appliance.

Figure 14.7 Positive/negative fanned flued room-sealed gas appliances

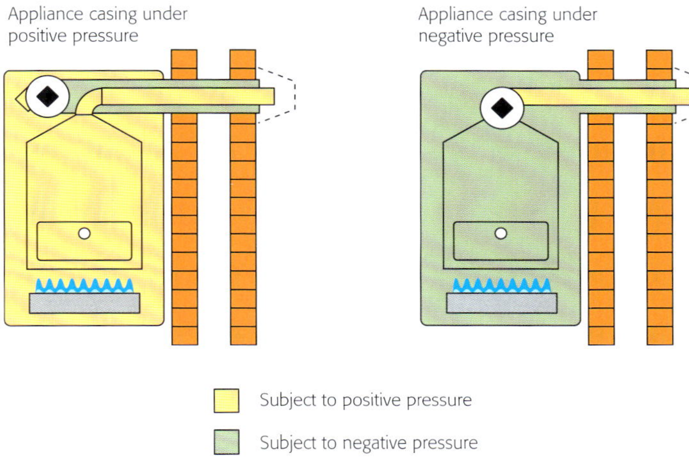

Why can positive pressure gas appliances be dangerous?

Historically, fanned draught room-sealed boilers were of the positive pressure type. For this type of appliance to operate safely, it is essential that the combustion chamber casing is firmly secured to the boiler chassis, as the manufacturer intended, with the correct seal in a good condition.

If not:

- there is a real risk that POC may escape into the room in which the appliance is installed
- due to the poor combustion that is likely to occur, high levels of CO could be produced, creating a dangerous environment

Figure 14.7 shows the differences between positive and negative pressure appliances.

How you test

Regulation 26(9) of the GSIUR requires you to test the effectiveness of any flue when you have worked on a gas appliance.

The Industry has developed the test method to help ensure that case seals of positive pressure gas appliances comply with the requirements of the GSIUR:

Step 1 – before you put the case back on the appliance

Do the following checks:

- are any water leaks evident?
- is the back plate or case corroded?
- where there is corrosion, is it likely to affect the integrity of the case, back plate, or case seal?

Note: Using a sharp instrument (e.g. a screwdriver), carefully check the extent of the corrosion. If the instrument does not perforate the corroded area, take this as acceptable – but you must advise the gas user of the problem and potential consequences if they do not repair it.

- are the combustion chamber insulation linings intact?
- is the back plate or the case distorted or damaged? Pay particular attention to the area where the case and seal meet. If so, it may have been caused by explosive ignition of the main burner
- is the case sealing material intact and in good condition? (e.g. pliable, free from discolouration, trapped debris, etc.). Will it continue to form an adequate seal between the case and the back plate?
- is anything trapped or likely to be trapped, when the case is put back on (e.g. wires, thermocouple capillaries, tubes, etc.)?
- are other gaskets and seals intact?
- is the pilot inspection glass undamaged?
- are the case fastenings and fixings (including fixing lugs) in good condition? (e.g. screws/nuts stripped)
- are there any signs of discolouration on or around the appliance, which may have been caused by leaks of POC from the appliance?

Rectify any defects you identify in Step 1 as necessary and proceed to Step 2.

Note: Where you identify defects, classify them using the following criteria in accordance with the current GIUSP:

1. Where there are inappropriate or missing case fittings or defective seals, which you cannot remedy, but there is no evidence of leakage, classify the appliance as 'AR'.
2. If there is evidence of actual leakage, class the appliance as 'ID'.
3. Where suitable replacement seals are no longer available class the appliance as 'ID' and regard it as obsolete.

Step 2 – when you have put case back on the appliance

Do the following checks:

- is the case fitted correctly?
- is a "mark" visible showing that the case had previously been fitted closer to the back plate?
- are all the case screws adequately tightened?
- is a bright area visible on the screw thread of any of the case securing screws, indicating that the screw was previously secured more tightly?
- is anything trapped and showing through the case seal?

Rectify any defects you identify in Step 2 as necessary. Proceed to Step 3.

Step 3 – Operate/light the appliance

- ensure that the main burner remains lit (i.e. set the appliance and room thermostats to their highest settings)
- check for possible leakage; initially this can be done by running your hands around the boiler casing and back plate
- then check for possible leakage etc. as in Step 4 where practicable

Step 4 – Check for possible leakage of combustion products from the appliance

- where joints have been disturbed, check with non-corrosive LDF to confirm that there are no gas escapes
- check for possible leakage of combustion products from the appliance using a taper, an ordinary match, or similar (use a taper to get into less accessible locations)

Note: Whilst you can use smoke tubes and smoke matches, the results may require further interpretation.

- light the taper/match and allow the flame to establish. Position the flame very close to the case seal or any possible leakage point (e.g. back panel)
- the flame will be blown quite easily by the draught caused by a leak. Move the taper around the entire seal and use fresh tapers, as required
- to investigate the seal at the bottom of the case – hold the lit taper between the bottom of the case and the appliance control panel. Does the flame flicker slowly or is it disturbed by leakage flowing from the case? Try the taper in several positions

Attention: DO NOT confuse natural convection with leakage. DO NOT look for a gas escape with this method.

Rectify any defects as necessary and re-check. If still unsure, seek expert advice

Note: When using this method, be careful not to set fire to surrounding fixtures/furnishings.

Update – Health and Safety Executive (HSE) advice on preferred smoke device

The HSE advise that a smoke producing device/wand is available that produces a steady thin stream of smoke. It can be used instead of a lighted taper/match and is the preferred method for you to use.

However, where you have any doubt regarding suitability of any smoke producing device/product (particularly with reference to COSHH requirements), follow the original procedure outlined.

Update on classifying the installation – new Appendix on testing positive pressure seals

The issue of testing positive pressure case seals is included in BS 5440-1: 2008 in an Appendix. There is also a 'Commentary and Recommendation' which states:

"The appliance certification criteria permit a limited amount of case and seal leakage due to manufacturing tolerances. It is therefore likely that some minor leakage might be identified on positive pressure fanned flue boilers, in particular where a thermocouple lead/thermostat capillary or ignition high tension lead etc. passes through a grommet/gasket, or where there is a metal fold/joint that forms a corner on the boiler case itself."

"In these instances it is necessary to assess whether the leakage is due to normal manufacturing tolerances or to a defect with the grommet/gasket taking into account any previous customer reports of fumes, signs of staining, condition of the grommet/gasket etc., before deciding that the leakage identified is due to normal manufacturing tolerances and whether the appliance is safe to leave in operation."

"If there is any concern as to whether the level of leakage is significant and providing the point of leakage is not due to a defect in the main boiler case seal, e.g. around a grommet or gasket, it might be possible to effect a permanent repair by supplementing the original grommet/gasket with high temperature silicone sealant."

"It is essential that perforations in the case material due to corrosion are not temporarily repaired and any defective main boiler case seals requiring replacement should only be replaced with the manufacturer's supplied or authorised component".

It is essential that:

Where you determine that any leakage is greater than normal manufacturing tolerances and you cannot rectify this at that time, you classify the appliance as 'ID' and make it safe in accordance with the current GIUSP.

Table 14.1 contains a list of old room-sealed fan assisted positive pressure gas appliances. This list is not exhaustive and many of those appliance identified will no longer be found in use, but use it as guidance to appliances that are believed to operate under positive pressure.

Warning: what to do where you find spillage – 'Immediately Dangerous' installations

It is essential that in ALL cases where you identify spillage:

- you make the appliance safe when you cannot rectify the fault

In the case of a new installation this includes:

- disconnecting and sealing the appliance from the gas supply immediately; and
- labelling the appliance accordingly

With an existing installation this includes:

- seeking permission from the gas user to seal the supply
- if permission is refused, turning the appliance off; and
- labelling the appliance accordingly

In both instances:

- adhere to the current GIUSP

Table 14.1 Room-sealed fan assisted positive pressure gas appliances

Manufacturer and model	Manufacturer and model
Alde International (UK) Ltd Alde 2927 Slimline **Brassware Sales Ltd** Ferrolli 76 FF* Ferrolli 77 FF* **Crosslee (JLB) (Pyrocraft)** AWB 23. 09 WT Combi **Crosslee (Trisave Boilers Ltd)** Trisave Turbo T45* Trisave Turbo T60* Trisave Turbo 30* Trisave Turbo 22* **Glow Worm Ltd** Economy 30F* Economy 40F* Economy 50F* Glow Worm Fuelsaver 35F* Glow Worm Fuelsaver 45F* Glow Worm Fuelsaver 55F* Glow Worm Fuelsaver 65F* Glow Worm Fuelsaver 80F* Glow Worm Fuelsaver 100F*	**Glynwed Domestic & Heating Appliances Ltd** AGA A50 AGA A50 A AGA A50 NG AGA A50 SS AGA A50 ANG AGA A60 AGA A60 NG AGA A75 NG Hi-light P50 Hi-light P50A SC Hi-light P50S Hi-light P50SS Hi-light P50S/A Hi-light P50S/A GLC Hi-light P50S/A SC Hi-light P50/A Hi-light P70 Hi-light P70S Hi-light P70SS **Halstead Heating & Engineering Ltd** Halstead 45F* Halstead 55F* Halstead 65F* Wickes 45F* Wickes 65F* Barlo Balmoral 45F* Barlo Balmoral 55F* Barlo Balmoral 65F* **Harvey Habridge Ltd** Impala MK 11 Impala MK 11 Ridgeseal Impala Super 2 (HF) Impala Super 2 (VF)
* Boilers where spares relevant to case seal problems are still available, based on information provided by manufacturers.	

Table 14.1 Room-sealed fan assisted positive pressure gas appliances (continued)

Manufacturer and model	Manufacturer and model
Potterton Myson Ltd	**Stelrad Group Ltd**
Myson (Thorn) Olympic 20/35F ‡	Ideal Elan 2 40F*
Myson (Thorn) Olympic 38/50F ‡	Ideal Elan 2 50F*
Myson (Thorn) Apollo Fanfare 15/30	Ideal Elan 2 60F*
Myson (Thorn) Apollo Fanfare 30/50	Ideal Elan 2 80F*
Supaheat 50/15 with 'A' control	Ideal Excel 30F*
Supaheat GC 50/15	Ideal Excel 40F*
Netaheat MK 1 10/16	Ideal Excel 50F*
Netaheat MK 1 16/22 BF	Ideal Excel 60F*
Netaheat MK 11 10/16 BF	Ideal Sprint 80F*
Netaheat MK 11 16/22 BF	Ideal W2000 30F*
Netaheat MK 11F 10-16 BF	Ideal W2000 40F*
Netaheat MK 11F 16-22 BF	Ideal W2000 50F*
Netaheat Electronic 6/10	Ideal W2000 60F*
Netaheat Electronic 10/16	**Worcester Bosch**
Netaheat Electronic 16/22	Heatslave 9.24 RSF*
Netaheat Electronic 10/16e	Worcester 9.24 Electronic RSF*
Netaheat Electronic 16/22e	Worcester 9.24 Electronic RSF 'S'*
Netaheat Profile 30e	
Netaheat Profile 40e	
Netaheat Profile 50e	
Netaheat Profile 60e	
Netaheat Profile 80e	
Netaheat Profile 100e	

‡ A safety enhancement kit has been designed for these appliances and is available from Potterton Myson Ltd.

* Boilers where spares relevant to case seal problems are still available, based on information provided by manufacturers.

Re-establishing gas supply/re-lighting appliances – 15

15 – Re-establishing gas supply/re-lighting appliances

Figures

Tables

Introduction

Information in this Part is relevant to domestic installations and gas appliances of rated input not exceeding 70kW (net) for 2nd and 3rd family gases and pipework up to and including 28mm copper/25mm steel (LPG) and 35mm copper/32mm steel (NG).

Mandatory requirements you must follow

You may be called upon to connect an installation pipe to a NG or LPG meter, LPG bulk storage vessel or cylinder, or to turn off the gas supply to carry out work on an installation. As a result, air may remain within a gas system or be admitted into it.

It is an essential requirement of the GSIUR that when you complete the gas related work you immediately:

- check for gas escapes
- make safe or, where necessary, purge the installation of air

What happens if air is not removed from the system?

If air is not completely removed, an explosive air/gas mixture or air pocket may remain. This can present a problem to appliances furthest from the meter, bulk storage vessel or cylinder(s).

Initially, the gas may be ignited at the appliance but it may subsequently extinguish when the air or air/gas mixture reaches the burner (see **Part 3 – Characteristics of combustion – Flammability limits – when fuel gases will burn in air**).

Purge installation pipework/fittings of air when you work

- if you carry out 'work' relating to a gas meter or gas pipework, you must ensure that you purge all installation pipework and fittings of air.
- a gas tightness test must be carried out immediately prior to purging and introducing the fuel gas.

For the average domestic sized installation using NG through a U6, G4, E6, G10 or U16 gas meter with installation pipework up to 35mm in diameter (with an installation volume of no greater than 0.02m^3 (NG); See **Purging volume – NG: to ensure safe condition** in this Part), you can generally purge the meter and pipework into the building, if you have taken the safety precautions detailed below:

1. During purging, take precautions to ensure gas is not allowed to accumulate in any confined space and:
   - do not turn on an electrical switch or operate an appliance
   - ensure there is no smoking or naked lights
   - ensure adequate ventilation is provided, by opening doors and windows etc. within the vicinity of the purge point
   - ensure where appropriate, that the responsible person for the premises has been informed
2. For NG installations where the total installation volume is greater than 0.02m^3 (this includes all U16/G10 meter installations), gas will need to be ignited as soon as possible at an appropriate gas appliance e.g. a gas cooker, or if a suitable appliance is not available, then you will need to connect a temporarily installed burner to the gas supply for this purpose.

Table 15.1 Calculating purge volumes

E6, G4, U6 meter and pipework ≤ 28mm	U6, G4 meter and pipework > 28mm to ≤ 35mm	E6 meter and pipework > 28mm to ≤ 35mm	G10, U16 meter and pipework ≤ 35mm
$0.01m^3$ ($0.35ft^3$)	1.5 x Installation Volume (meter/pipework*/fittings)	1.5 x Installation Volume (meter/pipework*/fittings)	1.5 x Installation Volume (meter/pipework*/fittings)
Gas meter volumes	(IV_m) $0.008m^3$	(IV_m) $0.0024m^3$	(IV_m) $0.025m^3$

≤ is less than or equal to > is greater than

* For details of pipework volumes see Table 15.2.

See 'Example' in 'Calculating purge volumes – NG' in this Part.

Table 15.2 Pipework volumes

Tubing sizes	Volume of 1m length of tube (m^3)
Steel or stainless steel	
15mm ½"	0.00024
20mm ¾"	0.00046
25mm 1"	0.00064
32mm 1¼"	0.0011
Copper tube	
15mm	0.00014
22mm	0.00032
28mm	0.00054
35mm	0.00084
PE pipe	
20mm	0.00019
25mm	0.00033
32mm	0.00053

3. When LPG is the fuel gas, there is a danger during purging that gas may fail to disperse (due to the fact that LPG is heavier than air and may accumulate at low level).

   Therefore carry out the purge through the appliance, applying a source of ignition (e.g. a lighted match) to the burner head until the purge is complete and the flame(s) stable. (See also **Part 18 – LPG – General requirements – Purging LPG gas installations – precautions you need to take**).

Purging volume – NG: to ensure safe condition

The purge volume is the total volume that needs to be passed through the gas installation to ensure that the installation is left in a safe condition and ready for use.

For purge volumes for particular gas installations with gas meters from U6/G4/E6 up to U16/G10 (see Figures 15.1 and 15.2 for typical meter indexes/dials) and pipework up to 35mm copper/32mm steel, refer to Table 15.1 and calculate where appropriate.

Figure 15.1 U6 meter dials

Figure 15.2 E6 meter indexes

Calculating purge volumes – NG

The purge volume (m^3) is calculated from the installation volume (IV), which is the sum of the meter volume (IV_m) + the pipework volume (IV_p) + the fitting volume (IV_f), if known.

Where the fitting volume isn't known, add 10% of the pipework volume.

Once the total installation volume is known, multiply the figure by 1.5 to obtain the required purge volume.

Table 15.1 and 15.2 provide installation volumes for gas meters and pipework

Example

A gas installation with a U6 gas meter, 10m of 35mm copper pipework, 5m of 22mm copper pipework and a further 5m of 15mm copper pipework – what would be the estimated purge volume of the complete installation?

Answer:

Purge volume (m^3) = 1.5 x total IV

Therefore:

- Gas meter (IV_m) = 0.008m^3 (see Table 15.1)
- Pipework (IV_p) = 10m x 0.00084 (35mm Cu) + 5m x 0.00032 (22mm Cu) + 5m x 0.00014 (15mm Cu)

  = 0.0084 + 0.0016 + 0.0007

  = 0.0107m^3
- Fittings (IV_f) = 10% of pipework volume

  = 0.00107m^3
- Total IV = 0.008 (IV_m) + 0.0107 (IV_p) + 0.00107 (IV_f)

  = 0.0197m^3
- Purge volume = 0.0197 x 1.5

  = 0.029m^3

The purge volume as calculated, is the volume of gas that may need to pass through the gas meter to ensure a complete purge (see **Purging procedure** in this Part)

Purging procedure

When purging a small domestic sized gas installation, it is recommended that you:

1. Extinguish any possible sources of ignition and where appropriate, advise the responsible person for the premises.
2. Make sure that all gas appliances are turned off and open doors and windows.
3. Turn on the gas supply and note the position of the test dial on an imperial meter – or that of the test drum/electronic index on a metric meter.
4. Select a gas point that is furthest from the gas meter to use as the purge point. This could be a gas appliance, or an appropriate fitting (e.g. an appliance isolation valve, see Note).

Note: Where the instillation volume is less than 0.02m^3 you may use an appropriate fitting to purge the installation.

5. For NG installations where the total installation volume is greater than 0.02m^3, or for LPG installations where there is a risk of gas collecting at low level, the gas will need to be ignited as soon as possible at an appropriate gas appliance e.g. a gas cooker. If a suitable appliance is not available, then you need to connect a temporarily installed burner to the gas supply for this purpose.
6. Open an appliance control tap, or loosen an appropriate fitting for a short time or, where necessary, ignite as soon as possible the gas at a gas cooker, gas space heater etc. (or a temporarily installed burner) and then turn off the control tap/tighten fitting.
7. Confirm the presence of gas e.g. by smell or a stable flame picture, as appropriate.

8. Return to the gas meter and note the volume of gas that has passed through it.
9. Where the appropriate volume has not passed through the gas meter, continue with steps 6 to 8 until the correct volume has been achieved.
10. Purge all branches of pipework and ensure a stable flame picture is achieved at each appliance within the premises.

Note: If you identify a non-commissioned appliance, disconnect this from the gas supply and label it appropriately. (See Part 10 – Emergency notices, warning labels and forms).

Purging through existing appliances – you have a duty of care

When purging through an existing gas appliance(s), do note that, although this is classed as 'work' on a gas appliance and therefore falls under Regulation 26(9) of the GSIUR, you are not required to examine:

1. The effectiveness of any flue.
2. The supply of combustion air.
3. Its operating pressure, heat input or both.
4. Its operation so as to ensure its safe functioning.

...where the purge has no adverse affect on the safety of the gas appliance or associated pipework and fittings (Regulation 26(10)(b)).

However and it's important to stress, you do have a 'duty of care' when purging and in the case of doubt, or where safety is likely to be affected (see **Visual risk assessment – to ensure gas appliance(s) are not a danger** in this Part) that you conduct your 26(9) checks bulleted above. See also **Re-lighting appliances – is purging necessary as well as duty of care?**

Re-lighting appliances – is purging necessary as well as Duty of Care?

When gas has been turned off at the meter, bulk storage vessel or cylinder, should the installation be purged after re-lighting – in addition to a gas tightness test?

Purging is not normally necessary provided:

- the system is an existing one
- the meter/pipework has not been disconnected or rearranged
- air has not been admitted into the system

However you do have a 'duty of care' when re-lighting an appliance(s) after a gas tightness test, to carry out a visual inspection to ensure that the appliance(s) is in a safe condition for further use (see **Visual risk assessment – to ensure gas appliance(s) are not a danger** in this Part).

Visual risk assessment – to ensure gas appliance(s) are not a danger

This guidance is for gas operatives with certificates of core gas competency

It describes the minimum criteria involved, when you encounter a gas appliance(s) not subject to the operational checks needed when 'work' has been undertaken, as required by the GSIUR.

The following trigger points outlined relate to you, if you hold core gas safety competency (e.g. CCN1, CMA1, etc.) – but they are not exhaustive.

Assess the situation using your experience, knowledge and judgement relevant to your level of competency – and note that the points are likely to lead to you making other specific visual assessments, depending on the appliance(s).

What is your minimum responsibility when making a visual risk assessment?

It is to ensure that the appliance(s) does not constitute a danger.

Only use the trigger points discussed here where you have not carried out gas work but have encountered the appliance(s):

- either directly whilst other gas work has been carried out, or
- as part of a 'check and re-light' procedure following the interruption of the gas supply (e.g. following a gas tightness test) and you need to re-establish the gas supply

Note: Where you have carried out gas work on a particular appliance(s), also complete the checks required by Regulation 26(9) of the GSIUR.

Visual risk assessment where no interruption of the gas supply has occurred

Figure 15.3 shows the 5 main trigger points that you need to consider when carrying out a visual risk assessment of an existing gas appliance(s) where no other gas work on that particular appliance(s) has been undertaken (e.g. whilst servicing a central heating boiler, a gas cooker is installed in the same room. Undertake a visual assessment of the gas cooker).

Follow the trigger points outlined in Figure 15.3 in any order and as necessary, looking for:

Location

Is the gas appliance installed in a suitable room/space as required by the GSIUR?

For example, an open-flued appliance installed in a bathroom or shower room, or a flueless appliance installed in an undersized room.

Flueing

If the gas appliance is flued (either open-flued or room-sealed), is there provision for adequate methods for the removal of the POC to atmosphere?

Ventilation

Where appropriate, is there provision for the supply of adequate ventilation for the appliance to operate safely?

For example, is there evidence of purpose-provided ventilation for an open-flued boiler?

Signs of distress

Are there any signs of distress on the gas appliance and/or the surrounding area?

For example, check for signs of discolouration and heat damage – such as scorching or finished surfaces becoming detached from worktops etc.

Stable and secure

Is the appliance installation both stable and secure?

Assess to ensure that under normal conditions, the appliance will remain fixed/installed in a manner that will not result in the appliance becoming unstable (e.g. free-standing appliances with damaged or missing supports).

Visual risk assessment following temporary interruption of gas supply

Figure 15.4 shows the 6 main trigger points you need to consider when carrying out a visual risk assessment of an existing gas appliance(s) where the gas supply has been temporarily interrupted – for example, if a replacement gas meter has been installed.

Figure 15.3 Visual risk assessment (no interruption of the gas supply has occurred)

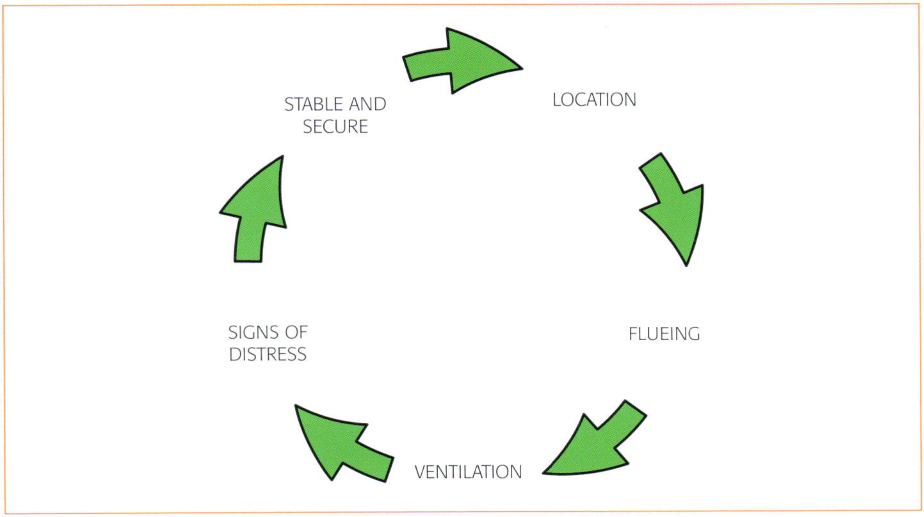

Figure 15.4 Visual risk assessment (following temporary interruption of gas supply)

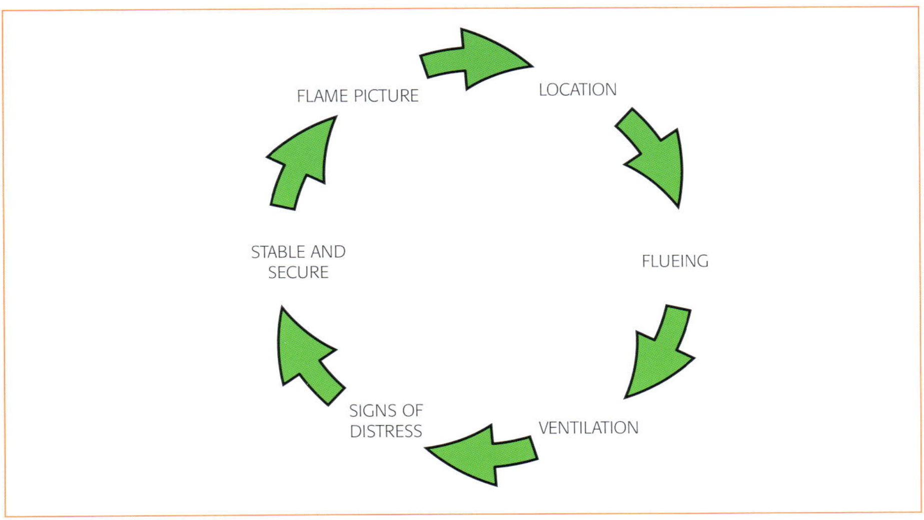

The 5 trigger points outlined in Figure 15.3 also apply to Figure 15.4 – with the addition of an assessment of an appropriate flame picture for the appliance type (e.g. live fuel effect gas fires and decorative fuel effect gas appliances are designed to produce a luminous flame).

Results of the visual risk assessment – what must you record/do?

You are not specifically required to record the results of a visual risk assessment – but you are advised to positively record such assessments. CORGI*direct* has developed a simple form to meet this need, see Figure 15.1.

However, where as a result of the assessment you have identified, or suspect, an unsafe situation, you must implement the current GIUSP and complete and issue the appropriate actions and warning notices/labels (see **Part 10 – Emergency notices, warning labels and forms** and also **Part 8 – Gas Industry Unsafe Situations Procedure**).

Figure 15.5 Visual Risk Assessment of Gas Appliances (CP9)

To confirm the validity of the gas operative please contact Gas Safe Register on Tel: 0800 408 5500

VISUAL RISK ASSESSMENT OF GAS APPLIANCES

The purpose of this form is to provide registered gas operatives a mechanism for the recording of visual risk assessment ONLY of gas appliances that they come in to contact with whilst engaged in other gas related work. The appliances documented HAVE NOT been tested/serviced or otherwise worked on, but have been visually assessed in accordance with Regulation 26 of The Gas Safety (Installation and Use) Regulations (GSIUR).

Registered Business Details ____________ (REG NO)
Company: ____________
Gas Operative Licence No: ____________
Address: ____________
____
Postcode: ____________ Tel No: ____________

Landlord/Managing Agent (where appropriate)
Name/Company: ____________
Address: ____________
____
____
Postcode: ____________ Tel No: ____________

Job Address
Name (Mr/Mrs/Miss/Ms): ____________
Address: ____________
____
____
Postcode: ____________ Tel No: ____________

Description of Gas Work Undertaken –
____
____
____
____
____

Gas Appliances Installed

Whilst engaged in gas work described above, how many other gas appliances installed in the property have you encountered/required to visually examine for safety as per Regulation 26 of The Gas Safety (Installation & Use) Regulations: ☐ (complete Appliance Details below)

Appliance Details

	Type	Make	Mode	Flue Type (OF/RS/FL)
1.				
2.				
3.				
4.				

Visual Inspection Details

	Flueing Satisfactory? (Yes/No/NA)	Ventilation Satisfactory? (Yes/No)	Signs of Distress? (Yes/No)	Stable & Secure? (Yes/No)	Location Satisfactory? (Yes/No)	Flame Picture Satisfactory? (Yes/No/NA)
1.						
2.						
3.						
4.						

Observations/Comments (as and where appropriate)		If Warning/Advice Notice issued insert serial No.*
	1.	
	2.	
	3.	
	4.	

I confirm that during the course of other gas works described on this form that I have visually inspected **ONLY** those gas appliances documented for safety as required by the GSIUR.

Print name: ____________
Gas operative signature: ____________
Date: ____________

Key: Top Copy – Gas User (PTO). Green Copy – Gas Operative

To re-order quote Ref. CP9

Domestic gas meters – 16

16 – Domestic gas meters

Figures

16 – Domestic gas meters

Tables

Introduction

Information in this Part is relevant to the installation and operation of domestic gas meters (NG and LPG) with a rating capacity up to and including 6m^3/hr.

It provides guidance for the installation and exchange of primary, secondary and prepayment domestic diaphragm gas meters that are manufactured to conform to BS EN 1359, or the previous Standard of BS 4161, Parts 3 and 5 (which are withdrawn) and BS EN 14236 for Ultrasonic domestic gas meters.

These gas meters will normally supply NG through a pre-set regulator to provide a nominal operating pressure (OP) of 21mbar at the meter outlet, or LPG (Propane) where the supply regulator has been set to provide a nominal operating pressure (OP) of 37mbar.

Working on gas meters

Guidance in the form of a Technical Bulletin (TB) issued by Gas Safe Register, which has been developed by IGEM on behalf of Meter Asset Managers Code of Practice (MAMCOP) Management Board provides clarification on competency requirements for gas meters.

TB 127 'Gas industry guidance on work on meter installations' applies to existing low-pressure-fed gas meter installations (operating pressure not exceeding 75mbar) up to 16m^3/hr.

The TB was issued September 2012 and is available via the engineer portal of Gas Safe Register's website - https://engineers.gassaferegister.co.uk

The guidance applies to existing meter installations where work is required to be undertaken, but not the replacement of a meter with a different meter.

There are three categories of work activity, them being:

A. Temporarily removing and refitting an existing meter i.e. where hot work is involved in close proximity to the gas meter

B. Repositioning an existing meter using same fittings i.e. relocating within the same space/cupboard using existing fittings

C. Relocating a meter, or repositioning a meter using different fittings

For category A work, operatives need to be Gas Safe Registered and hold core competences CCN1 or CMA1.

For category B work, operatives need to be Gas Safe Registered and hold core competences CCN1 or CMA1 and/or CMA1 + either MET1 or MET2, as appropriate.

For category C work and in addition to the requirement to being Gas Safe Registered, operatives will also require OAMI registration (Ofgem Approved Meter Installer).

Standards for installation you must follow

Install all gas meters and their fittings:

- in accordance with the GSIUR
- Building Regulations and those regulations appropriate to the geographical region in which they are installed
- BS 6400: 'Specification for installation, exchange, relocation and removal of gas meters with a maximum capacity not exceeding 6m^3/hr' -
    - Part 1: 2016 Low pressure (2nd family gases)
    - Part 2: 2006 Medium pressure (2nd family gases)
    - Part 3: 2007 Low and medium pressure (3rd family gases)

- IGEM/GM/6 (Edition 2) 'Non-domestic meter installations. Standard designs'
- IGE/GM/8 'Non-domestic meter installations. Flow rate exceeding $6m^3$/hr and inlet pressure not exceeding 38mbar'. Parts 1-5
- BS 7671: 2008 (Incorporating Amendment No.3) 2015 'Requirements for electrical installations. IET Wiring Regulations Seventeenth Edition'; and
- following manufacturer's installation instructions

The different types of modern gas meter

The most common domestic gas meter is the 'U6' (the designation 'U' meaning unit construction) diaphragm type, but this is now being replaced by either:

- the 'G4' (which is a diaphragm meter like the U6 but measures gas in m^3); or
- the next generation 'E' (electronic) meter which has a smaller case size, equivalent volume and gives a more accurate reading than the 'U6' type

Accurate reading meters – so gas users are charged fairly

Under the Gas Act, gas users are charged fairly for the gas they use by accurate reading meters that are not adversely affected by atmospheric or weather conditions. Smart meters are the next generation of gas and electricity meters and offer a range of intelligent functions.

Most of the smart meters that are being installed today use mobile phone-type signals to send meter readings to your supplier, and other wireless technologies to send information to the in-home display.

Smart meters bring a wide range of benefits. For example:

- smart meters give near real time information on energy use - expressed in pounds and pence
- be able to better manage your energy use, save money and reduce emissions
- smart meters will bring an end to estimated billing - you will only be billed for the energy you actually use.
- easier switching - smoother and faster to switch suppliers to get the best deals

Under current arrangements you pay for the cost of your meter and its maintenance through your energy bills, and this will be the same for smart meters.

Smart meters can work in prepayment or credit mode.

You must install U6, G4 and E6 gas meters as follows:

1. Primary meter – the first meter downstream connected to the gas service pipe.
2. Secondary meter – a meter that is supplied with gas via a primary meter (generally the property of the owner of the building).
3. Prepayment meter – a primary or secondary meter with an integral mechanism that allows a token (mechanical) or card (electronic) to purchase a predetermined volume of gas.

Attention: A pre-payment primary meter must not supply a secondary meter. This also includes 'Smart' meters used in pre-payment mode.

Figure 16.1 Basic components of a U6/G4 gas meter

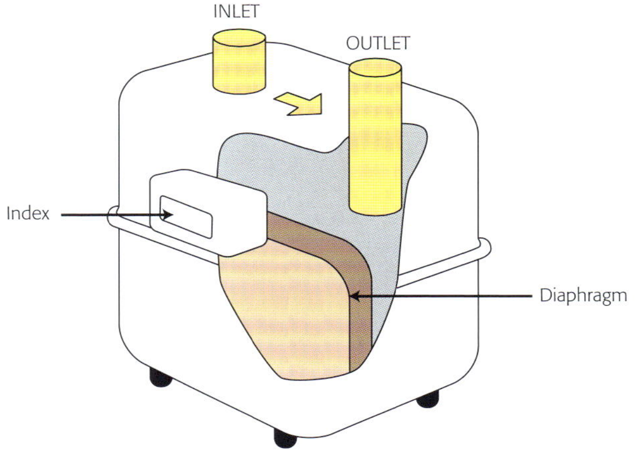

How a meter operates and measures gas

Diaphragm type

A U6 or G4 diaphragm meter case is divided into three main compartments:

- the top section contains the gas inlet and outlet, the valve assembly and the meters index (reading) mechanism, which can be either ft^3/hr or m^3/hr
- the lower section is divided into two equally sized sealed chambers. Each chamber contains a central flexible diaphragm. The gas passes through these chambers allowing an accurate measured volume to be recorded by the index. Two valves control gas flow to and from each chamber.

How this works

Gas flows from the meter outlet as soon as a burner is turned on, or flows continually if a pilot light is operating.

The gas (now under a greater differential pressure in the upper chamber) enters one of the lower chambers on one side of the diaphragm via an open valve. This expands the volume on that side of the compartment and forces the gas on the other side of the diaphragm into the outlet. As the diaphragm expands one way, it moves a lever known as a flag, which is linked to the valves and the meter index.

Before full movement (one way) of the diaphragm has taken place, the inlet/outlet valves move so that the second chamber receives gas, allowing the process to be repeated.

The main components of a U6 or G4 diaphragm meter are shown in Figure 16.1.

Figure 16.2 Typical E6 gas meter

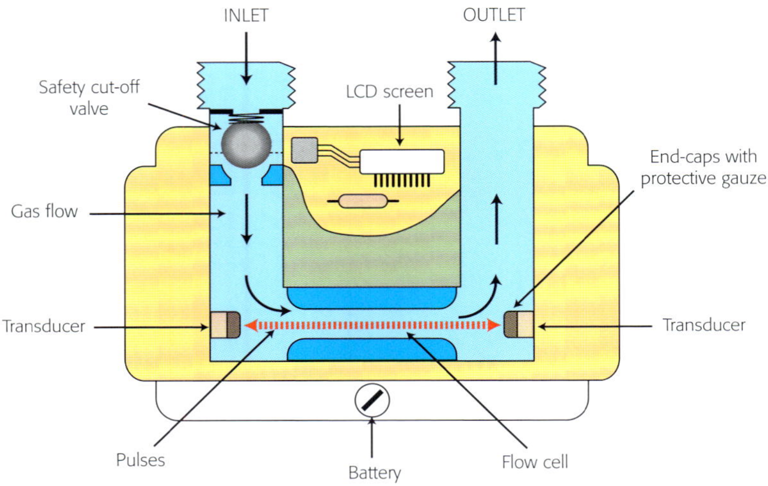

R5 Index – provides an automatic meter reading system (AMR)

Modern diaphragm gas meters may also have an R5 index. It provides the meter with an AMR system by means of a magnet fixed to one of the gear wheels which rotates to activate a read switch, creating a pulse.

The pulse is then transmitted to a circuit connected to a socket outlet. This allows remote recording to be carried out of the gas meter index.

Electronic type – measures very accurately

The electronic E6 meter (see Figure 16.2) is an alternative to the U6. Generally constructed of mild steel or from suitable plastic based materials, it has a Liquid Crystal Display (LCD).

The meter case size varies between manufacturers, but is usually the size of an average house brick.

- the E6 meter measures gas very accurately and continually checks for meter malfunction, interference, tampering and service requirements
- the LCD display indicates any fault etc.
- you can use compatible electronic equipment to allow meter readings to be made remote from the building in which the meter is installed

How it works

The meter has a battery (typically lasting 10-years) and measures gas flow by using ultrasound pulses at 2-second intervals. There are no moving parts within the meter.

Gas flows through it due to demand and gas pressure. The pulses move with and against the gas flow within the meter and are recorded by two electrical transducers. The time the pulses are 'in flight' is measured and calculated to record gas volume flow in m^3/hr on the LCD display.

This has an eight-digit reading: five digits before the decimal point, followed by a further diagnostic display.

Note: A transducer is an electrical device which can convert an electrical signal into a mechanical signal i.e. it can act as a transmitter and receiver.

You install an E6 meter similarly to a U6 or G4 meter and the inlet and outlet union threads are the same size. The meter also utilises the standard meter wall bracket.

Other differences re: E6/U6/G4 meters that you need to know

There are significant differences in the characteristics of the E6/U6/G4 meters that affect both permissible leakage rate (pressure loss) and gas rate testing:

1. The permitted gas leakage tolerance for NG allowed for by standards and procedures for installations with appliances connected is different for each meter type (providing there is no smell of gas).

   For further guidance on testing for gas tightness, see **Part 6 – Testing for gas tightness (Natural gas)**.

2. The gas rating test for an E6/G4 meter requires a comparison of gas flow over 2 minutes, but with a U6 meter, this is achieved by recording the time taken to burn 1ft^3 of gas.

   For further guidance on gas rating, see **Part 11 – Checking/setting burner pressures/gas rates**.

Note: The reason for the difference in tolerance is due to the reduced meter volume of the E6 meter compared with that of the U6/G4 meters.

Figure 16.3 Typical electronic token meter

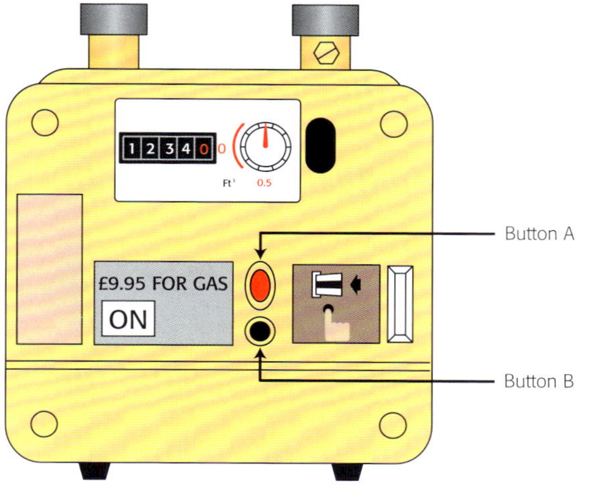

Electronic token meter (ETM) – current prepayment technology

The ETM uses a 'Gascard' to pay for the gas used (see Figure 16.3). The meter is U6 in size and volume, with an index dial reading window and an LCD display panel showing credit available and details of any faults.

The ETMs currently installed form part of the 'Quantum' System consisting of:

1. A meter complete with integral 'shut off' valve and electronic module incorporating a card slot;
2. A 'Customer' Gascard – used by the gas user to purchase credit;
3. A 'Charger Unit' – found in Post Offices and other retail outlets (Paypoint);
4. A 'Servicecard' – used by the owner of the meter (and their authorised agents) to commission and maintain the meter.

How it works

- when a Gascard (which has been initialised in the meter) is placed in a charger unit (Paypoint), the gas user can purchase credit
- when the Gascard is inserted into the card slot of the meter, the gas user can transfer the purchased credit to the meter itself
- when credit has been transferred, the meter re-calculates the monetary value of the gas available and also deducts a daily standing charge
- it can also deduct payments against any outstanding debt

Safety advice

With all prepayment meters, the gas user must turn off all taps (including pilot lights) on appliances served by the meter – before reinstating the gas supply.

The meter also has an anti-tamper device which, when activated by shock or movement, will close the valve. The anti-tamper device can only be re-set by the owner of the meter (or their authorised agent), using a Servicecard.

Attention: Due to the anti-tamper device, you must exercise great care when you disconnect, handle or work on, or near this type of meter.

The meter inlet and outlet gas connections are of standard thread size. To fit the ETM follow the installation requirements for U6/G4 meters.

Meter installation design – what you need to consider

Select the right fittings/components in each case. Particularly ensure they are suitable for the lowest operating gas pressure up to the maximum incidental pressure (MIP).

When you design the gas meter installation, you must also take into account:

- the building design
- the potential gas load
- the gas pressure available within the gas network e.g. whether the installation will be designed for a low pressure-fed, medium pressure-fed, NG or a LPG meter installation

Hazardous area classification

Domestic installations are exempt from The Dangerous Substances and Explosive Atmospheres Regulations 2002 (DSEAR) . For the purposes of design, meters installed in domestic premises, in either an internal location or in a meter housing, are deemed to be in a Zone 2 hazardous area (an area in which an explosive air/gas mixture is not likely to occur in normal operation and, if it occurs, it will exist only for a short time).

However, DSEAR does apply to non-domestic installations, confirmation shall be obtained from the responsible person for the site of any existing hazardous area classification for the proposed location.

The meter and any equipment electrically connected to it shall be suitable for use in a Zone 2 area or be suitable for use in the zone in which it is to be used, whichever is the more onerous requirement. Any other equipment fitted inside the meter housing shall be suitable for use in the zone in which it is to be used and shall be connected in accordance with IGEM/GM/7A .

Restricted locations – essential advice

For new meter installations in buildings with two or more floors above the ground floor, do not install a meter (and its associated controls) on or under a stairway or in any other location, where the stairway or other location provides the only means of escape in the event of a fire.

Gas meter installations in multi-occupancy buildings i.e. flats (see Note) should also comply with the requirements of IGEM/G/5 (Edition 2) 'Gas in flats and other multi-dwelling buildings' – which provides further clarification/guidance on issues relating to 'means of egress'.

Note 1: IGEM/G/5 (Edition 2) defines within its scope –

These procedures cover the design, installation, operation and maintenance of gas supplies to, and installations within, any building containing multiple individual dwellings and within the dwellings themselves (see Note 2).

This embraces buildings containing solely domestic dwellings, and dual-purpose buildings containing both commercial properties such as shops and offices and individual dwellings

Note 2: Each individual dwelling within a building would be independently occupied, with a dedicated gas supply whether that be a gas service or installation pipework. This means that, unless a building does not contain a domestic dwelling, the following arrangements are covered by IGEM/G/5: a building containing or served by more than one primary meter; a building containing, or whose contents are served by, more than one secondary meter that is used for billing/charging purposes; a building containing or served by just one primary meter but where more than one additional emergency control valve (AECV) has to be installed in compliance with GS(I&U)R Regulation 9(1) and, in general, a building comprising more than one individual dwelling i.e. is multi-tenancy and/or multi-owner occupancy.

For existing gas meter installations – including buildings with less than two floors above the ground floor and replacement meters in buildings with two or more floors above the ground floor – the installation should comply with the requirements for new installations (see **Gas meter installations** in this Part).

Pantry/food store

Do not site a new meter where food is, or is likely to be stored.

Damp or corrosive conditions

Do not site a new meter in an area where it may be subject to damp (unless it is a semi-concealed meter installation) or corrosive conditions.

Basements and cellars

Do not install LPG meters in a room or internal space below ground level e.g. a basement or cellar (see also **Gas meter installation – Location and siting** in this Part).

You can install such meters into rooms which are basements with respect to one side of the building, but open to ground level on the opposite side.

Medium pressure-fed meter installations – additional requirements

Do not locate medium pressure meter regulators and relief valves inside buildings. Do not locate multiple medium pressure-fed meter installations within a single compartment, housing or building.

Semi-concealed meter boxes

Do not site these types of boxes in areas that are prone to flooding.

General installation details

The emergency control valve: what you need to know

All gas meters must have an ECV fitted at the inlet to all primary and secondary meters:

- to allow the gas to be turned off in the event of an emergency, and
- to allow work to be carried out on the gas installation

You must ensure it is accessible and clearly identify its operation (see **Part 9 – Emergency controls and isolation valves** and **Part 10 – Emergency notices, warning labels and forms**).

Identify all ECVs (AECVs) to make the gas user aware why they are there and how they operate.

Where you install a meter in any premises at a distance of more than 2m from the nearest upstream ECV in the premises (see **Working on gas meters** in this Part), you may not supply or provide gas for the first time through that meter unless you prominently display a suitably worded permanent notice on or near the meter, indicating the position of the control (see **Part 10 – Emergency notices, warning labels and forms**).

Note: If you install an ECV remote from a gas meter, you must label the valve to comply with these requirements.

The GSIUR also require all gas meters to have a suitably worded label/notice, in permanent form, prominently displayed on or near the meter on the actions to take in an emergency – and when you install a gas meter you are responsible for fitting this notice (see **Part 10 – Emergency notices, warning labels and forms**).

Meter regulators – set for constant operating pressure

You must set meter regulators to maintain a constant operating pressure. The nominal operating pressure for NG is 21mbar and for LPG (Propane) 37mbar (see **Low pressure-fed gas meter installations** and **Pressure absorption** in this Part).

Check the regulator and adjust (see Note) if necessary (see **Part 7 – Checking and/or setting regulators** for further guidance).

Note: Only a person authorised to do so by the gas transporter or supplier may adjust a meter regulator.

Industry meter regulator check device

The device comprises a fixed drilled orifice, vented to atmosphere at one end and a union for attaching to the meter outlet at the other end.

The drilled orifice is sized to pass a known quantity of NG and is used to ensure that the meter regulator is operating correctly, where no gas appliances are connected.

Non-return valve

Where an installation uses a gas compressor, premixed blown, compressed air or extraneous gas in conjunction with the gas supply, the gas meter and incoming gas supply shall be protected by a non-return valve approved by the gas transporter installed downstream of the meter insulation.

Table 16.1 Diversity factors of gas appliances burning NG or LPG

Appliance Type	Diversity factor
Central heating boiler (not combination boiler)	1.0
Circulator	1.0
Convector heater	1.0
Combination boiler	0.8
Instantaneous water heater (multi-point)	0.8
Room heater (gas fire)	0.6
Oven	0.6
Hob (hotplate)	0.6
Sink water heater	0.6
Tumble dryer	0.6
Wash boiler	0.6
Cooker	0.4
Refrigerator	0.0

Table 16.2 Gas meter designation and rating

Meter designation	Maximum flow rate (capacity) m^3/hr
U 2.5*	2.5
U 4*	4.0
U 6	6.0
G 4	6.0
E 6	6.0

Note: *U 2.5 and U 4 meters are only for use with LPG

Meter sizing – and 'diversity of use factor'

Before you connect a gas appliance to an installation, you must ascertain the gas meter is capable of passing sufficient gas for the new and existing gas appliances.

To calculate the size (capacity) for a gas meter, you need to determine the maximum combined heat input rating (gross) of all appliances connected to the installation, to calculate the maximum demand at any one time.

However, as all appliances are unlikely to be in use at the same time a 'diversity of use factor' is introduced into the calculation. This reflects the time that individual appliances can be expected to be in use (see Table 16.1).

Table 16.3 Example

Appliance Type	Maximum heat input kW (gross)	Diversity factor	Load or rating
Central heating boiler	22.0	x 1.0	= 22.0
Cooker	21.5	x 0.4	= 8.6
Room heater	5.0	x 0.6	= 3.0
Tumble dryer	3.0	x 0.6	= 1.8
			Total 35.4kW

Convert this total figure (35.4kW) to MJ/hr (35.4 x 3.6 = 127.44 MJ/hr)

Divide this figure (127.44 MJ/hr) by the calorific value of the gas to convert to m^3/hr

NG	Propane gas
$127.44 \div 38.76(MJ/m^3) = 3.28m^3/hr$	$127.44 \div 95(MJ/m^3) = 1.34m^3/hr$

Conclusion: An E6, G4 or U6 meter for NG should be used and a U 2.5 meter for Propane gas (see Table 16.2).

It does not necessarily depend on the maximum heat input rating of the appliance – only on the time it may be in use.

Note: You normally identify gas meters by their badged capacity (maximum flow rate) i.e. U6, G4 or E6 = 6m³/hr (see Table 16.2).

How to calculate the size of meter required

- you multiply the maximum heat input rating (gross) of each individual appliance in kW by its diversity factor
- add these figures together and convert to Megajoules (1kW = 3.6MJ/hr)
- divide this total by the calorific value (CV) of the gas being used e.g. averaged $38.76MJ/m^3$ for NG and $95MJ/m^3$ for Propane

Note: The CV's are an averaged value and may differ to that stated by the gas supplier.

The calculation in Table 16.3 is an example for each gas type.

Protective bonding conductor

A gas installation within a property with an electrical supply shall have a main protective bonding conductor connecting the pipework to the electrical installation's main earth terminal, as specified in BS 7671

See **Part 5 – Installation of pipework and fittings – Equipotential bonding – minimises hazards from electrical shock** for further guidance.

When to carry out tightness testing and purging

- when you install any new meter, exchange an existing meter, or restore a gas supply to the meter, you must test the meter, together with any installation pipework and appliances for gas tightness
- also, where you carry out work on a gas meter or any installation it serves, you must ensure that the meter and all its installation pipework and fittings are purged of air

See **Part 6 – Testing for gas tightness (Natural gas)** and **Part 15 – Re-establishing gas supply/re-lighting appliances** for further guidance. For LPG installations, see **Part 18 – LPG – General requirements** and/or CORGI*directs* Gas Installer Manual Series, 'LPG – Domestic, Including Permanent Dwellings, Leisure Accommodation Vehicles, Residential Park Homes and Boats' (see **Part 22 – CORGI*direct* Publications**).

Relocating a gas meter/meter installation – options available

You can relocate some gas meter/meter installations – but before you commence, you must inspect the installation to determine the most appropriate method.

Depending on circumstances, there are usually three options for you to consider:

1. Alter the position of the gas service pipe and the complete gas meter installation.
2. Relocate the gas meter only.
3. Relocate the gas meter and regulator.

Before you agree to relocate the gas meter, you need the correct competency for the activity (see **Working on gas meters** in this Part) and importantly, permission from relevant parties, to include:

- the owner of the meter/meter installation
- the owner of the property, and
- the gas transporter (GT)

If after inspecting the gas meter installation, you feel the most appropriate method is to alter the gas service pipe and the complete gas meter installation, then contact the relevant gas supplier/transporter – so that they can carry out the necessary work.

You must not relocate medium pressure-fed gas meter installations to inside a property.

Note: Where work is required to alter the upstream gas supply, only the GT, or those persons authorised by the GT may carry this work out.

Where you propose to relocate the meter or meter installation, then additional pressure losses need to be considered/calculated – follow the recommendations outlined in the relevant part of BS 6400 for meters up to 6m^3/hr and IGEM/GM/6 (Edition 2) or IGE/GM/8, as appropriate for meters up to 16m^3/hr.

When an electrical isolation joint is fitted downstream of the ECV (in a legacy installation) it shall only be removed with the consent of the Network Operator

Gas rating check

This check determines the volume of gas being burnt by an appliance. It converts the time it takes to pass a known volume of gas through a gas meter in order to determine the appliance gross heat input. You can then compare this figure to the appliance's data plate.

You can carry out this check in a matter of minutes with the use of a stopwatch and calculator, or by using a stopwatch and a pre-calculated gas rating chart (see **Part 11 – Checking/setting burner pressures/gas rates** for further guidance).

Gas meter installations

Location and siting

The building contractor and the gas supplier or GT normally discuss where to site the gas meter.

Primary gas meters are generally and preferably installed externally in purpose-made meter boxes, semi-concealed into adjacent ground, or located on, or built into the building's external structure. Alternatively, they can be sited into boxes, housings or compounds at the property boundary.

Where a number of primary meters serving different parts of the building are grouped together e.g. in large premises converted into flats or in flats over business premises, the instalation shall be carried out such that:

- the meters are situated in an area to which there is access at all times
- the meters are enclosed in a single lockable housing or in individual lockable meter boxes
- a suitable labelled key for the meter housing is provided for each occupant of the dwelling each meter serves
- each meter or individual meter box is clearly marked to indicate the dwelling it serves
- meter housings are installed, marked or labelled to indicate the dwelling to which each meter relates and that the marking or labelling is permanent and visible when the compartment is closed
- each individual dwelling (e.g. each house, flat or maisonette) using a supply of gas is provided with an AECV whether or not that premises contains a gas meter which is situated as near as is reasonably practicable to the point where the pipe supplying gas enters the premises.

If an external position is not possible, you can install a meter within a building, as long as the requirements of **Restricted locations** are complied with i.e. the meter and any associated pipework and fittings, will not cause an obstruction or be a danger to any person (see also **Internal meter position in multi-occupancy dwellings** in this Part).

Only install coin/token operated prepayment gas meters within a building accessible only to authorised persons – and conveniently located for easy use and access to the coin/token mechanism.

When a diaphragm meter is used, it shall be installed upright and level such that it cannot be readily tilted.

Note: Where you locate a gas meter for a timber/light steel frame building, refer also to CORGI*directs* gas installer manual 'Gas Installations in Timber/Light Steel Frame Buildings' from the current Gas Installer Manual Series, see Part 22 – 'CORGI*direct* Publications'.

Internal meter position

When fitted inside a building, ensure that the meter, pipework and fittings are in an adequately ventilated area that allows easy access for:

1. Use of the ECV.
2. Installation and maintenance (setting/re-setting of the regulator and maintaining the ECV).
3. Exchange or removal of the meter and regulator and components.
4. Meter reading.
5. Operating any prepayment mechanism and emptying the collection box.

- ensure that the meter is not subjected to damage from damp or corrosive conditions e.g. either at floor level or wall mounted, or where chemical cleaners could come into contact with and attack the meter case
- The meter shall be secured such that it is not in direct contact with any wall and shall be protected either by design, installation or position from direct contact with any cement and/or cement composition and any floor that might be wetted
- ensure the meter is not installed where there is a significant risk it will be submerged by flooding
- ensure that it is not subject to heat or extremes of temperature; nor in close proximity to electrical wiring, meters or switch-gear, etc. (see Note)

Note: A gas meter, together with its associated controls and pipework needs to be installed at least 150mm away from any electricity meter/switchgear or 25mm from the electric supply and distribution cables.

Where a gas meter and its associated controls are located within 150mm of an electricity meter/switchgear or 25mm from the electric supply and distribution cables, fit a non-combustible partition (constructed of electrical insulating material) between the meter and other electrical equipment.

Pliable connections

BS 6400-1 advises that only one pliable connection should be used within a meter installation (the meter installation is regarded to be from the ECV to the outlet of the meter, with the exception of semi-concealed meter installations).

The reason given for this are many, but primarily it is to prohibit tampering of the meter installation.

Therefore any flexible pipework connected to the outlet of the gas meter would be deemed to be part of the installation pipework and subject to being in accordance with BS 6891: 2015 Specification for the installation and maintenance of low pressure gas installation pipework of up to 35mm (R1$^1/_4$) on premises.

BS 6891 states "Unless the meter is securely restrained (e.g. by a meter bracket), the connection to the meter installation shall be in securely fixed rigid pipe for at least the first 600 mm".

Internal meter position for multi-occupancy dwellings

In addition to complying with the requirements of BS 6400-1, 2 and 3 (as appropriate), gas meters installed in multi-occupancy dwellings i.e. flats need also to comply with the requirements of IGEM/G/5 (Edition 2).

As discussed previously in this Part, do not install a meter (and its associated controls) under a stairway or in any other location, where the stairway or other location provides the only means of escape in the event of a fire.

If however, there are more than one means of escape from the premises and an external installation is not possible, a gas meter may be installed in such a location where:

- a risk assessment confirms the risk is mitigated; and
- the common, alternative means of escape

is completely independent from the other means of escape i.e. a fire or explosion in one means of escape will not hinder escape by the occupants in the other; or

- the escape, in which the meter is installed is largely exposed to the external air; or
- an excess flow valve and thermal cut-out is installed complying with a recognised standard, the meter is located within a housing having at least 30 minute fire resistance and the housing has a self-closing door

If the meter is to be located within the sole means of escape, within an individual dwelling (inside a flat), then the meter installation needs:

- to incorporate a thermal cut-out immediately upstream of the meter; or
- the meter is enclosed in a compartment/box/cupboard having at least 30 minutes fire resistance and having a self-closing door

External meter position

Purpose designed boxes, housings and compounds for external meter installations provide protection from the weather and from vandalism. The installer shall ensure that any meter housing is labelled to indicate:

- the meter housing manufacture's details and details of the housing including the manufacturer's name address and telephone number and the model number of the box (if the meter housing is a box), ventilation details and the classification of the hazardous zone deemed to exist within the housing
- the gas family and MOP for which the housing is intended:
- adequately ventilate housings suitable for the size of the meter and associated fittings and controls
- design them to prevent any escape of gas into a wall cavity or other part of the building, via pipe outlets, ducts or sleeves
- meter boxes, housings or compounds should only be accessible by a suitably marked key, which the gas user should have in their possession
- outhouses, garages etc. should only be accessible by a suitably marked key in the possession of the gas user
- ensure the meter is not installed where there is a significant risk it will be submerged by flooding.

Provide easy access for:

1. Use of the ECV.
2. Installation and maintenance (setting/re-setting of the regulator and maintaining the ECV).
3. Exchange or removal of the gas meter, regulator and components.
4. Meter reading.

- ensure that the meter is not subjected to damage from damp or corrosive conditions e.g. at either floor level or wall mounted, or where chemical cleaners could come into contact with and attack the meter case
- The meter shall be secured such that it is not in direct contact with any wall and shall be protected either by design, installation or position from direct contact with any cement and/or cement composition and any floor that might be wetted
- ensure the meter is not installed where there is a significant risk it will be submerged by flooding
- ensure that it is not subject to heat or extremes of temperature; nor in close proximity to electrical wiring, meters or switch-gear, etc. (see Note to **Internal meter position** in this Part).

Meter boxes – types and positioning

The builder and the gas supplier, or GT will normally decide upon the position of a gas meter box. Generally, four purpose-designed boxes are used:

1. Built-in.
2. Surface-mounted.
3. Semi-concealed.
4. Universal.

These boxes should comply with BS 8499: 2009 'Specification for domestic gas meter boxes and meter bracket'.

The installer shall ensure that any meter housing is labelled to indicate:

- the meter housing manufacturer's details and details of the housing
- the gas family and MOP for which the housing is intended.

Attention: For medium pressure-fed NG/LPG meter installations, any meter boxes must not permit any pipework or Protective bonding conductor to pass directly from the box into the building. For example, the gas pipework/cable should exit the bottom front of the box before entering the building through a gas tight sleeve. (see Figure 16.8).

Figure 16.4 Typical built-in meter box

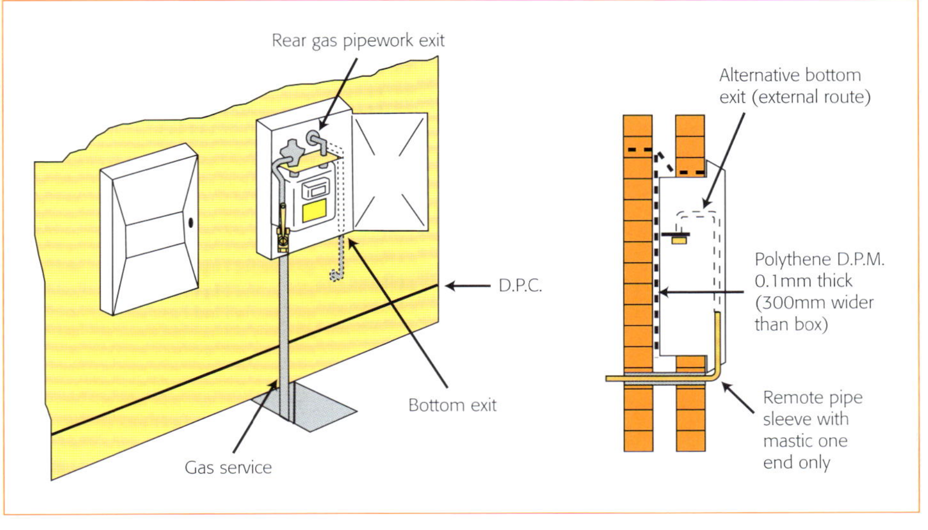

Ventilation required for meter boxes

Meter boxes need to incorporate non-closable ventilation as follows:

Built-in and surface mounted boxes – require a minimum of 2% of the floor area, equally distributed at both high and low level.

Semi-concealed boxes – require a minimum of 6% of the floor area, equally distributed around the box lid.

Universal boxes – require a minimum of 2% of the floor area, equally distributed at both high and low level, or a minimum of 6% of the floor area, equally distributed around the box lid.

Built-in

The builder normally installs the meter box (see Figure 16.4) into an external wall leaf. The box normally enters the wall cavity with the base of the box above the finished ground level and damp-proof course/membrane.

- secure the box to the building, using all the manufacturer's fixing positions and seal to the wall surface to prevent rain getting in
- where a meter box is in contact with the internal wall, protect that wall from damp penetration by a suitable damp-proof membrane, installed as shown in Figure 16.4
- for low pressure-fed gas meter installations where you plan a rear exit gas installation pipework supply, you must use the wall spigot sleeve supplied with the meter box (see 'Attention page 19').

  For further guidance on the installation of sleeves, see **Part 5 – Installation of pipework and fittings**
- the spigot sleeve must be continuous between the inner finished face of the building and the meter box and secured by the screws and flange provided
- if you are installing the installation pipework from the meter, you must seal the spigot sleeve within the meter box. Use an appropriate non-setting mastic, or similar
- do not make any holes in the box (the only ones must be the manufacturer's pre-formed openings). Any Protective bonding conductor may only enter the box via these openings

The box is normally supplied with a meter bracket to secure the meter in position.

For example, the gas pipework/cable should exit the bottom front of the box before entering the building through a gas-tight sleeve (see Figure 16.4, Alternative bottom exit).

Surface-mounted

- the meter box (see Figure 16.5) is designed to be located on an external wall and you site it in relation to the gas service pipe, or its intended position
- ensure that the chosen site does not restrict the removal of the box front cover, or cause an obstruction to any persons or vehicles if it is sited on a pathway or drive
- fix it to the wall by a back plate, to which is secured the meter fixing bracket
- use all the manufacturer's securing points with suitable sized screws or wall bolts. The box is designed with pre-prepared exit holes for the gas pipework outlet supply, providing a rear or bottom option
- where the gas pipework enters the building, you must enclose it in a gas tight sleeve (see 'Attention page 19')

If you fit the box to a timber frame building, you will find further information in 'Gas Installations in Timber/Light Steel Frame Buildings' from CORGI*directs* current Gas Installer manual series, see **Part 22 – CORGI*direct* Publications**.

- do not make any holes other than ones in the box prepared by the manufacturer. Any

Figure 16.5 Typical surface-mounted meter box

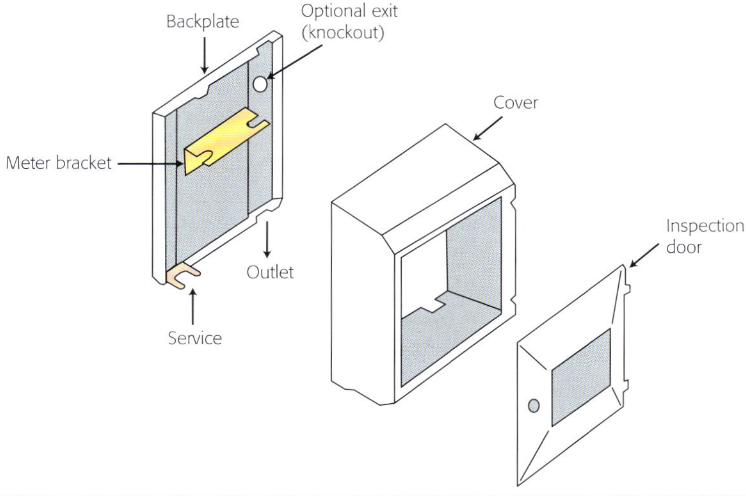

Protective bonding conductor must only enter the box via these openings

- you normally site the box with its base between 500mm and 1500mm from finished ground level. It should not bridge a damp-proof course or membrane.

Semi-concealed

The meter box (see Figure 16.6) is designed to be located against an outside wall and is partially buried in the ground. A builder, or the GT, normally install it and site it in relation to the gas service pipe.

The chosen site must not restrict the lifting up of the box top cover nor, if sited on a pathway or drive, must it obstruct any person or vehicle.

If you are installing the box, follow the meter box manufacturer's installation instructions.

- excavate a suitable hole and mark on the box the depth in relation to the finished ground line
- use all the manufacturer's fixing points, with suitable screws or wall bolts and the box spacers
- ensure that the box fixings do not penetrate any damp-proof course or membrane

Figure 16.6 Typical semi-concealed meter box

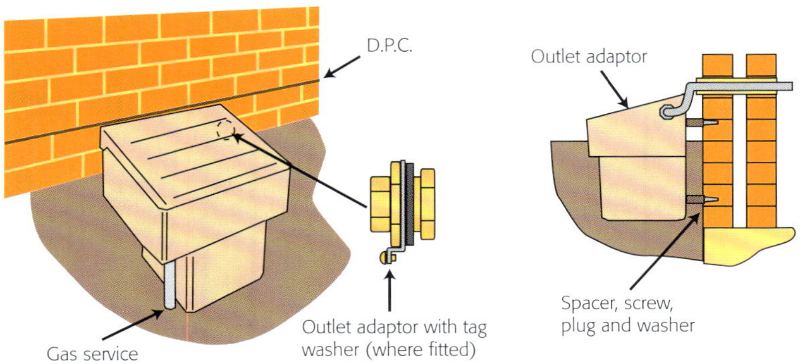

Generally, the gas meter ('U', 'G' or 'E' type), the associated controls and fittings compatible with this type of box are unique.

The gas installation pipework must only exit the box via the prepared hole and outlet adaptor and enter the building above any damp-proof course, or membrane enclosed in a suitable gas tight sleeve.

For further guidance on the installation of sleeves, see **Part 5 – Installation of pipework and fittings**.

Do not make any holes in the box other than those prepared by the meter box manufacturer. If necessary, any Protective bonding conductor should only enter the box via these openings and it should be connected to the installation pipework or tag, where fitted.

Note: Where you install a meter for a LPG installation into a semi-concealed meter box, refer to the meter box manufacturer's installation instructions regarding where to site it.

Universal

The meter box (see Figure 16.7) is designed to be located against an outside wall and is installed either:

- wholly above ground, or
- partially below ground (see Note)

Note: The box will have a level marked on it indicating the ground level for partially below ground installations.

Do not make any holes in the box other than those prepared by the meter box manufacturer. If necessary, any Protective bonding conductor should only enter the box via these openings.

Figure 16.7 Typical universal meter box

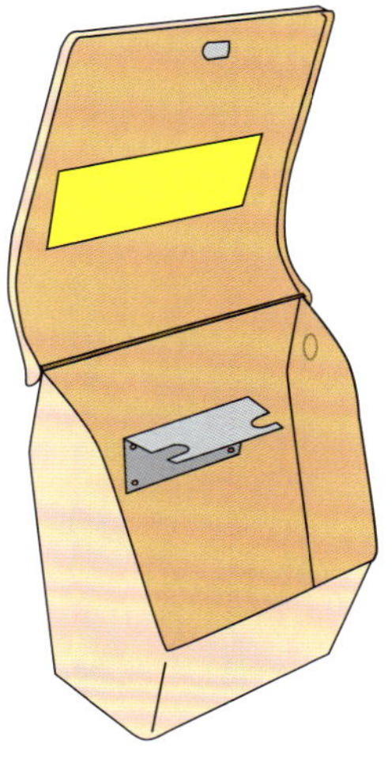

Meter boxes housing medium pressure-fed meter installations – additional restrictions

Medium pressure-fed meter installations within meter boxes have additional restrictions placed upon them.

These installations have special regulators that operate at inlet pressures normally in excess of 75mbar must have a pressure relief valve (PRV) and a vent pipe terminating to a safe position direct to outside air.

- consider any electrical equipment or openings into buildings such as openable windows, doors, airbricks, flue terminals or any other similar opening (see Tables 16.4, 16.5 and Note)

Note: Where the relief valve vent pipe needs to be extended (more than 150mm up to a maximum of 1500mm), to meet the requirements of Tables 16.1 and 16.2, you will need to increase the diameter of the pipe.

This is to maintain maximum relief valve discharge should a fault occur. The pipe will need to be greater than 10mm internal diameter e.g. 12mm or 15mm copper tubing.

- built-in/surface-mounted meter boxes must not allow any pipe or cable to pass directly from the box into the building (see Figure 16.8 for the correct method of routing installation pipework and bonding cable)
- ensure that built-in meter boxes utilised for medium pressure-fed meter installations also have a purpose designed non-adjustable louvered vent at high level and sufficient ventilation at low level to supply good all round ventilation.

Figure 16.8 Correct method for routing installation pipework for a medium pressure-fed meter installation

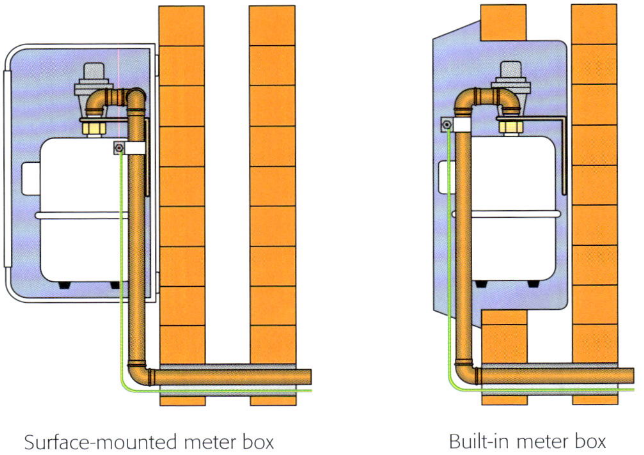

Table 16.4 Clearances for relief valve vent pipe tips and meter housings (NG) – medium pressure-fed meter installations

Relief valve discharge (m^3/hr(std))	A (m)	B (m)	C (m)	D (m)
> 0 ≤ 1	0.18	0.33	0.57	0.85
> 1 ≤ 2	0.18	0.33	0.8	1.3
> 2 ≤ 2.25	0.18	0.33	1.0	1.55

A. Minimum distance of a meter box from any opening such as an openable window, door, air brick, appliance flue terminal, other breaches etc. into the property.

B. Minimum distance above or at the side of a meter box to any electrical equipment.

C. Minimum distance from any relief valve vent pipe tip to any opening into the property.

D. Minimum distance from any relief valve vent pipe tip to any electrical equipment.

Table 16.5 Clearances for relief valve vent pipe tips and meter housings (Propane) – medium pressure-fed meter installations

A. Semi-concealed meter box to opening	B. Meter housing other than semi-concealed to opening	C. Meter housing to electrical equipment	D. Vent tip to opening	E. Vent tip to electrical equipment
1.0m	0.18m	0.33m	0.57m	0.85m

A. The minimum distance from the meter box to any un-trapped drain, gully, balanced flue terminal or low level opening into the building e.g. doors or air bricks which are below 250mm from ground level.

B. The minimum distance from the meter housing to any opening into the building e.g. opening windows, doors, air bricks and balanced flues.

C. The minimum distance from any meter housing to any electrical equipment.

D. The minimum distance from any relief valve vent pipe tip to any opening into the property or un-trapped drain or gully.

E. The minimum distance from the relief valve vent pipe tip to any electrical equipment.

Installing 'U', 'G' and 'E' series gas meters

- transport and handle all gas meters with great care, to avoid mechanical damage or shock
- always keep diaphragm meters upright
- install all new 'U', 'G' and 'E' series meters level and with adequate support, generally provided by a purpose-made meter bracket or shelf
- for semi-concealed boxes, only fit meters specially designed for the purpose and use an appropriate fixing kit (see Figure 16.6)

Note: Where existing lead connections are encountered, it would be good practice to replace with an appropriate alternative.

However providing the existing lead pipework is suitably sized and in good condition and not modified i.e. re-jointed it may be re-used (see Part 5 – 'Installation of pipework and fittings – Lead (lead composition) pipework – not to be used for new installations').

Checks prior to installing the gas meter

Before the gas meter is installed, you must do certain basic checks, including:

- confirm that the pressure within the upstream network is compatible with the proposed installation and components e.g. low pressure-fed, medium pressure-fed (see 'Attention') NG or LPG
- confirm the gas service pipe and ECV is appropriately labelled
- confirm the ECV operates correctly
- confirm the proposed location for the gas meter installation is acceptable, including ventilation requirements
- confirm where appropriate, the meter housing is suitable for the purpose

Figure 16.9 Standard 'U', 'G' and 'E' series meter installation (NG)

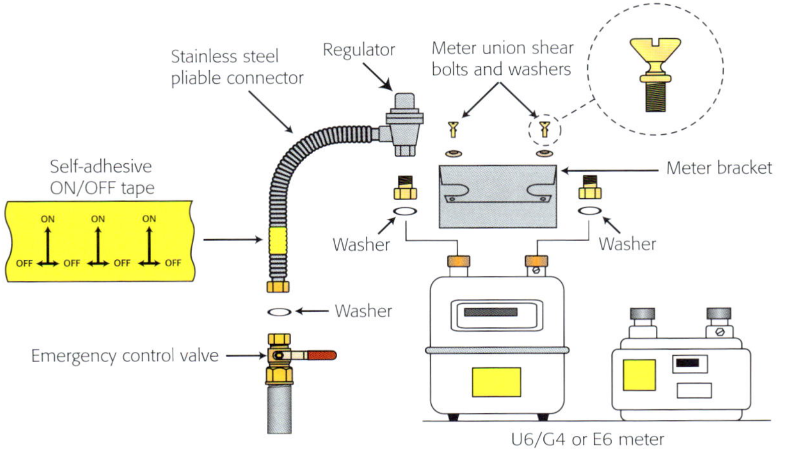

- confirm where there is an internal gas installation supply in place – a temporary continuity bond will need to be connected between the gas service and the internal gas supply
- prior to undertaking any meter work the installer shall carry out a site specific risk assessment. Appropriate mitigating action shall be taken for any hazards identified. The installer shall be alert for any other hazards that arise during meter work and where necessary, update the site-specific risk assessment. As part of this risk assessment the installer shall ensure that any exposed metal work is checked for stray electrical voltage with an appropriate voltage detector.

Note: The use of a non-contact voltage detector (single pole) capable of indicating voltages of 50 V or greater (see Figure 16.10 for an example) on all exposed metalwork in the work area assists in detection of stray voltage that might be harmful. It should be noted that such voltage detectors are not designed to respond to current flow and should be used in accordance with the manufacturer's instructions. Where any doubt about the electrical installation exists the advise of a competent electrician should be sought.

Attention: For medium pressure-fed gas meter installations, you need to carry out additional tests on the regulator/safety system, in accordance with the manufacturer's instructions, or relevant industry standard.

Figure 16.10 Non-contact voltage detector (single pole)

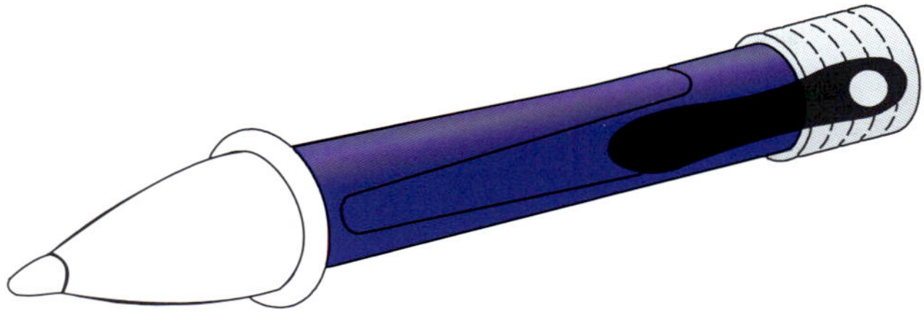

Primary meters

A primary meter installation commences at the outlet of the ECV and terminates at one of the following:

- the outlet of the meter
- the outlet of the meter adaptor, (where fitted); or
- for a semi-concealed meter installation with a pliable connection downstream of the meter, the outlet of the meter box adaptor

General guidance to help you

Assemble meter and associated fittings in the correct order. Figure 16.9 shows a typical NG meter installation. You must fit any regulator and filter (if applicable) to the meter inlet connection. The meter and regulator assembly can be supported and installed in a manner that minimizes strain being placed on any connections and allows easy removal and refitting of the meter and regulator assembly, restrict movement and reduce the likelihood of tampering.

A meter that is to be installed in an above ground meter housing can be fitted to a meter bracket that has been fitted within the meter housing.

A meter that is to be installed in a semi-concealed meter box can be restrained from movement by either the design of the box or the use of a meter bracket that has been fitted, where a diaphragm meter is used it can only be installed upright and level such that it cannot be readily tilted.

The meter and its regulation assembly can be secured such that they are not in direct contact with any wall and shall be protected either by design, installation or position from direct contact with any cement and/or cement composition and any floor that might be wetted.

Note: This can be achieved by raising the meter above the floor using the purpose designed meter support or using a meter manufactured with feet.

Where a meter bracket is used the union connections shall be fitted to the bracket such that they cannot be removed without the use

1. of a special tool or key.
2. Only use approved fittings.

Attention: The meter inlet and outlet connections have threads manufactured to BS 746. Never force incompatible threads onto these special meter union fittings.

3. Always use the correct meter washers with the meter union fittings.

4. The preferred method of fixing a meter is to use one rigid (copper or mild steel) connection for the outlet supply and one pliable (flexible stainless steel) connection for the inlet supply. The bending radius for pliable connections must not be less than two pipe diameters and shall not be bent so that the corrugations are close enough for condensation to bridge the gap, thereby providing a potential risk of corrosion.

Note: For a semi-concealed meter installation, the inlet connection is likely to be rigid and the outlet connection pliable. However where the meter is restrained in position by the meter box or bracket, two pliable connections can be used provided a meter box outlet adapter is fitted and the pressure absorptions of the primary meter installation does not exceed 4mbar.

5. When you fit the meter to an existing or new installation, you must immediately test the whole installation for gas tightness and purge it of air. See **Parts 6 – Testing for gas tightness (Natural gas)** and **Part 15 – Re-establishing gas supply/re-lighting appliances** for further guidance.

You will find additional information on LPG tightness testing in **Part 18 – LPG – General requirements**.

6. If the meter is not connected immediately to the gas installation pipework, you must test the meter and associated fittings for gas tightness and purge of air.

   You must then seal the outlet connection with a suitable fitting. Leave the gas supply ECV in the 'off' position.

7 If you fit the meter to a gas installation with an appliance(s) connected, check the regulator for correct operation. See **Part 7 – Checking and/or setting regulators** for further guidance (also see Attention).

8. You must correctly label the meter and the ECV. See **Medium pressure-fed gas meter installations (NG)** in this Part and **Part 10 – Emergency notices, warning labels and forms** for further guidance.

Attention: If you install the meter at a property prior to the completion of the internal gas installation/appliances, then you can ascertain the accuracy of the regulator by using an Industry meter regulator check device. See 'Industry meter regulator check device' in this Part.

Note: The meter, associated pipework and fittings may require a Protective bonding conductor. See 'Equipotential bonding – minimises hazards' in Part 5 - Installation of pipework and fittings.

Secondary meters

- you must only supply a secondary meter through a primary credit meter – and you must label the primary meter, indicating the number and position of all the secondary meters it supplies
- all secondary meters must have a label stating 'Secondary meter' (see **Part 10 – Emergency notices, warning labels and forms**)
- the GSIUR require that a line diagram in permanent form is displayed on or near the primary meter or gas storage vessel and or near all emergency controls connected to the primary meter. The diagram must show the configuration of all meters, installation pipework and emergency controls

Note: If you change the configuration of any meter, installation pipework or ECV and this affects the accuracy of the line diagram, you must amend it to show the altered configuration.

- you must fit an ECV as close as possible to the meter inlet connection
- you do not normally need to fit a regulator to the secondary meter, as the regulator connected to the primary meter should be adequate to control pressure for the gas installation
- you must notify the supplier of gas to the premises that a secondary meter has been fitted.
- it is a requirement of BS6891 that the maximum pressure absorptions across the installation pipework is a maximum of 1mbar. Where there is a secondary meter, this requirement applies to the pipework including the secondary meter.

For general installation guidance, see **Primary meters** in this Part.

Note: The meter and associated pipework and fittings may require a Protective bonding conductor. See 'Equipotential bonding – minimises hazards' in Part 5 - Installation of pipework and fittings.

Multiple meter installations – follow these guidelines

Separate primary meters serving individual dwellings within a large building such as blocks of flats can be installed in a communal compound.

Where this is the case, ensure that meter installations comply with the following guidelines:

1. Multiple meter installations need to be of the low pressure-fed type only.
2. Meters should be accessible at all times.
3. Meters should be enclosed in a single lockable compartment, or within individual lockable meter boxes.
4. Each separate meter box or compartment should be clearly and permanently marked to indicate the dwelling that it serves.
5. A suitably marked key to the compound should be held by the responsible person for each dwelling.

Note: Further guidance for multiple meter installations in multi-occupancy dwellings can be found in IGEM/G/5 (Edition 2).

16 – Domestic gas meters

Disconnecting a gas meter – what you need to do

You may be required to disconnect and remove a meter for a short period of time e.g. when making a soldered joint on the installation pipework (category A work as discussed in TB 127 – see **Working on gas meters** in this Part).

Take great care when handling any meter, in particular ETMs, see **Electronic token meters (ETM)** in this Part.

- once disconnected, seal the inlet and outlet meter connections with dust caps or similar, to prevent the ingress of dirt or the release of an explosive gas/air mixture, and place in a safe, ventilated position
- when refitted, you must test the meter and the gas installation for gas tightness and purge of air. See **Part 6 – Testing for gas tightness (natural gas)** and **Part 15 – Re-establishing gas supply/re-lighting appliances** for further guidance

You will find additional information on LPG tightness testing in **Part 18 – LPG – General requirements** and CORGI*directs* Gas Installer Manual 'LPG – Domestic, Including Permanent Dwellings, Leisure Accommodation Vehicles, Residential Park Homes and Boats' (see **Part 22 – CORGI*direct* Publications**).

Attention: To comply with GSIUR regarding protecting persons and property, you may need to fit a temporary continuity bond across the installation when disconnecting a gas meter.

See Part 5 – 'Installation of pipework and fittings – Temporary continuity bond – use for your own safety before you start work'.

Removing a gas meter – steps to follow (after you get permission)

Before you remove a gas meter, you need permission from either or all: the gas meter owner, the gas supplier or the owner of the property (see **Working on gas meters** in this Part).

When you remove a gas meter you must:

- turn off the gas at the ECV and safely release the pressure from the installation e.g. at an appliance
- seal the ECV and the internal gas supply with an appropriate gas fitting
- clearly mark any live gas pipework
- where necessary, fit a permanent Protective bonding conductor between the upstream gas service pipe and the internal downstream gas installation (see Note)

Note: Where the gas service pipe and internal gas installation are still in contact with each other via a meter bracket, you may not need to fit a permanent Protective bonding conductor. A permanent continuity bond shall not be fitted such that it bridges any insulation joint fitted upstream of the ECV.

See Part 5 – 'Installation of pipework and fittings – Temporary continuity bond – use for your own safety before you start work'.

Maintenance

During any work that necessitates connection or disconnection of any meter or associated gas fittings, a temporary continuity bond shall be fitted to avoid electrical danger, whether or not the permanent protective bonding conductor has been established. The temporary continuity bond shall remain in position until the work is completed.

Metering equipment shall be maintained in accordance with the manufacturers' instructions.

Maintenance shall be carried out such that when the maintenance engineer leaves the site (the following list is non-exhaustive):

1. the meter installation:
   a) is installed correctly
   b) is in sound mechanical condition
   c) operates at the intended set points
   d) operates in a safe manner
   e) has appropriate notices affixed
   f) has, where fitted, a functional data transmission system
2. any official metrological meter seal is intact
3. the regulator seal is intact and has been sealed by the manufacturer or by an OAMI
4. the general condition and security of the installation is satisfactory
5. the meter housing/box is not damaged and its integrity is such as not to allow gas to enter any adjacent cavity or property
6. the ECV operable and is accessible to the consumer
7. the flow rates and operating pressures for the installation are within the design constraints
8. the meter is not noisy
9. the index movement has smooth operation and is not jerky
10. the metering pressure is as intended, reset if necessary and, if the meter regulator seal has to be broken, resealed by an OAMI
11. the meter installation has the appropriate ventilation
12. any meter or electrical apparatus is suitable for use in any hazardous area associated with the meter
13. the electrical separation is in satisfactory
14. the protective bonding conductor is present
15. there are no diagnostic indicators displayed
16. any electronic display is operable
17. the installation has the appropriate notices affixed or any defects or non-conformities found have been rectified or made safe and reported to the appropriate body.

Battery replacement

Any replacement batteries shall meet the meter manufacturer's specification. Batteries shall be replaced in accordance with the meter manufacturer's instructions.

Low pressure-fed gas meter installations

Low pressure-fed NG/LPG (Propane) gas meter installations are supplied with gas at a pressure of up to 75mbar. This pressure needs to be reduced to the domestic operating pressure of 21mbar (NG) and 37mbar (Propane) by a regulator fitted upstream of the meter.

This regulator protects the meter, its associated controls and the internal downstream installation against over-pressurisation (see **Part 7 – Checking and/or setting regulators** and **Part 12 – Checking of appliance gas safety devices/controls** for further guidance).

Attention: Meter regulators for low pressure-fed NG installations are set at the factory to deliver an operating pressure of 21mbar at an inlet pressure of 30mbar and a flow rate equivalent to 3.5m^3/hr of gas.

However, at low through-puts (typically less than 0.5m^3/hr – U6/G4/E6 meter) and high through-puts (typically greater than 3.5m^3/hr – U6/G4/E6 meter), the outlet pressure provided by the meter regulator will be in the range of 19 to 23mbar

When you check meter regulator settings on site, it is important you obtain sufficient flow rate.

You can do this by operating an appliance – or you can use a suitable industry meter regulator check device to provide a minimum flow rate of 0.5m^3/hr (see **Industry meter regulator check device** in this Part).

Pressure absorption

It should be borne in mind when checking the meter supply pressure that conditions upstream of the meter, i.e. in the distribution main/service, can in certain areas and at certain times of the day/night restrict the pressure available at the meter.

Figure 16.11 Typical medium pressure-fed gas meter installation (NG)

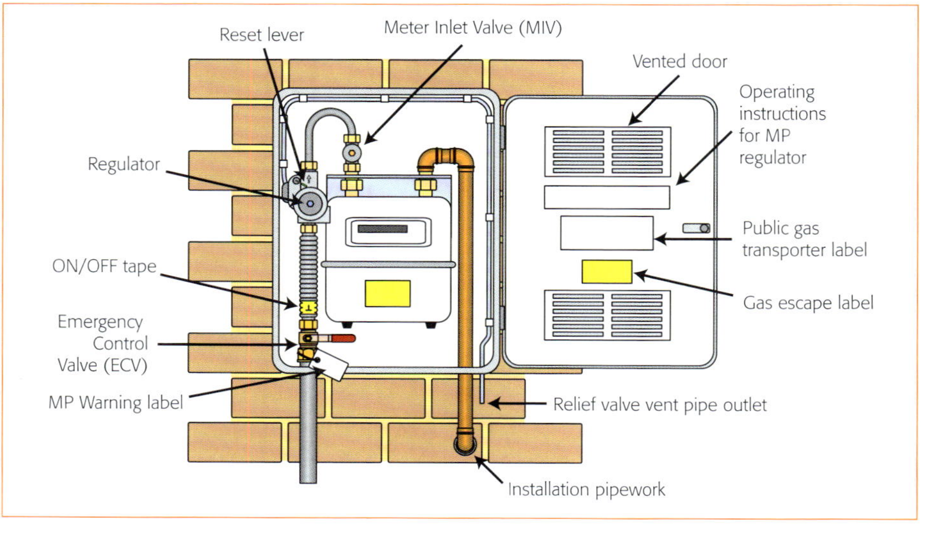

BS 6400-1 specifies that the pressure absorption (i.e. pressure loss) across a NG meter installation when it is at full capacity shall not exceed 4mbar.

Therefore, should the supply conditions ever drop as low as 19mbar on a meter installation that operates at full capacity, it may be possible that the operating pressure at the outlet of the meter is restricted to 15mbar.

Medium pressure-fed gas meter installations

Medium pressure-fed NG/Propane gas meter installations are supplied with gas at a pressure not exceeding 2bar. This pressure needs to be reduced to an operating pressure of 21mbar (NG) and 37mbar (Propane) by a 'medium to low pressure' regulator fitted upstream of the meter.

This regulator protects the meter, its associated controls and the internal downstream installation against over-pressurisation.

These special regulators that operate at inlet pressures normally in excess of 75mbar must have a pressure relief valve (PRV) and a vent terminating to a safe position direct to outside air.

1. Do not install medium pressure-fed gas meter installations inside a dwelling or building (see Figure 16.11 for details of a typical (NG) control and fittings arrangement).
2. You must also site a 'Warning Notice' adjacent to the meter, to provide important information about medium pressure-fed gas meter installations, including:
    - a warning that the installation is supplied with gas at a pressure of up to 2bar
    - clear instructions on how to re-set the safety trip for the pressure control and safety device
    - identification that there is an excess flow valve fitted in the gas service pipe upstream of the meter installation, where this applies
    - clearly indicating the ECV, with instructions on how to operate it SLOWLY when turning on the gas supply
    - a label on the meter inlet valve (MIV)
    - a label on built-in and surface-mounted meter boxes, stating that the outlet pipework must not exit via the rear of the box

Follow the general installation guidance for **Primary meters** in this Part.

Note: See Part 6 – 'Testing for gas tightness (Natural gas)' for further guidance.

Attention: Meter regulators for medium pressure-fed NG/Propane gas installations are set at the factory to deliver an operating pressure of 21mbar (NG) and 37mbar (Propane gas).

When you check meter regulator settings on site, it is important that you obtain a sufficient flow rate.

Figure 16.12 Typical LPG meter installation

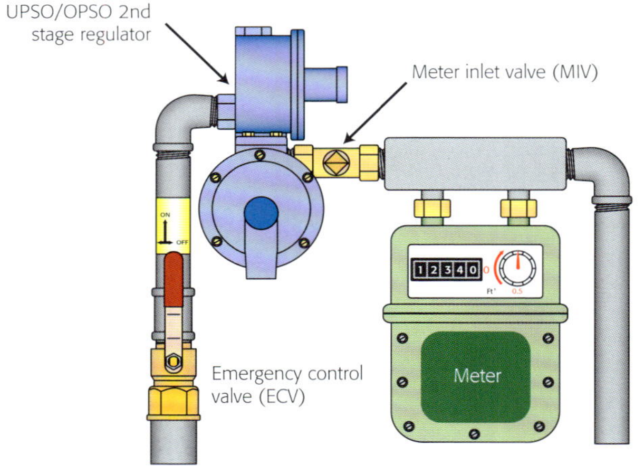

LPG gas meter installations (Propane) – additional guidance

You normally install an LPG meter in an external meter box and it will generally be supplied from a bulk storage vessel. Pressure reduction from the storage vessel to the meter will be in two stages:

1. First stage regulation should be at the bulk storage vessel reducing the pressure to 75mbar.
2. Second stage regulation can be either at the bulk storage vessel (downstream of the 1st stage regulator), or be at the inlet to a meter, reducing the pressure to an operating pressure of 37mbar for Propane (Figure 16.12 shows a typical LPG gas meter installation).

The second stage regulator must include protection against under/over pressure and will normally be in the form of an under-pressure shut off/over-pressure shut off (UPSO/OPSO) valve. See **Part 18 – LPG – General requirements – Operation and checking of gas safety devices and controls (LPG)** for further guidance.

The UPSO/OPSO valve must include manual re-set devices and a PRV vented direct to the outside air.

Follow the general installation guidance for **Primary meters** in this Part, but you must ensure that all fittings, controls, union washers and any joining material are compatible for use with LPG.

Leisure Accommodation Vehicles (LAVs) and Residential Park Homes (RPHs)

Gas supplies to LAVs (caravan holiday homes) and RPHs (residential park homes) may be by either NG or LPG and may be supplied by one of the following methods:

1. Via a single primary meter or bulk storage vessel usually sited on the park boundary, or
2. A primary meter/bulk storage vessel with secondary meters to each home, or
3. Individual primary meters to each home.

For meter installations, follow the guidance given in **Primary meters** or **Secondary meters** in this Part. You may also find the following guidelines useful:

- fit a primary meter serving the whole park in a permanent, sturdy, ventilated, enclosure or box, accessible by a suitably marked key (in the possession of a responsible person for the park)
- a permanent line diagram must be prominently displayed on or near to the primary meter detailing the configuration of all secondary meters, installation pipework and ECVs.

  If you change the configuration of any meter, installation pipework or ECV, you must amend the diagram to reflect this change
- a bulk storage vessel that supplies secondary meters must have a line diagram, as specified above
- fit individual primary or secondary meters in a permanent, sturdy, ventilated box adjacent to the home, accessible by a suitably marked key (in the possession of the gas user for the home)

Note: For further guidance on the installation of gas pipework from the meter position to the home, refer to the guidelines in CORGI*directs* Gas Installer Manual Series 'LPG – Domestic, Including Permanent Dwellings, Leisure Accommodation Vehicles, Residential Park Homes and Boats' (see Part 22 – 'CORGI*direct* Publications').

Permanently moored boats

Gas can be supplied to permanently moored boats in a similar method as for permanently sited LAVs and RPHs, via primary or secondary meters.

Ensure that:

1. The meter is located on the bank above normal flooding level, or
2. If you fit individual secondary meters, they are capable of isolation from a position which is unaffected by flood conditions i.e. a floating pontoon or the bank.

For further information on the gas installation pipework from the meter position to the boat, see CORGI*directs* 'LPG – Domestic, Including Permanent Dwellings, Leisure Accommodation Vehicles, Residential Park Homes and Boats'.

Landlord's Gas Safety Record – 17

17 – Landlord's Gas Safety Record

Figures

Responsibilities of landlords/managing agents – and premises involved

The GSIUR place duties on landlords – and their managing agents – in 'relevant premises' in both the private and public sectors, to ensure that all gas installation pipework, gas appliances and chimney systems owned by them are maintained in a safe condition.

This duty was placed on landlords/managing agents to protect tenants from suffering due to faulty and poorly maintained gas appliances.

'Relevant premises' include:

- residential premises provided for rent by local authorities, housing associations, private sector landlords, housing co-operatives, hostels, colleges and universities
- rooms let in bed-sit accommodation, private households including bed and breakfast accommodation
- rented holiday accommodation such as chalets, cottages, flats, caravans and narrow boats on inland waterways

Note: The list is quite extensive. Landlords may need to seek further advice to ensure that they comply with the Regulations.

Which appliances must a landlord maintain safely?

- any chimney system which serves any 'relevant gas fitting' – which the tenant would not be entitled to remove from the premises, once they vacate

Note: Pipework and appliances are defined as relevant gas fittings.

- appliances e.g. central heating boilers, where the appliance is located remotely from residential areas but provides heat to the residential areas of the property (this applies even though the tenant may not have access to the appliance or chimney system)
- areas such as bars, restaurants or sitting rooms in a hotel – if these are served by, for example, a gas fire for the residents' benefit

Are any appliances not a landlord's responsibility?

- appliances owned by the tenant – though the landlord or managing agent still have a duty of care regarding maintaining the chimney system safely

What is this landlord's/managing agents' 'duty of care'?

Under Section 3 of the Health and Safety at Work etc. Act (HSWA), their duty of care is to maintain a chimney system serving a gas appliance owned by the tenant.

The landlord's obligation is to ensure that any chimney system that serves a tenant's own appliance is maintained to effectively remove POC safely.

This might mean implementing a programme of routine maintenance.

Note: Gas appliances left in the property by the out-going tenant (e.g. gas fires, cookers, cabinet heaters, etc.), after the tenancy agreement has finished, become 'relevant gas fittings' as defined by the GSIUR, if left for a new tenant to use. They will then need to be maintained and safety checked.

Under the GSIUR, what must landlords/managing agents do?

1. They must ensure that installation pipework, appliances and, where applicable, their chimney systems, are checked for safety – at intervals of not more than 12 months.

Note: A newly installed appliance does not need a Landlord's Gas Safety Record for the first 12 months following installation.

However, it would be deemed good practice to include the appliance on any record you issue during the first 12 months of installation, especially in the case of RPHs, LAVs and boats, which are hired out in the course of a business e.g. holiday homes.

2. As well as keeping written records of these checks for two years, they must:
   - provide a copy of the record to the tenant within 28 days of the check taking place
   - ensure they provide a copy of the last current Landlord's Gas Safety Record to a new tenant – before the tenant occupies those premises

Note: Where premises are for 'short let' (i.e. less than 28 days) they may instead prominently display a copy of the current record form within the premises e.g. holiday homes.

Be aware of your responsibilities as a gas operative

If you have reason to suspect that an appliance/installation is not safe to use (e.g. 'ID' or 'AR', follow the current GIUSP. See **Part 8 – Gas Industry Unsafe Situations Procedure** and **Part 10 – Emergency notices, warning labels and forms**) for further guidance.

In addition, you should have sufficient equipment to undertake all necessary tests (e.g. smoke pellets, smoke matches, non-corrosive LDF, ECGA, etc.).

Information that is required on record

When a landlord/managing agent instructs you to carry out a Landlord's Gas Safety Record check, you need to produce legible documentation (a record) of the gas safety check you've undertaken - including any associated paperwork such as a Warning Notice, where faulty appliance/installation is encountered.

Failure to do so will prevent the landlord/managing agent from proving compliance with the GSIUR

The check must include items 1 to 9 as follows

1. The date you checked the pipework, appliance or chimney.
2. The address of the premises at which the pipework, appliance or chimney is installed.
3. The name and address of the landlord (or, where appropriate, managing agent) of the premises at which the pipework, appliance or chimney is installed.
4. Location details and a description of each appliance or chimney you check.
5. Any defect you identify.
6. Any remedial action you took.

7. Confirmation that the check undertaken complies with the requirements of the GSIUR i.e. "Where a person performs work on a gas appliance, he shall immediately thereafter examine:
   a) The effectiveness of any flue.
   b) The supply of combustion air.
   c) Its operating pressure or heat input or, where necessary, both.
   d) Its operation so as to ensure its safe functioning".
8. The name and signature of the individual carrying out the check. See also **Electronic records and signatures – what you need to know** in this Part
9. The registration number of the individual, or their employer.

CORGI*directs* 'Landlord/Home Owner Gas Safety Record' Form (Order Ref: CP12, see Figure 17.1 overleaf) meets the requirements of the GSIUR.

Figure 17.1 has also been populated with dummy information to illustrate how the form should be completed – note that for the boiler 'NA' has been entered in to the combustion analyser fields; manufacturer's instructions state that combustion analysis is not required as it is set by the manufacturer. Always be guided by the appliance manufacturer.

CORGI*direct* provides these in 'no carbon required' (NCR) paper pads of 50 forms, with three copies:

- WHITE copy – Landlord/Managing Agent/Home Owner
- GREEN copy – Tenant
- YELLOW copy – Registered Business

The cover to the pad incorporates notes for satisfactory completion of the form.

The forms are serial numbered so that they can be easily traced and filed.

How to complete a Landlord/Home Owner Gas Safety Record (CP12)

It is recommended that you complete this form as you progress through the safety checks.

Complete it fully and accurately, ensuring that the response is relevant to the question.

The safety check needs to include, but should not be limited to, an examination of the matters referred to in the GSIUR.

1. Record all details of the check accurately.
2. If there are more than four appliances at an address, use an additional form.
3. Use a new form for each tenant and postal address (where applicable).
4. If you identify faults and remedial work is required but has not been carried out, still complete the record and present it to the home owner or landlord/managing agent, with a copy to the tenant (where applicable).

Note: For rented accommodation, the landlord/managing agent is responsible for having the remedial work completed.

5. If you have rectified faults identified, complete the section entitled Remedial Action Taken in the appropriate appliance line.

Registered business (trading title) details:

- the registration number, name of the registered company, the registered gas operative carrying out the safety check and the individual's licence number (found on your ID card), the address and telephone number are also required

Note: If work is subcontracted by a registered business to another registered Installer business, then do ensure the record shows the registration number of the Installer business who actually carried out the work.

Figure 17.1 Landlord/Home Owner Gas Safety Record (CP12)

To confirm the validity of the gas operative please contact Gas Safe Register on Tel: 0800 408 5500

Serial No. B 123456

LANDLORD/HOME OWNER GAS SAFETY RECORD

This form allows the recording of the results of the required checks as defined by the Gas Safety (Installation and Use) Regulations.
The information recorded on this form does not confirm that the installation was installed by a Registered Installer or that the installation complies with any relevant Building Regulations.
Chimney systems were inspected visually and checked for satisfactory evacuation of products of combustion, a detailed internal inspection of the chimney system has not been carried out.

Registered Business Details REG NO 123456
Gas operative Jo Bloggs (Print name)
Operative licence No. 987654
Company Bloggs & Son
Address 123 Average Avenue
Average Town, Averagington
Postcode AV12 3ER Tel No. 01234 567890

Job Address
Name (Mr/Mrs/Miss/Ms)
Address 16 The Street
Average Town
Averagington
Postcode AV2 4TT
Tel No. 01234 000200

Landlord (or where appropriate their agent)
Name (Mr/Mrs/Miss/Ms) Mr. Nice
Address 22a The Towers
Average Town
Averagington
Postcode AV0 8IT Tel No. 01234 000200
Number of appliances tested 2

APPLIANCE DETAILS

	Location	Appliance type	Make	Model	Landlord's appliance Yes/No/NA	Appliance inspected Yes/No	Flue type OF/RS/FL
1	Kitchen	Cooker	Leisure	LGV65	Yes	Yes	FL
2	Utility Room	Boiler	Baxi	Duo-tec 28 HE	Yes	Yes	RS
3							
4							

INSPECTION DETAILS / **AUDIBLE CO ALARM**

	Operating pressure in mbar or heat input in kW	Initial combustion analyser reading (if applicable)	Final combustion analyser reading (if applicable)	Safety device(s) correct operation Yes/No/NA	Ventilation provision satisfactory Yes/No	Visual condition of chimney/termination satisfactory Yes/No/NA	Flue performance checks Pass/Fail/NA	Appliance serviced Yes/No	**Appliance safe to use Yes/No**	Approved CO alarm fitted Yes/No/NA	Is CO alarm in date Yes/No/NA	Testing of CO alarm satisfactory Yes/No/NA
1	12.7 kW	0.005	0.005	Yes	Yes	NA	Pass	No	Yes	NA	NA	NA
2	24.7 kW	NA	NA	Yes	Yes	Yes	Pass	No	Yes	NA	NA	NA
3												
4												

DEFECT(S) IDENTIFIED

	Defect(s) identified	If Warning/Advice Notice issued insert serial No*
1		
2		
3		
4		

REMEDIAL ACTION TAKEN

	Remedial action taken
1	
2	
3	
4	

Gas installation pipework satisfactory visual inspection Yes/No	Yes
Emergency Control Valve (ECV) accessible Yes/No	Yes
Satisfactory gas tightness test Yes/No/NA	Yes
Protective equipotential bonding satisfactory Yes/No	Yes

NEXT SAFETY CHECK DUE WITHIN 12 MONTHS

This Safety Record issued by: Signed
Print Name: Jo Bloggs
Received by: Signed Tenant/Landlord/Agent/Home Owner
Date appliance(s)/chimney(s) checked: 1/3/2012

Key: **Top Copy** – Landlord/Managing Agent/Home Owner **Green Copy** – Tenant **Yellow Copy** – Registered Business
* Refer to separate Warning/Advice Notice
To re-order quote Ref. CP12

- the form must be signed and dated by the registered gas operative who carried out the safety check

Note: If two or more operatives are safety checking the same installation i.e. large property with multiple appliances, then either one operative takes overall responsibility for the safety checks and therefore signs and dates the landlord record form(s), or each operative involved issues their own landlord record check form containing their signature and date for their part of the safety check.

Tenant/Home Owner (job address) details:

- enter here the name, address and telephone number of the property where the inspection is being carried out

Landlord/agent details:

- enter here the name, address and telephone number details of the landlord, or where appropriate, their managing agent
- **Number of appliances tested** – enter in the box the total number of appliances tested

Appliance details:

- **Location** – room/area in the property in which the appliance is installed
- **Appliance type** – boiler, cooker, water heater, etc.
- **Make** – the manufacturer's name
- **Model** – the name of the appliance, e.g. Classic 240ff
- **Landlord's appliance** – indicate YES, NO or N/A (Not Applicable)
- **Appliance inspected** – carry out an inspection on all appliances owned by the landlord or home owner. Indicate YES or NO

  Record any appliance(s) not owned by the landlord (i.e. tenant's own) in the APPLIANCE DETAILS section; enter NO in the 'Inspected' box. Draw a line through the correspondingly numbered INSPECTION DETAILS line (each line is numbered 1-4 in both the APPLIANCE and INSPECTION DETAILS fields)

Note: Record any details of faults found on tenants' own appliances (those not owned by the landlord/managing agent) in the DETAILS OF ANY FAULTS section.

Take the appropriate action i.e. issue a Warning Label and Warning notice, advise the tenant and landlord of the issues and with permission of the tenant/landlord, as appropriate, either turn off the offending appliance or isolate it from the gas supply (see the current GIUSP and Part 10 – 'Emergency notices, warning labels and forms').

- **Flue type** – open-flued (OF), room-sealed (RS) or flueless (FL)

Inspection details:

- **Operating pressure in mbars or heat input in kW** – the appliance operating pressure (mbar) or the heat input rating (kW)
- **Initial combustion analyser reading and Final combustion analyser reading** – where applicable (refer to manufacturer's instructions), using a portable electronic combustion gas analyser (ECGA) – often refered to as a flue gas analyser (FGA) – measure and record details of the combustion performance readings.

  The combustion performance test should be carried out paying due regard to appliance manufacturer's instructions and/or where manufacturer's instructions are not available, in accordance with BS 7967.

Initial combustion analyser reading – this applies as follows:

1. An ECGA can be used as a diagnostic instrument to determine the need for remedial work e.g. a service or maintenance work. Enter the "as found" reading.
2. Where a decision is made to service an appliance the initial combustion analyser reading can be used to determine whether or not a full strip down service is undertaken.

Final combustion analyser reading – following a service in accordance with manufacturer's instructions and/or BS 7967, the final combustion analyser reading can be used to determine whether the appliance is safe for continued use.

- **Safety device(s) correct operation** – correct operation of any FSD, ASDs, etc. Indicate YES, NO or N/A, as appropriate

Note: Test flame supervision devices (FSDs) to ensure they shut off the gas supply positively within the maximum fail safe period. Test atmosphere sensing devices (ASDs) and spillage monitoring devices (TTBs) in accordance with appliance manufacturer's installation/maintenance instructions
(see Part 12 – 'Checking of appliance gas safety devices/controls – Flame protection controls: stop gas flow' for further guidance).

- **Ventilation provision satisfactory** – indicate YES or NO

Note: If purpose provided ventilation is not required, it will be considered as satisfactory, so enter Yes in the box.

- **Visual condition of chimney and termination satisfactory** – carry out a visual inspection of the accessible areas of the chimney system. It is very important that you inspect chimneys fitted in, or running through, loft areas, voids (see Note 1), etc. (where applicable). Indicate Pass, Fail or N/A, as appropriate

Note 1: Chimney systems (flues) in voids is of particular concern and are covered in detail in Part 13 - 'Chimney standards – Room-sealed fanned-draught chimney systems concealed within voids'.

Note 2: Remove gas fires and inspect catchment spaces. Other appliances that are connected to unlined brick/masonry chimney systems require the catchment space to be checked where there is an inspection opening (see Part 14 – 'Flue testing – Flue checks – for open-flued appliances' for further guidance).

- **Flue performance checks** – Flueless appliance – N/A. For Open-flued gas appliances, you must carry out both a flue flow check and an appliance spillage check – Pass or Fail. Room-sealed gas appliances: check condition of case seals etc – Pass or Fail

Note: In the absence of the appliance manufacturer's installation instructions, see Part 14 – 'Flue testing – Flue checks – for open-flued appliances' for flue flow and spillage testing procedures.

- **Appliance serviced** – indicate YES or NO

Note: A service is a separate activity to a Landlord Gas Safety Record. Where a service is identified or desired, this should be agreed between the gas operative and the customer.

- **Appliance safe to use** – indicate YES or NO

  Where 'No' is indicated, complete 'Defects identified' field and where appropriate, complete a Warning Notice, entering its serial number in the space provided.

Audible CO alarm

The following checks with regards to CO alarms are not prescribed by the GSIUR, but are regarded as good practice offering another level of safety to gas users.

- **Approved CO alarm fitted** – indicate YES, NO or N/A, as appropriate. The CO alarm, where fitted, should comply with BS EN 50291 and carry a CE mark. CO alarms will also be marked with either a BSi Kitemark or LPCB logo to show independent testing and certification.
- **Is CO alarm in date** – indicate YES, NO or N/A, as appropriate. Check that the CO alarm is within its service life; typical life span ranges from 5 to 10 years. The service life will be indicated within the CO alarm instructions and typically on the product itself.
- **Testing of CO alarm satisfactory** – indicate YES, NO or N/A, as appropriate. Confirm correct positioning and operation of the CO alarm by using the test button function incorporated within the CO alarm.

Note: Should an approved CO alarm not be present and/or checks/tests show the installed CO alarm to be faulty and/or out of date, advise the Landlord and Tenant to have an approved CO alarm installed, or where due to a fault, replaced as necessary.

Defects identified and remedial action taken

- **Details of any faults/remedial action taken** – note details of any faults and rectification work for the appliances tested and enter these in the corresponding line in these sections (e.g. if you find a fault on the No 3 appliance, note details in the No 3 Row)
- **Label and warning notice issued** – any unsafe gas appliances/installations identified should have a Warning Notice completed and Warning Label affixed, in accordance with the GSIUR and the current GIUSP. CORGI*directs* Warning Notice's are serial numbered and this number should be entered in to the field.

Gas pipework and controls

- **Gas installation pipework satisfactory visual inspection** – carry out a visual inspection of exposed areas of pipework. A gas tightness test is also recommended. Details should be noted. State YES or NO
- **Emergency Control accessible** – state YES or NO
- **Satisfactory gas tightness test** – where undertaken, state YES, NO or N/A, as appropriate
- **Protective equipotential bonding satisfactory** – look for its presence, mechanical connection to pipework and correct sizing (where unsure, advice tenant/landlord to have the PEB checked by a suitably competent person). State YES or NO

Sign-off

- **This safety record is issued by** – the gas operative carrying out the safety check needs to sign and print their name and date the record.

  Wherever possible, obtain a signature from the responsible person, e.g. landlord/managing agent, tenant, home owner, etc.

Specific checks for a Landlord's Gas Safety Record – the essentials

We give some details here to help on specific checks on appliances/installations – but the list is not exhaustive. Do use your own competencies to carry out a full and comprehensive inspection.

The range of appliances listed just covers the most common ones. You may find others.

You must carry out a safety inspection on every relevant appliance/installation – without exception. This must include:

- checking the satisfactory operation of all flame supervision devices
- checking the flame picture
- checking the burner pressure and/or the rated heat input against manufacturer's data
- checking physical stability and the presence and effectiveness of any stability devices that are fitted (where appropriate)
- checking that there is a satisfactory chimney system. Check the chimney throughout its length, (including inspection of catchment space, loft/roof areas, voids etc. where appropriate)
- conducting flue tests to ensure removal of the POC (flue flow and spillage tests)
- checking there is satisfactory provision of all necessary ventilation, including compartment ventilation
- identifying any appliance or installation defects and taking the appropriate actions
- investigating any evidence of unsafe operation and reporting this to the responsible person(s)

Gas meter and installation pipework – the landlord's duty

Although not specifically required to check gas meter/installation pipework, a landlord does have a duty to maintain the gas installation pipework in a safe condition.

So take the opportunity to carry out some basic checks as follows:

1. Carry out a visual inspection of exposed areas of pipework.
2. It is recommended that you carry out a gas tightness test and note the details on the record form.
3. Bring any concerns you identify – either with the gas meter or its ancillary controls – to the attention of the Gas Transporter.
4. Inspect the location and condition of the gas meter/ancillary controls, and the installation pipework.

Checklist:

- is the meter correctly sited and accessible?
- are the meter inlet and outlet connections in a satisfactory condition?
- is the gas ECV accessible to all users and suitably labelled? (see **Part 10 – Emergency notices, warning labels and forms**)
- is the lever attached to the ECV and can it be easily operated in the correct direction (moving downwards one quarter of a turn to turn the gas supply off)?
- is there evidence that the installation pipework has appropriate PEB?
- is the installation pipework of adequate size?
- is the installation pipework adequately protected against corrosion and sufficiently supported?

Restricted locations – what you can or cannot do

Basements and cellars

- do not locate an appliance fitted with an automatic means of ignition for use with LPG in a room, or internal space below ground level, such as a basement or cellar
- such appliances can be installed into rooms which are basements with respect to one side of the building – if they are open to ground level on the opposite side

Bath or shower rooms

- you must not install open-flued or flueless appliances in a room or internal space containing a bath or shower (this includes any cupboard/compartment or space (e.g. cubicle), which has an air path or connecting door opening into the bath or shower room)
- you may only install room-sealed appliances in such spaces. But only consider this location if there is no alternative – and then only if allowed by the appliance manufacturer.

Covered passageways/ginnells

- do not site natural-draught/fanned-draught room-sealed and open-flued fanned-draught flue terminals within a covered passageway/ginnell between properties e.g. terraced properties.
- in the case of room-sealed appliances, flues terminated in these positions may cause the POC to re-enter the air inlet duct, creating poor combustion at the burner. Flues terminated in these positions will produce combustion products, which could contain CO that could accumulate, entering habitable areas above the passageway (see the current GIUSP for further guidance)

Additional requirements for gas-fired water heaters (all rooms)

All open-flued and flueless instantaneous water heaters, must incorporate a safety device designed to shut down the appliance before there is a dangerous quantity of POC in the room concerned.

This may be in the form of an ASD. This includes any previously used or second-hand open-flued or flueless water heaters.

Bedroom/bed-sitting rooms

Whilst a bedroom/bed-sitting room is not an ideal location for a gas cooker or a gas tumble dryer, you may install them in this location provided that:

- for gas cookers – the room concerned is greater than 20m^3 in volume
- for tumble dryers – the room volume is at least 7m^3/kW(net) of the appliance heat input

Note: Refer also to the particular appliance manufacturer's installation instructions.

For gas appliances installed in a room used or intended to be used as sleeping accommodation then the following applies:

1. Gas appliances of greater than 12.7kW net (14kW gross) rated heat input, installed in a room used or intended to be used as sleeping accommodation must be room-sealed. This includes installations into any cupboard/compartment or space (e.g. cubicle) that has an air path or connecting door opening into the bedroom/bedsitting room.
2. Gas appliances of 12.7kW net (14kW gross) rated heat input or less may be room-sealed, or they must incorporate a safety control designed to shut down the appliance before there is a dangerous quantity of POC in the room concerned. This device should be in the form of an ASD.

Flueless, or non-room-sealed space heating, or water heating appliance over 14kW heat input (gross), or under 14kW heat input (gross) without a built-in atmosphere sensing device, installed in a bedroom or bed-sitting room, under the Gas Industry Unsafe Situations Procedure (GIUSP) is classified as At Risk.

Alterations to rooms in rented properties

Where the sleeping arrangements have been altered as described, landlords have a duty of care under Section 3 of the HSWA (see **Part 1 – Gas safety legislation – The Health and Safety at Work Etc. Act (HSWA) – the wide ranging duties explained**), to take reasonably practical steps to ensure the safety of their tenants.

They could therefore consider:

1. Replacing/removing the appliance(s).
2. Re-housing the tenant into other suitable accommodation.

Any existing appliances that do not comply with the current requirements must be removed from any room where it is proposed to convert that room into sleeping accommodation.

Temporary change of use of a room to sleeping accommodation

A temporary change of use of a room is not recognised under the regulations and therefore, any gas appliance that is found not to comply should either be:

- replaced with an appliance that meets the minimum requirements; or
- classified as 'At Risk' in accordance with the procedure detailed in the current GIUSP and turned off with the gas user's permission until the appliance(s) is either replaced, or until such times as the consumer/tenant is able to stop using the space as sleeping accommodation. Alternatively, consideration should be given to re-housing the tenant into other suitable accommodation.

Note: Owner-occupied properties

Unless the householder believes or has been advised by a registered engineer that the appliance is dangerous (in which case he/she should not be using the appliance), there is no legislation that prevents private householders from using a living room as sleeping accommodation where gas appliances which do not meet the requirements of the regulations have already been installed.

Although there is no legislation in place to prevent private householders from using an appliance in this manner, it should still be classified as 'AR' in accordance with the procedure detailed in the current GIUSP and turned off with the consumer's permission until such time that the appliance(s) is replaced.

The primary intention of the Regulations is to prevent landlords deliberately changing rooms from, for example, lounges into bedrooms:

- if there were gas appliances already installed in them that were not room-sealed; or
- not fitted with safety devices that will shut down the appliance before a dangerous quantity of POC can build up in the room concerned.

Unsafe situations – what you must do in this context

When you undertake any gas work, you may discover an unsafe installation/appliance – for example, gas escapes outside of tightness test tolerances. You have a duty of care under the GSIUR to take the appropriate action in accordance with the current GIUSP (see **Part 8 – Gas Industry Unsafe Situations Procedure** and **Part 10 – Emergency notices, warning labels and forms**).

In addition to recording this on the Landlord Gas Safety Record, you need to:

- issue appropriate documentation – CORGI*direct* produces a suitable Warning Notice (Order Ref: CP14)
- label the appliance/installation – CORGI*direct* produce 'Danger Do Not Use' labels (Order ref: WL1)
- take further appropriate action as necessary (see **Part 8 – Gas Industry Unsafe Situations Procedure** and **Part 10 – Emergency notices, warning labels and forms**)

Landlords cannot delegate any of their duties to their tenants and they need to take all 'reasonable steps' to comply with the requirements of the GSIUR.

When you are denied access to carry out safety checks, you must keep records to show that a landlord has demonstrated their commitment to satisfy the requirements of the GSIUR.

Note: This does not include using force to gain entry to the property. Proper action through the courts may be appropriate.

Gas appliances – checks on each type must comply with the GSIUR

Guidance and checklists to help you

These regulations (Regulation 26(9)) require that when you perform work on a gas appliance including completing a Landlord's Gas Safety Record, you must immediately thereafter examine the following:

- the effectiveness of any flue;
- the supply of combustion air;
- its operating pressure or heat input, or where necessary, both; and
- its operation, so as to ensure safe functioning

Space heating appliances

Here is guidance when checking gas fires, convector heaters, Inset Live Fuel Effect (ILFE) gas fires and Decorative Fuel Effect (DFE) gas appliances supplied with either Natural gas or LPG.

Central heating and/or hot water systems connected to a combined appliance are dealt with under the **Central heating boilers** section in this Part.

Checklist:

- is the location acceptable? (see **Restricted locations – what you can or cannot do** in this Part)
- is the ventilation correct? (see **Part 4 – Ventilation**)
- check the general operation of the appliance
- is the gas user aware of any faults?
- check the operation of the control tap
- check the flame picture
- check the correct positioning and condition of fuel effect components or radiants (advise renewal where appropriate)

- are there any signs of spillage on the appliance or in the area around the appliance?
- if the appliance is wall-mounted, is it positioned at the correct height above any combustible material?
- if the appliance is fitted to a surround, is the surround manufactured from a suitable material and sealed to the wall correctly?
- remove any closure plate and check that the catchment space behind the fire is of adequate size and structurally sound
- where a gas flue box or flue gas collector has been utilised, ensure that it is in good condition and is suitable for the appliance type to which it is connected (see also **Part 13 – Chimney standards – Chimney materials – must conform to relevant British Standard(s) – Gas flue boxes – why and where to install** and **Flue gas collector boxes – when and how to use**)
- visually inspect the chimney system throughout its length and carry out a flue flow test. Is the termination satisfactory? Are there any chimney pot inserts fitted? Are there any signs of smoke escaping from anywhere other than the correct termination point? Is the chimney the correct size for the appliance?
- re-seal any closure plate in position on all four sides using suitable sealing material
- where an appliance has been removed, re-fit and check all joints for gas tightness using non-corrosive LDF
- check burner operating pressure and/or appliance heat input rating
- carry out a spillage test as detailed in the appliance manufacturer's installation instructions
- check the operation of the flame supervision device (where appropriate)
- check the condition of the appliance, e.g. heat exchanger, seals, etc. (where appropriate) (see also **Part 14 – Flue testing**)

Note: For further guidance see CORGI*directs* manual entitled 'Gas Fires and Space Heaters' – from the current Gas Installer Manual Series – Domestic.

Cookers

Be aware that any room containing a cooker needs an opening window, door or other means of ventilation, opening direct to outside air. If purpose-provided ventilation is required, fit this direct to outside air.

When you assess ventilation, consider other appliances fitted in the same room, even if they are solid fuel or oil burning appliances. Refer to **Part 4 – Ventilation**.

Where extract fans or cooker hoods are fitted, you need to check whether their use affects the safe operation of open-flued appliances in the room or adjoining area(s) (see also **Restricted locations – what you can or cannot do** in this Part).

Checklist:

- is the location acceptable? (see **Restricted locations – what you can or cannot do** in this Part)
- is the ventilation correct? (see **Part 4 – Ventilation**)
- check the general operation of the appliance
- is the gas user aware of any faults?
- check the operation of all control taps
- check the flame picture on all burners
- check burner operating pressure and/or heat input rating
- check the operation of flame supervision device(s), where appropriate

- check the operation of the safety shut-off valve (SSOV), where appropriate, i.e. drop down glass lid
- check that the oven's thermostat shuts down to a by-pass rate when hot. Is the flame still stable?
- check the condition of the appliance, e.g. grill frets (see Note), oven door seals, etc.
- check that the flexible gas connection is fitted correctly (looped downwards), is under no stress and is suitable for the gas being used i.e. hoses suitable for LPG are normally identified by a red band
- check that a stability device is fitted and that it is secure (where appropriate)

Note: Research shows that cooker grills have been responsible for a number of CO poisoning incidents. So take extra care when you check the operation of the grill.

- ensure that the grill fret is in good working order and that it is not buckled, out of shape, damaged or out of line with the burner
- remove any lint from the burner injector to ensure a good flame picture

Note: For further guidance see CORGI*directs* manual entitled 'Gas Cookers and Ranges' – from the current Gas Installer Manual Series – Domestic.

Water heaters

This section gives guidance on safety checks on all major types of domestic gas-fired water heating systems including instantaneous water heaters, storage systems and circulators (see also **Restricted locations – what you can or cannot do** in this Part)

Checklist:

- is the location acceptable? (see **Restricted locations – what you can or cannot do** in this Part)
- is the room size adequate? (see Note)

Note: Existing water heaters that are located in rooms intended for sleeping accommodation and which are less than 20m^3 in volume must be room-sealed.

For further guidance see CORGI*directs* manual entitled 'Water Heaters' – from the current Gas Installer Manual Series – Domestic.

- is the room/compartment ventilation correct? (see **Part 4 – Ventilation**). This is very important where an older open-flued appliance is installed in a bathroom or shower room
- check the general operation of the appliance
- is the gas user aware of any faults?
- check the flame picture
- are there any signs of spillage on the appliance or in the surrounding area?
- check the burner operating pressure or heat input rating (see **Part 11 – Checking/setting burner pressures/gas rates**)
- visually inspect the chimney system, including roof spaces, where appropriate
- if the appliance is open-flued, carry out a flue flow test
- if the appliance is open-flued, carry out a spillage test
- if the appliance is room-sealed, is the termination satisfactory? Does it require a terminal guard?
- check the operation of all FSDs
- check the condition of the appliance, e.g. heat exchanger, case seals, etc. (see **Part 14 – Flue testing** for further guidance)

- where necessary, are the appropriate warning labels fitted to warn the gas user about:
  - 5 minute limitation on the use of an instantaneous flueless water heater
  - the need to turn off a large instantaneous open-flued water heater before entering a bath
  - restrictions on the use of a compartment

Central heating boilers

This section gives guidance for checking free standing/wall mounted central heating boilers, BBUs (see Note) and combination boilers (see also **Restricted locations – what you can or cannot do** in this Part).

Note: When checking a combined appliance record as two separate appliances on the landlords gas safety record form.

Checklist:

- is the location acceptable? (see **Restricted locations – what you can or cannot do** in this Part)
- check the room/compartment ventilation is correct (see **Part 4 – Ventilation**)
- check the general operation of the appliance
- is the gas user aware of any faults?
- check the flame picture
- if the appliance is open-flued, are there any signs of spillage?
- visually inspect all visible parts of the chimney system (especially the roof spaces, voids (see **Part 13 – Chimney standards – Room-sealed fanned-draught chimney systems concealed within voids**), where appropriate); is it correctly fitted and installed?
- if the appliance is room-sealed, is the termination satisfactory?
  Does it require a terminal guard (see **Part 13 – Chimney standards – Terminal guards - when to use**)?
- check the burner operating pressure and/or heat input rating
- carry out a combustion analysis (where appropriate)
- carry out a flue flow test (where appropriate)
- carry out a spillage test (where appropriate)
- check the operation of all FSDs
- check the condition of the appliance, e.g. heat exchanger, seals, etc.
  (see also **Part 14 – Flue testing**)

Note: For further guidance, see CORGI*directs* manual entitled 'Central Heating – Wet and Dry' – from the current Gas Installer Manual Series – Domestic.

Guidance on checking both categories of warm air heaters:

1. Natural convection warm air heater.

This type of appliance is not generally suitable for a ducted warm air system. It is designed for location within a centrally-sited convection chamber, which is a purpose-built compartment constructed of brick, block or prefabricated panels. Low level air inlet grilles allow cold air to be drawn into the chamber, where it is warmed by the appliance heat exchanger and then discharged, by natural convection, through grilles or registers fitted at the top of the compartment.

2. Fan assisted warm air heater.

Warm air is fan assisted through a network of ducts throughout the building. Types normally found are up flow, down flow and horizontal (cross) flow.

Either type may be combined with an integral circulator for heating domestic hot water.

Further guidance is given in **Water heaters** in this Part (see also **Restricted locations – what you can or cannot do** in this Part).

More detailed guidance relating to warm air units can be found in CORGI*directs* 'Central Heating – Wet and Dry' manual (see **Part 22 – CORGI*direct* Publications**).

Checklist:

- is the location acceptable?
(See **Restricted locations** in this Part)
- check that the room/compartment ventilation is correct (see **Part 4 – Ventilation**)
- check the general operation of the appliance
- is the gas user aware of any faults?
- check the flame picture
- if the appliance is open-flued; are there any signs of spillage?
- visually inspect all visible parts of the chimney system (especially the loft area, voids (see **Part 13 – Chimney standards – Room-sealed fanned-draught chimney systems concealed within voids**), where appropriate). Is it correctly fitted and installed?
- if the appliance is room-sealed, is the termination satisfactory?
Does it require a terminal guard
(see **Part 13 – Chimney standards – Terminal guards - when to use**)?
- is there a full and unobstructed return air path from all heated rooms – except the kitchen, bathroom, shower room and toilet?

  On open-flued appliances, it is essential to ensure that the return air arrangements avoid interference with the operation of the flue by the air circulation fan.

  Where you install a common return air duct in a collection area (e.g. the hall), the return air grille from the collection area needs to be positively connected – by ducting directly to the return air inlet on the appliance

Note: Classify existing open-flued warm air heater installations installed in compartments, which have no positive return air connection as 'AR' in accordance with the current GIUSP.

Where you find a domestic open-flued warm air heater with fanned warm air circulation, with no provision for a positive return air arrangement, then consult the appliance manufacturer for the particular requirements for each appliance type.

- where the heater is mounted on a plenum box (connection between the heater and the ductwork system), check that the seal/gasket is fitted between the heater and plenum box and that it is in good condition
- check the burner operating pressure and/or heat input rating
- if the appliance is open-flued, carry out a flue flow test
- carry out a spillage test (where applicable) paying particular attention to other open-flued appliances in the same or adjoining room, to ensure that any fan does not affect the operation of their flue(s)
- check the operation of all FSDs
- check the condition of the appliance, e.g. heat exchanger, seals, etc.
(see also **Part 14 – Flue testing**)

Note: For further guidance see CORGI*directs* manual entitled 'Central Heating – Wet and Dry' – from the current Gas Installer Manual Series – Domestic (see Part 22 – CORGI*direct* Publications).

Leisure accommodation vehicles and boats

This section gives guidance for carrying out safety checks on LAVs, including caravan holiday homes (see **Part 19 – Definitions**) and boats supplied with LPG from bulk storage vessels or cylinders.

- if the accommodation is supplied with NG, carry out the safety checks previously described for individual appliances

The GSIUR require that in respect of any tenant whose right to occupy the accommodation is for a period not exceeding 28 days, then a copy of the record may instead be displayed in those premises.

Leisure accommodation vehicles and boats are often hired out as holiday accommodation.

- as they are normally for short term tenancy (occupancy less than 28 days), a copy of a Landlord's Gas Safety Record may be displayed within the accommodation
- tenants occupying premises exceeding 28 days must be provided with a copy of the relevant record

A specially designed leisure industry gas safety record form has been developed for the leisure industry (see Figure 17.2).

Checklist

Ensure that:

1. The cylinder/final connection to the LAV/boat is satisfactory.
2. You have visually inspected the gas installation pipework.
3. The gas installation is checked to ensure gas tightness.
4. The flexible connection from the second stage regulator to the LAV/boat inlet point is in good condition, with no significant cracks, is properly secured, etc.
5. The regulator is operating correctly.
6. If fitted, the UPSO/OPSO (see **Part 19 – Definitions**) is operating correctly.
7. A sufficient permanent supply of air for the safe combustion of gas is available in the room/space where the appliance(s) is located, through suitably sized and located ventilators.
8. Gas dispersal holes (commonly known as 'drop holes') are suitably sized and located, where appropriate.
9. The LPG operating pressure and/or heat input of all appliances is correct.
10. You inspect chimney systems serving appliances to ensure they:
    - are constructed of suitable materials
    - are suitably sized
    - are suitably located/routed
    - are in good condition
    - pass a satisfactory flue flow test
11. Where appropriate, the appliance passes a satisfactory spillage test.

Note: See Part 14 – 'Flue testing' and the manual entitled 'LPG – Domestic Including Permanent Dwellings, Leisure Accommodation Vehicles, Residential Park Homes and Boats', from CORGI*directs* Gas Installer Manual Series – Domestic for further guidance (see Part 22 – CORGI*direct* Publications).

12. You carry out a visual check of the 'flame picture' (i.e. appearance) to ensure that it is burning safely, or take a measurement of the combustion performance of the appliance, using a suitable ECGA where appropriate (see also **Part 3 – Characteristics of combustion – Electronic portable combustion gas analysers**).

Figure 17.2 Leisure Industry landlord's Gas Safety Record (CP2)

To confirm the validity of the gas operative please contact the gas registration body.

Serial No. 123456

LEISURE INDUSTRY LANDLORD'S GAS SAFETY RECORD

NEXT SAFETY CHECK DUE WITHIN 12 MONTHS

GAS safe REGISTER

This inspection is for gas safety purposes only in accordance with The Gas Safety (Installation and Use) Regulations. Flues were inspected visually and checked for satisfactory evacuation of products of combustion. A detailed internal inspection of the flue integrity, construction and lining has not been carried out.

Registered Business Details REG NO

Gas operative ______ (Print name)

Operative licence No. ______

Company ______

Address ______

Postcode ______ Tel No. ______

Job Address

LAV serial No./Boat Name ______

Pitch/Mooring Number ______

Park/Marina ______

Address ______

Postcode ______ Tel No. ______

Landlord (or where appropriate their agent)

Name (Mr/Mrs/Miss/Ms) ______

Address ______

Postcode ______ Tel No. ______

No. of appliances tested ______

APPLIANCE DETAILS

	Location	Appliance type	Make	Model	Landlord's appliance Yes/No/NA	Appliance inspected Yes/No	Flue type OF/RS/FL
1							
2							
3							
4							

INSPECTION DETAILS / **AUDIBLE CO ALARM**

	Operating pressure in mbar or heat input in kW	Initial combustion analyser reading (if applicable)	Final combustion analyser reading (if applicable)	Safety device(s) correct operation Yes/No/NA	Ventilation provision satisfactory Yes/No	Visual condition of chimney/termination satisfactory Yes/No/NA	Flue performance checks Pass/Fail/NA	Appliance serviced Yes/No	**Appliance safe to use Yes/No**	Approved CO alarm fitted Yes/No/NA	Is CO alarm in date Yes/No/NA	Testing of CO alarm satisfactory Yes/No/NA
1												
2												
3												
4												

DEFECT(S) IDENTIFIED		If Warning/Advice Notice issued insert serial No.*
1		
2		
3		
4		

REMEDIAL ACTION TAKEN	
1	
2	
3	
4	

Cylinder/final connection hoses to LAV/boat satisfactory Yes/No ☐

Gas installation pipework (visual inspection) satisfactory Yes/No ☐

Gas tightness test satisfactory Yes/No/NA ☐

ECV accessible and operable Yes/No ☐

LPG regulator operating pressure (mbar) ☐

LPG regulator lock-up pressure (mbar) ☐

Safety Record issued by: ______

Print Name: ______

Received by: ______ Tenant/Landlord/Agent

Date appliance(s)/flue(s) checked: ______

Key: **Top Copy** – Landlord/Managing Agent **Green Copy** – Tenant **Yellow Copy** – Registered Business * Refer to separate Warning/Advice Notice To re-order quote Ref. **CP2**

13. The appliance and associated gas fittings operate safely.
14. The gas tightness of the appliances is checked.
15. The appliances are physically stable, securely fitted and correctly connected to other fittings.
16. You record any faults found and take remedial actions.

On completion of the checks:

Classify any faults found on the gas installation as described in the current GIUSP, label and document accordingly (see also **Part 8 – Gas Industry Unsafe Situations Procedure and Part 10 – Emergency notices, warning labels and forms** for further guidance).

Following a statutory annual safety check:

Complete and affix a Leisure Industry Gas Safety Record form in a suitable position, i.e. inside a cupboard door.

Safety check or service?

It is important to understand that the duties placed on landlords in the GSIUR require a safety inspection to be carried out. The checklists in this Part should satisfy the requirements.

When a service is required, the specific operations will normally be detailed in the appliance manufacturer's installation instructions, but they will not necessarily define the operations required to service the appliance.

In the absence of manufacturer's servicing instructions, general gas appliance servicing details on individual appliances can be found in the appropriate manuals from CORGI*directs* Gas Installer Manual Series – Domestic.

Painted appliance cases – things to consider

You will occasionally find gas appliance cases that have been re-painted, in an effort to give the appliance a new 'lease of life'.

It is important you remember that appliance cases can be both functional and decorative. As they are normally close to the heat produced by the appliance itself, the type of paint and the way they are re-painted can be critical.

The original type of paint or finish used would have been part of the design of the appliance and the appropriate type of covering required specified by the manufacturer.

If a gas user wants to paint an appliance casing, then they/you need to seek specialist advice from the appliance manufacturer as to whether this is acceptable.

Note: Gas Safe Register have a Technical Bulletin on painted cases – TB 058 Painted appliance cases – which can be accessed at https://engineers.gassaferegister.co.uk

If so, what would be correct products to use – for example, a suitable substance to remove the original coating and appropriate coverings to use when the case is re-painted. The appliance manufacturer's advice should be followed at all times.

Before the case is re-painted, it should be removed from the appliance. This will normally involve the assistance of a registered gas engineer. This is especially true if the appliance is a room-sealed appliance and the case forms part of the room-sealed arrangement.

Note: The appliance should not be used until the finish has had sufficient time to dry; the appliance or the paint manufacturer will give advice on suitable drying times.

If you find an appliance that has been painted with a product not approved by the manufacturer, if there are signs of distress apparent, caused by the re-painting of the appliance case:

- regard the appliance as 'AR'
- follow the current GIUSP (see also **Part 8 – Gas Industry Unsafe Situations Procedure** and **Part 10 – Emergency notices, warning labels and forms** for further guidance)

Void properties – the checks required before re-let

When a tenant vacates a property, landlords need to ensure that gas fittings/appliances are safe before they re-let the property.

Note: Gas appliances left in the property by the out-going tenant (e.g. gas cookers, fires and cabinet heaters), after the tenancy agreement has finished, would become 'relevant gas fittings' as defined by the GSIUR if left for a new tenant to use and will need to be maintained and safety checked.

Before the property is re-let:

Make suitable checks and rectify or remove any unsafe equipment before any new tenancy can begin.

These checks should also include an inspection of the installation pipework and a gas tightness test. This check does not necessarily require a full safety check to the extent required annually.

Generally, the check should include:

1. A visual inspection of all gas appliances.
2. Checking for obvious signs of incomplete combustion or spillage of POC.
3. Any associated chimneys; and
4. Provision of ventilation, where necessary.

Note: It is strongly recommended that open-flued appliances have a spillage test, in accordance with the appliance manufacturer's installation instructions. Gas operatives should positively record the results of any tests carried out.

CORGI*directs* 'Gas Safety Inspection' form (Order Ref: CP4) or 'Landlords/Home Owner Gas Safety Record' (Order Ref: CP12) offer an appropriate means to record these results (see Part 10 – 'Emergency notices, warning labels and forms' for further guidance).

This however, is not a prescriptive list of checks. Both you – as the gas operative – and landlords/managing agents, should assess the risk on each individual basis.

Take into account, together, the type of property and appliances involved – and ensure all have taken all reasonable steps to discharge responsibilities under the GSIUR.

Electronic records and signatures – what you need to know

Information for landlords and gas installation businesses

The GSIUR require landlords:

- to arrange for annual safety checks of relevant gas appliances and flues
- to keep a record of that check for a period of 2 years
- to provide copies of the record to tenants, who have rights to inspect the original on request

The legislation sets out particulars to be included on the check record. These include the name and signature of the person carrying out the check and their registration number (this is mandatory). Both the record and signature can now be in electronic format.

Electronic solutions and security

Many gas installation businesses and landlords are adopting electronic systems. These offer advantages, particularly for work planning and monitoring.

However, the use of electronic landlord gas safety records may lead to an increased potential for forgery/alteration and therefore reduced safety assurance for tenants.

This guidance confirms a goal-setting approach (developed following discussions with industry and government) to allow the industry to develop electronic solutions compatible with necessary security.

Guidance:

1. You can provide landlords with electronic formats as follows:

   Gas installers may provide a landlord with a Landlord's Gas Safety Record in electronic format with an electronic signature. The landlord then keeps this as the original record. This record should:

   - be capable of being reproduced in hard copy format when required (e.g. for the tenant/HSE/housing department)
   - be secure from loss and interference
   - uniquely identify the gas operative who carried out the safety check. This could be, for example, an electronic representation of the operative's signature, a scanned signature, a payroll number unique to the operative, the name of the operative etc. The employer needs to have secure systems that link the individual operative to the unique identifier
   - these objectives should be achieved by management systems/appropriate technology in accordance with relevant British/European Standards

2. You can also send/give copy to tenant as follows:

   With the landlord's agreement, you can send or give a copy of the record directly to the tenant to provide additional assurance of its source. The record already includes your registration number as the operative.

3. Safeguard your position as follows:

   In case the landlord subsequently alters the record, you should keep your own secure copy to allow subsequent verification.

Some relevant legislation and standards (not an exhaustive list):

- The Electronic Signatures Regulations 2002
- Data Protection Act 1998
- BS ISO/IEC 27002: 2013 Information technology. Security techniques. Code of practice for information security management
- BS ISO/IEC 27001: 2013 (BS 7799-2: 2005) – Information techniques. Information security management systems. Requirements
- BS ISO 15489-1: 2016 – Information and documentation. Records management. General
- BIP 0025-1: 2002 – Effective Records Management. A management guide to the value of BS ISO 15489-1
- BIP 0008-1: 2014 Evidential weight and legal admissibility of information stored electronically. Code of Practice for the implementation of BS 10008
- BIP 0008-2: 2014 Evidential weight and legal admissibility of information transferred electronically. Code of Practice for the implementation of BS 10008
- BIP 0008-3: 2014 Evidential weight and legal admissibility of linking electronic identity to documents. Code of Practice for the implementation of BS 10008

LPG – General requirements – 18

18 – LPG – General requirements

Figures

Tables

Introduction

Information in this Part is specifically for LPG in Permanent Dwellings (PD).

Reference will be made to other parts within this manual where they are relevant to both NG and LPG. Additionally, with the revision to IGEM/UP/1B (Edition 3), LPG/Air mixtures are also covered when discussing tightness testing in this Part.

LPG installation in permanent dwellings

Many requirements are the same for the installation of LPG systems as for NG systems, but there are some important differences you need to recognise. This part covers essential requirements for LPG installations in PDs.

Most of the information in this Part is referenced from:

- BS 5482-1: 2005 Code of practice for domestic butane- and propane-gas-burning installations – Part 1: Installation at permanent dwellings, residential park homes and commercial premises, with installation pipework sizes not exceeding DN25 for steel and DN28 for corrugated stainless steel or copper *(partially replaced)*
- BS 6891: 2015 Specification for the installation and maintenance of low pressure gas installation pipework of up to 35 mm (R1¼) on premises
- IGEM/UP/1B (Edition 3) Tightness testing and direct purging of small Liquefied Petroleum Gas/Air, Natural Gas and Liquefied Petroleum Gas installations (Amendment October 2012)
- UKLPG CoP 22: 2011 Design, installation and testing of LPG piping systems
- UKLPG CoP 24: Pt 1 2006 Use of LPG cylinders at residential and similar premises.

For guidance on the installation of LPG in non-PDs (including caravan holiday homes, residential park homes and boats) refer to:

- CORGI*direct's* manual entitled 'LPG – Domestic Including Permanent Dwellings, Leisure Accommodation Vehicles, Residential Park Homes and Boats' from the current Gas Installer Manual Series – Domestic;
- UKLPG Code of Practice 21: 2011 Guidance for safety checks on LPG appliances in caravans (Incorporating Amendment 1: August 2000);
- UKLPG User Information Sheet (UIS) 014 Gas soundness testing of LPG service pipework, installation pipework and appliances in holiday homes, caravans and motor homes

Installation of pipework and fittings for LPG – sometimes different from NG

Gas operatives working on LPG installation pipework and fittings must be competent to do so.

Generally, materials used for NG are also suitable for use with LPG. This Part provides guidance to any exceptions and additional requirements for LPG installations.
For further information, refer to
Part 5 – Installation of pipework and fittings.

For LPG installations where a gas meter is fitted (e.g. a bulk storage vessel(s) feeding a metered estate) installation pipework is defined as any pipework or fitting on the outlet of the primary meter installation to points at which appliances are to be connected.

For LPG installations where a gas meter is not fitted (i.e. bulk storage vessel/cylinder(s) it is defined as any pipework or fitting on the outlet of the ECV to points at which appliances are to be connected (see Figures 18.1 to 18.4).

Figure 18.1 Typical single cylinder installations

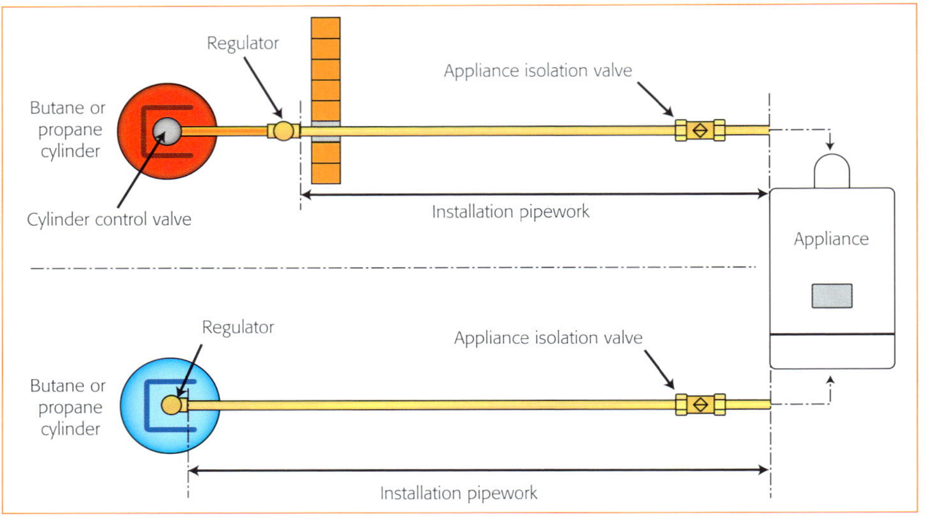

Figure 18.2 Typical multiple cylinder installation

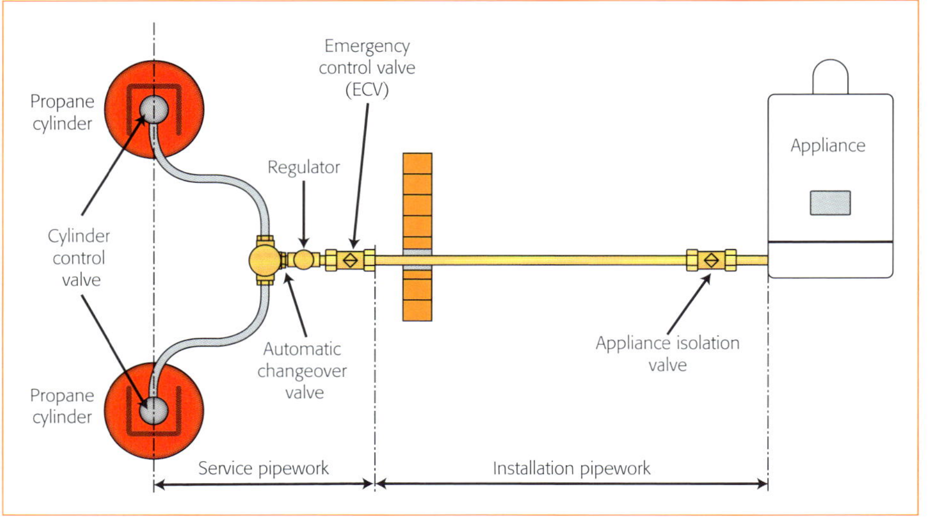

Figure 18.3 Typical bulk storage vessel installations - low pressure service pipework

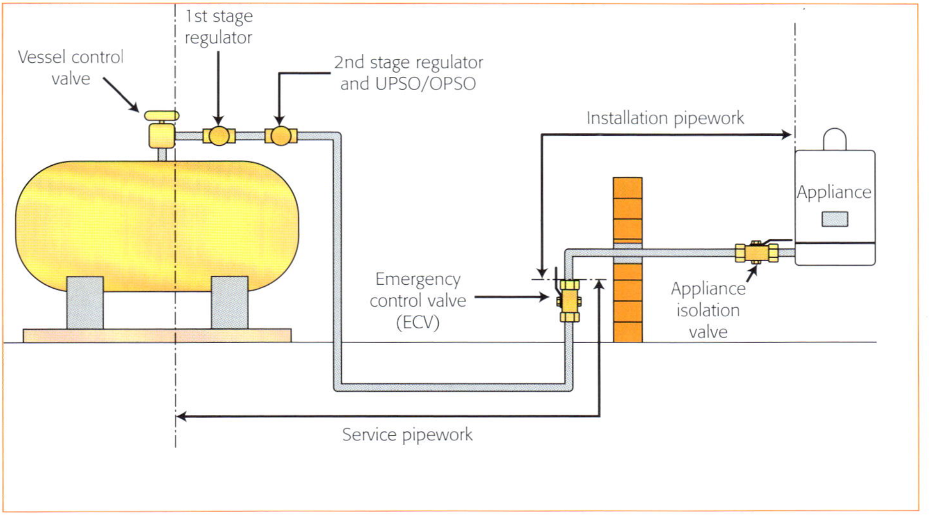

Figure 18.4 Typical bulk storage vessel installation - medium pressure service pipework

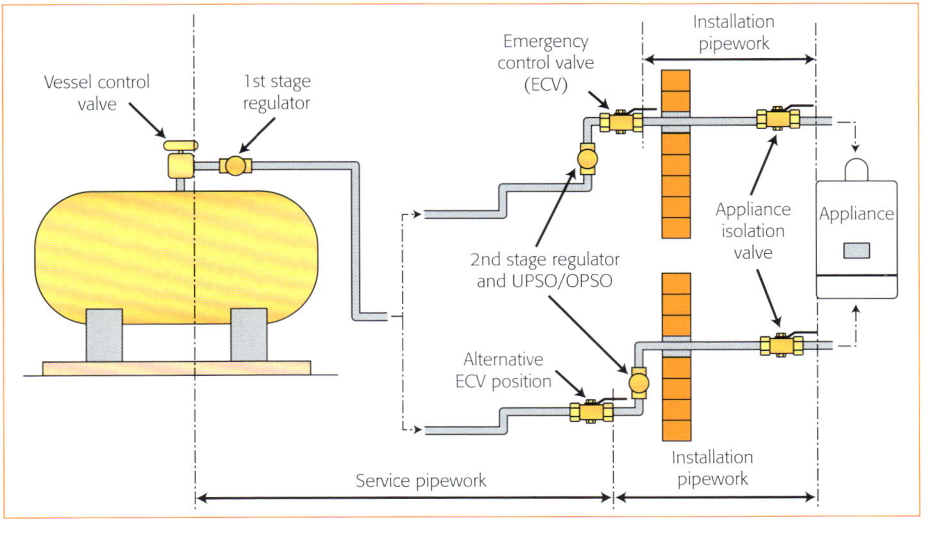

Note: Regard pipework installed between the bulk storage vessel/cylinder(s) and the ECV at the premises, as 'service pipework'.

Key differences between NG/LPG pipework/fittings

Although NG and LPG pipework and fittings are very similar in use, key differences affect the safety of the installation.

Here are guidelines to help you:

- LPG, due to its chemical composition, will find any weaknesses in joints – particularly threaded joints.
- LPG is aggressive to certain non-metallic materials (natural rubber and many plastics). So make sure that fittings and hoses are suitable for use with LPG.
- never use two parallel threads together on pipework or fittings.
- jointing compounds suitable for LPG (conforming to relevant part of BS EN 751) are required.
- the threaded valves used on LPG cylinders generally have a 'left hand' thread. Propane cylinders normally have a female left handed outlet, but 27mm 'clip-on' valves are now used for BBQ and patio heater applications. Butane cylinder valves normally have 21mm 'clip-on' connections, except for some 4.5kg cylinders, which has a left-handed male outlet.
- appliance taps and valves etc. need to be designed for use with LPG. Taper plug taps designed for NG are not suitable.
- Taps and plug valves should be of the spring loaded type for LPG use and lubricated with a special grease (this should contain 25% molybdenum disulphide base). Oil-based lubricants will be dissolved by LPG, causing the tap to become dry, stiff to turn and prone to gas escapes.
- ball valves using PTFE inserts that require no lubricant are generally suitable for use with LPG

Sizing gas supplies (LPG) – important considerations

General requirements

As with NG installations when you design LPG installations, take great care to ensure the pipe size you select is of adequate diameter to supply sufficient gas to all appliances at the same time.

Failure to do so on LPG installations may result in the same issue as have been described for NG installations. For further information, refer to **Part 5 – Installation of pipework and fittings**.

Initial design considerations

Pressure regulators fitted to LPG bulk storage vessel/cylinder(s) installations are designed and adjusted to ensure that the pressure is reduced from its storage pressure (under normal conditions this will range from approximately 2 bar – 9 bar) to operating pressure, irrespective of demand. The is sometimes done in a single stage, such as with the regulators found on many cylinder installations, or in multiple stages as is seen on bulk storage vessels.

Where the pressure is reduced in multiple stages there will be a first stage regulator, a second stage regulator and even on some installations a third stage regulator. The last regulator on the system is referred to as the final stage regulator and it is this regulator that delivers the required operating pressure to the installation.

All the regulators on an installation though need to be able to pass a sufficient capacity of gas for the installation.

Figure 18.5 Design pressure loss for a typical multiple cylinder installation

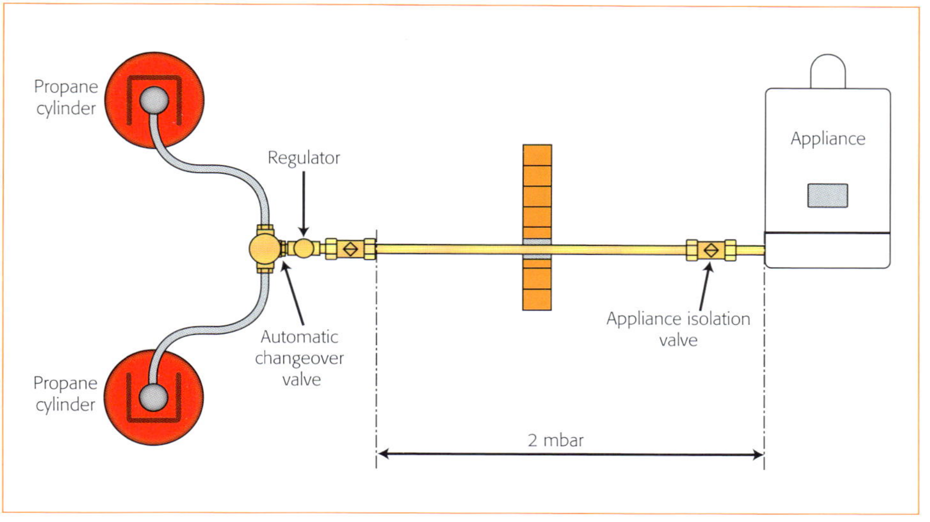

Figure 18.6 Design pressure loss for a typical bulk storage vessel installations - low pressure service pipework

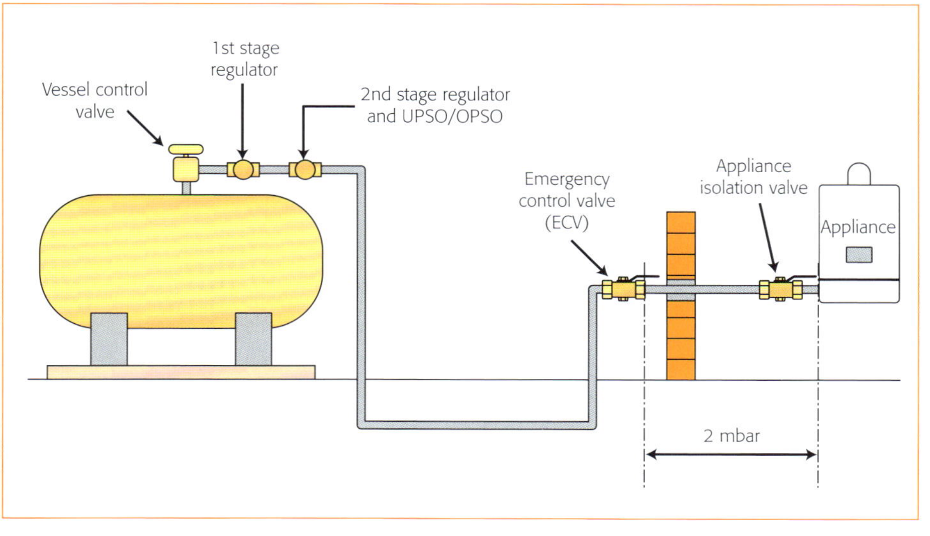

Figure 18.7 Design pressure loss for a typical bulk storage vessel installations - medium pressure service pipework with the final stage regulator after the ECV

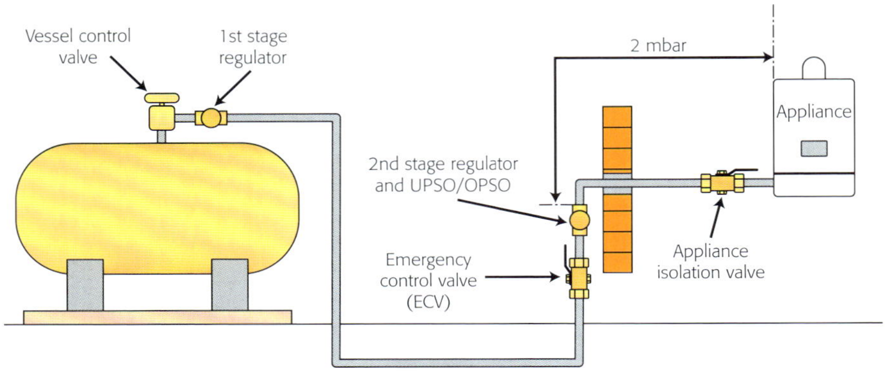

LPG appliances are designed to operate at different pressures than NG appliances and therfore the final stage regulator feeding the installation pipework is set at a different pressure to that of the meter regulator on an NG installation. The nominal outlet pressure of this regulator also differs dependent on the type of LPG supplying the installation as follows:

- Propane - 37 mbar
- Butane - 28 or 29 mbar depending on what type of regulator is fitted.r

The design pressure loss between the outlet of a primary meter installation, or where no meter is installed, the outlet of the ECV or the outlet of the final stage regulator when fitted after the ECV and the point to be connected on any appliance inlet shall not exceed 2 mbar at the design installation flow rate

Figures 18.5 to 18.7 show some typical LPG installation configurations, including regulator positions, and the section of the installation pipework that should not have greater than a 2 mbar pressure loss.

Due to the higher CV of LPG, smaller volumes of gas are required to provide the same heat input as NG. Pipe sizes that would be adequate for NG will generally be more than adequate for LPG.

Guide for pipe sizing (LPG)

The process for sizing LPG installation pipework is the same as that given for NG described in **Part 5 – Installation of pipework and fittings**, however there are a number of differences that are discussed in this Part.

One of the main differences is the tables that are used to determine the pressure loss data for the different types of pipe. Tables 18.1 and 18.2 show the pressure loss data for copper pipe and pressure loss in fittings for LPG installations.

Table 18.1 Approximate LPG pressure loss data (mbar/m) - copper pipe

Flow Rate	Heat Input		Nominal Pipe Size (mm) A)						
(m3/h)	Gross (kW)	Net (kW)	8 (6)	10 (8)	12 (10)	15 (13)	22 (19)	28 (25)	35 (32)
0.25	6.47	5.88	0.4413	0.1148	0.0405	0.0119	0.0021	0.0006	0.0002
0.50	12.93	11.76	1.4367	0.3705	0.1298	0.0379	0.0064	0.0018	0.0006
0.75	19.40	17.63		0.7421	0.2590	0.0753	0.0127	0.0035	0.0011
1.00	25.86	23.51		1.2194	0.4246	0.1231	0.0206	0.0057	0.0018
1.25	32.33	29.39		1.7959	0.6243	0.1807	0.0302	0.0083	0.0026
1.50	38.79	35.27			0.8565	0.2474	0.0412	0.0113	0.0035
1.75	45.26	41.14			1.1200	0.3231	0.0537	0.0147	0.0046
2.00	51.72	47.02			1.4137	0.4074	0.0676	0.0185	0.0058
2.25	58.19	52.90			1.7369	0.5001	0.0829	0.0227	0.0071
2.50	64.65	58.78				0.6009	0.0995	0.0272	0.0085
2.75	71.12	64.65				0.7097	0.1174	0.0320	0.0100
3.00	77.58	70.53				0.8264	0.1365	0.0372	0.0116
3.25	84.05	76.41				0.9508	0.1569	0.0428	0.0133
3.50	90.51	82.29				1.0827	0.1786	0.0486	0.0151
3.75	96.98	88.16				1.2222	0.2014	0.0548	0.0170
4.00	103.44	94.04				1.3690	0.2254	0.0613	0.0191
4.25	109.91	99.92				1.5230	0.2506	0.0682	0.0212
4.50	116.38	105.80				1.6843	0.2770	0.0753	0.0234
4.75	122.84	111.67				1.8527	0.3045	0.0827	0.0257
5.00	129.31	117.55					0.3332	0.0905	0.0281
5.25	135.77	123.43					0.3630	0.0985	0.0305
5.50	142.24	129.31					0.3939	0.1069	0.0331
5.75	148.70	135.18					0.4259	0.1155	0.0358

Table 18.2 Typical allowances for pressure loss in fittings

Nominal Pipe Size			Equivalent Length (m)				
			Bends		Fittings		
Steel		Copper	45^0	90^0	90^0 elbow	Tee A) (Entering Branch)	Tee A) (Exiting Branch)
≤ 15 mm	R $^1/_2$	≤ 15 mm	0.15	0.20	0.40	0.75	1.20
20 mm	R $^3/_4$	22 mm	0.20	0.30	0.60	1.20	1.80
25 mm	R 1	28 mm	0.25	0.40	0.80	1.50	2.30
32 mm	R $1^1/_4$	35 mm	0.30	0.50	1.00	2.00	3.00
A) Use the largest connection size on the fitting that is not necessarily the branch size							

Note: For pressure losses associated with press end fittings, CSST / PLT tube & fittings, flexible connections, secondary meters, check valves, regulators, etc. reference needs to be made to the manufacturer's data.

Figure 18.8 Pipe sizing example (LPG) of a typical copper pipe gas installation

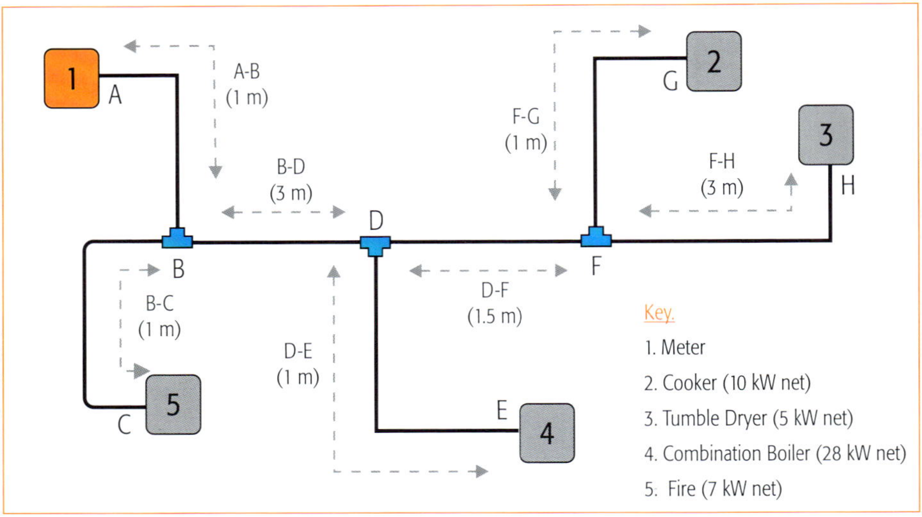

Figure 18.8 give the example of a typical copper pipe installation that was used in **Part 5 – Installation of pipework and fittings**.

This time the installation is to be fed by LPG rather than NG. As previously mentioned, due to the higher CV of LPG and greater pressure loss allowed that pipework sizes for NG would be more than adequate for LPG. However, as well as the additional work that may be involved in installing larger size pipes, this could result in increased costs to the customer.

Therefore the installation needs to be sized for LPG rather than NG using the tables and maximum pressure loss criteria for LPG.

The step by step process of deperming the size of pipework that is required is the same as that which was described in Part 5, so has not been repeated in this Part. The pipework though has been sized using Table 18.1 and 18.2 with the results shown in Table 18.3.

What can be seen from this table is that the main pipework run from point A to point F has been reduced in size. Sections A-B and B-D can satisfactorily be installed in 22 mm, whilst for NG this had worked out as 28 mm, and section D-F can be installed in 15 mm compared to 22 mm for NG.

Even with these reduced pipe sizes it can be seen that the total pressure loss to each appliance is well within the maximum 2 mbar allowed. In fact, with the exception of the pipework run to the boiler (point A to E), the pressure losses to all the other appliances are even below 1 mbar.

This means that it would be possible, should the designer / gas operative wish to, to reduce the pipe sizes even further and still deliver adequate pressure at the inlet of each appliance. Remember though to always check the manufacturers requirements for minimum pipe size for the appliance connection.

Table 18.3 Pipe sizing results (LPG)

1	2	3	4	5		6	7	8
Pipework Section	Maximum Heat Input A)	Estimated Pipe Diameter	Pipe Length	Fittings Allowance B)		Adjusted Length	Pressure Loss/Metre	Pressure loss C)
				Type	Equivalent Length			
	kW (net)	mm / DN	m		m	m	mbar /m	mbar
A-B	50	22	1	Elbow x1 Tee (Enter) x 1	0.6 1.2	2.8	0.0829	0.24
B-C	7	15	1	90⁰ Bend x 2	0.4	1.4	0.0379	0.06
B-D	43	22	3	N/A	0	3	0.0676	0.21
D-E	28	15	1	Tee (Exit) x 1 Elbow x 1	1.80 0.40	3.2	0.1807	0.58
D-F	15	15	1.5	N/A	0	1.5	0.0753	0.12
F-G	10	15	1	Tee (Exit) x 1 Elbow x 1	1.20 0.40	2.6	0.0379	0.10
F-H	5	15	3	Elbow x1	0.40	3.4	0.0119	0.05

Pressure loss to points to be connected for each appliance

Appliance	Maximum Heat Input kW (Net)	Pipework Sections and Associated Pressure Loss					Total Pressure Loss (mbar)	Satisfactory
Cooker	10	A-B	B-D	D-F	F-G		0.67	Yes
		0.24	0.21	0.12	0.10			
Tumble Dryer	5	A-B	B-D	D-F	F-H		0.62	Yes
		0.24	0.21	0.12	0.05			
Boiler	28	A-B	B-D	D-E			1.03	Yes
		0.24	0.21	0.58				
Fire	7	A-B	B-C				030	Yes
		0.24	0.06					

Flexible hoses, tubing and associated assemblies (LPG) – which to use and how to use

Use of hoses, tubing and associated assemblies

Flexible, hoses, tubes and associated assemblies may be used in a number of situations at a permanent dwelling (PD). These are the:

- high pressure connections between a cylinder and wall mounted regulator or changeover device;
- low pressure connection between a cylinder mounted regulator and fixed installation pipework; andr
- low pressure connections to moveable appliances.

They should not be used for other purposes.

Figure 18.9 Typical markings on flexible hoses and tubes used for LPG

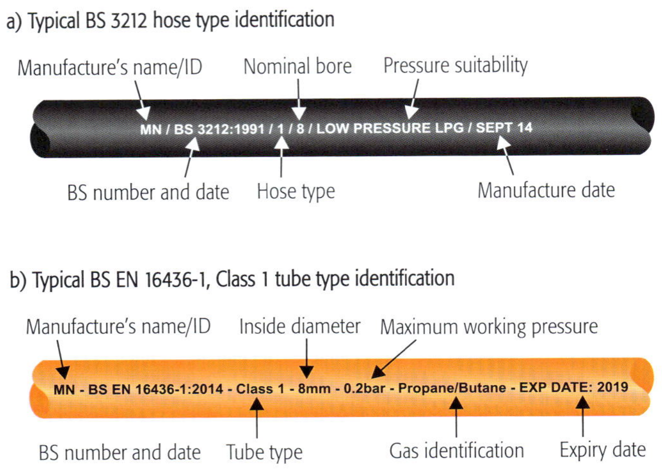

Suitable hoses, tubing and associated assemblies

In addition to metallic flexible pipework, non metallic flexible hoses, tubing and their assemblies may be used on LPG in the situations previously described. These non metallic hoses and tubes should conform to one of the following standards:

- BS 3212;
- BS EN 1763-1; or
- BS EN 16436-1.

All flexible hoses and tubes should be marked so that they are readily identifiable. Typical markings for hoses and tubes that are currently available are illustrated in Figure 18.9.

You will see that BS 3212 hoses are marked with a manufacture date marked on them (as are BS EN 1763-1, and BS EN 16436-1 class 2 and 3 hoses). Whilst they have no specific service life or exchange interval it is unlikely that there services life will exceed 10 years and therefore replacement at this age is recommended. BS EN 16436-1 class 1 tubes include an expiry date and should not be used after the declared date.

It should be remembered that these hoses are frequently outside and exposed to the elements as well as being constantly in contact with LPG, so whilst these hoses and tubes may be date sensitive, this should not override the fact that need to be physically inspected. If they show signs of physical damage, such as cuts, cracking, blistering or kinking, they should be replaced.

You will see from the markings that hoses and tubes are marked to indicate the pressures they are suitable to be used on. For example, with hoses manufactured to BS 3212 they are either:

- **Type 1**: Low pressure (up to 50 mbar); or
- **Type 2**: High pressure (up to 17.5 bar).

and are marked either 'high pressure' or 'low pressure'.

High pressure hoses may be used in low pressure situations but low pressure hoses must not be used on high pressure.

Generally low pressure hoses are black and high pressure hoses are orange, however other colours may be encountered, for example, high pressure hoses supplied as a complete assembly with crimped or swaged ferrules for use may have a black cover.

Other hoses may have a braided or armoured covers, for example where they are used in areas that may be prone to rodent attack, and these covers mask any markings and the colour. They will be identified though by other means such as a tag connected to the hose.

Hose and tube connections

Flexible hose and tube connections should preferably be made using prefabricated fittings. Where this is not appropriate, connections should only be made to suitable and correctly sized nozzles that are secured using approved crimp clips or worm drive clips (see Note).

Note: Hose or tubing with an internal diameter of less than 8mm and hoses operated at a pressure exceeding 50 mbar should always be secured by crimp clips or swaged fittings. Worm drive clips should not be used.

Other general installation requirements

There are a number of other installation considerations that need to be taken into account that are explained below.

Length

All flexible pipework should be in a single continuous length, being kept as short as possible, though long enough to provide the necessary flexibility, without excessive strain on the hose or the end fittings.

For high pressure connections between the cylinder vapour valve and a wall mounted regulator or change over device the maximum recommended length per cylinder is 1 m.

For low pressure connections between cylinders and fixed pipework installations, the maximum recommended length is 500 mm.

Hose Route

Flexible pipework connections shall not pass through walls, floors, ceilings or other partitions, nor should they pass between rooms. Where necessary the connection should be supported.

Temperature

Do not use flexible hoses, tubing and associated assemblies where they may be subject to temperatures above 50°C. unless designed for higher temperatures.

Electrical continuity

Where necessary, electrical continuity shall be maintained across any flexible pipework..

High pressure connections between a cylinder and wall mounted regulator or changeover device

High pressure flexible connections are used to connect a cylinder to the regulator or changeover device and facilitate the exchanging of cylinders. They are normally prefabricated connections and if fitted to a changeover device may incorporate a non return valve.

Note: The flexible connections between a cylinder valve and a regulator/changeover device are generally known as 'pigtails'

Low pressure connection between a cylinder mounted regulator and fixed installation pipework

Low pressure flexible connections are used to connect a cylinder mounted regulator to the fixed pipework installation and facilitate the exchanging of cylinders. ed rigid internal installation gas pipework.

Traditional nozzles used on LPG installations, such as the "Fulham Nozzle", are known to have high-pressure losses and these pressure losses along with the internal diameter of any flexible hose or tube need to be considered when designing the installation.

Low pressure connections to moveable appliances

Flexible hoses can be used to connect flueless moveable appliances, e.g. cookers, gas rings, Bunsen burners and barbecues.

Use the minimum practicable length. They will be provided with integral taper threaded metal ends. Alternatively, secure them at each end to a nozzle, using a suitable hose clip.

Appliances connected to micro-points will be supplied with a right angle bayonet-type plug-in connection, which keeps the hose close to the wall and is self-sealing when disconnected.

Where you use a flexible hose connection with a domestic cooker, a proprietary cooker hose and bayonet should be used. It is generally identified with a red band or stripe, indicating its suitability for use with LPG. Other rubber hoses and tubes must not be used

As with other hoses and tubes, those connected to flueless moveable appliances also need inspecting on a regular basis (see Note).

Note: Gas Safe Register has produced a Technical Bulletin – TB137 'Deterioration of LPG flexible connectors' – addressing this issue. The TB can be downloaded at https://engineers.gassaferegister.co.uk

Tightness testing permanent dwellings, caravan holiday homes and residential park homes

Criteria to meet and instruments to use

The introduction of the IGEM/UP/1B (Edition 3) and the expansion of its scope to include the tightness testing of LPG/air mixtures and LPG (Propane and Butane), in addition to NG installations means that the previous 'soundness' testing procedure covered by BS 5482-1 is now superseded.

Therefore, the tightness testing procedures discussed in this Part for LPG/air mixtures and LPG installations (Propane and Butane) is based on two Standards – depending on whether the pipework to be tested is 'Installation pipework' or 'Service pipework':

Installation pipework

IGEM/UP/1B (Edition 3) 'Tightness testing and directing purging of small Liquefied Petroleum Gas/Air, Natural gas and Liquefied Petroleum Gas installations' – Amendment October 2012 – applies to gas pipework downstream of the outlet of the ECV/AECV/supply control valve/final stage regulator, as appropriate, including a gas meter (where installed) for installations with a maximum operating pressure of 2bar at the outlet of the ECV and:

- pipework no greater than 35mm diameter
- meter, where installed, no greater than 16m^3/h
- installation volume not exceeding 0.035m^3
- for LPG/air mixtures – operating pressure at the outlet of the meter not exceeding 21mbar
- for LPG – operating pressure at the final stage regulator not exceeding 37mbar

(See Figures 18.10a, 18.10b and 18.10c)

Service pipework

UKLPG CoP 22 Design, Installation and Testing of LPG piping systems: 2011 – applies to gas pipework upstream of the ECV and:

- an outside diameter of service pipework not exceeding 63mm; and
- a volume of service pipework not exceeding $0.02m^3$

(See Figures 18.10a, 18.10b and 18.10c)

Both procedures utilise a pressure gauge for testing pressure loss in the pipework and appliance system when under pressure, with all inlets and outlets sealed.

This is usually the most practical method of testing, but alternative procedures are possible.

Pipework tightness testing is expected to meet the 'No discernible pressure drop' criteria based on the use of:

- a 'U' gauge; or
- an electronic/digital gauge

See **Definition of 'No discernible pressure drop'** in this Part.

The selected gauge needs to be used in accordance with manufacturer's instructions and, in the case of electronic/digital pressure gauges, be calibrated annually or as the manufacturer advises.

Gas appliances are permitted a very small leakage rate for practical reasons. This may occasionally create a discernible pressure drop when you test complete gas installations with appliances connected, particularly if the pipework and appliance internal volume is small.

The allowable pressure drops will be discussed within the procedures outlined in this Part.

Definition of 'no discernible pressure drop'

For **Installation pipework** – a pressure drop (movement) of 0.25mbar or less on a 'U' gauge is considered to be not discernible. As 0.25mbar is allowable for 'U' gauge it follows that it is an acceptable pressure drop if measured by more sensitive instrumentation such as electronic/digital gauges, provided the gauge is regularly calibrated.

For **Service pipework** – a pressure drop (movement) of 0.5mbar or less on a 'U' gauge is considered to be not discernible. For electronic/digital gauges, 0.3mbar for low pressure and 5mbar for medium & high pressure is considered not discernible.

Note: The use of a Bourdon gauge is no longer acceptable for Service pipework.

Pressure gauge accuracy

For **Installation pipework** – a 'U' gauge needs to be capable of being read to an accuracy of 0.5mbar and an electronic/digital gauge, 0.1mbar or better,

For **Service pipework** – a 'U' gauge needs to be capable of being read to an accuracy of 0.5mbar with a range of 0–100mbar. For electronic/digital gauge measuring pressures below 100mbar, an accuracy of 0.3mbar and above 100mbar, an accuracy of 1mbar or better.

Note: All gauges need to be regularly re-calibrated for accuracy in accordance with the manufacturer's instructions.

Figure 18.10a Single cylinder – 'Installation' and 'Service' pipework configurations

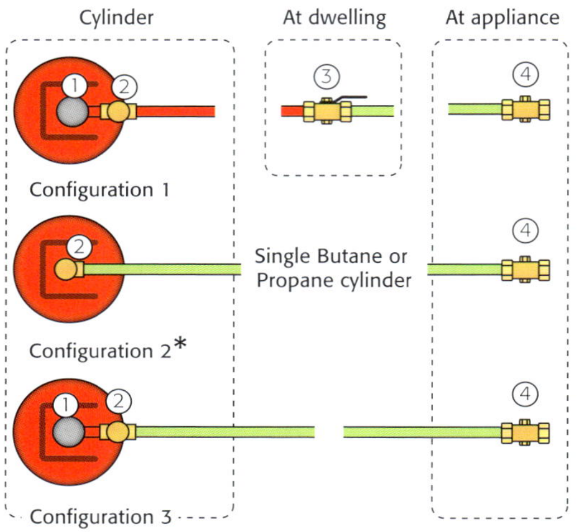

Service pipework – tightness tested to UKlpg CoP 22

Installation pipework – tightness tested to IGEM/UP/1B (Ed 3)

① Cylinder control valve

② Single stage regulator

*Single stage regulator on single Propane or Butane cylinder also acts as the ECV

③ ECV

④ Appliance isolation valve

Note: Service pipework which is accessible and under 3m in length can be tested using non-corrosive LDF – see Note to 'Tightness testing low pressure service pipework – new and existing installations – using air/inert gas or LPG'.

Figure 18.10b Multiple cylinders – 'Installation' and 'Service' pipework configurations

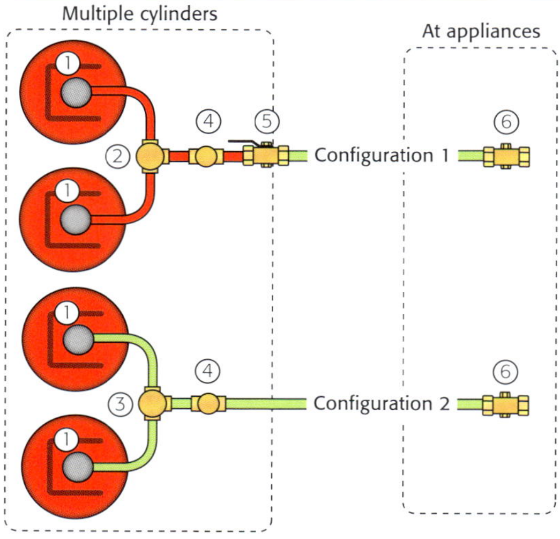

Service pipework – tightness tested to UKlpg CoP 22

Installation pipework – tightness tested to IGEM/UP/1B (Ed 3)

① Cylinder control valve

② Manual or automatic changeover device

③ Manual changeover device

④ 1st, 2nd or single stage regulator

⑤ ECV

⑥ Appliance isolation valve

Note: Service pipework which is accessible and under 3m in length can be tested using non-corrosive LDF – see Note to 'Tightness testing low pressure service pipework – new and existing installations – using air/inert gas or LPG'.

Figure 18.10c Bulk storage vessel – 'Installation' and 'Service' pipework configurations

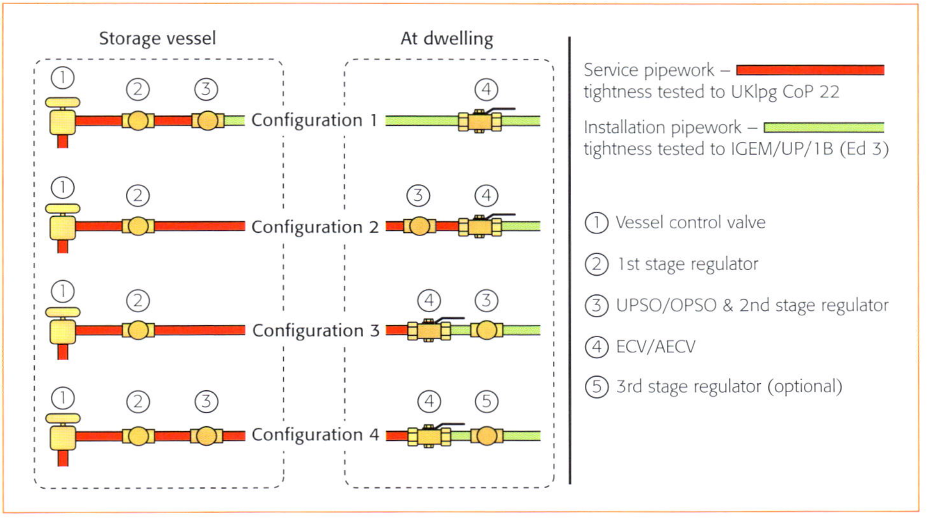

What these tightness testing procedures cover

1. Permanent domestic installations
2. Residential Park Homes (RPH's)
3. Leisure Accommodation Vehicles (LAV's)
4. Small permanent commercial installations*.

While these testing procedures encompass commercial installations, these installations are not covered in this manual – reference should be made to CORGI*directs* '
Essential Gas Safety – Non Domestic'
manual for further guidance.

See Figures 18.1 – 18.4 for examples of typical LPG installations.

You must carry out a test to ensure gas tightness:

- before gas charging and commissioning of newly installed pipework and before extensions to existing pipework
- whenever any work is carried out on an installation that might affect gas tightness
- if a gas escape is suspected
- if there is a smell of gas
- before re-establishing an existing gas supply

The test procedures discussed in this Part are split in to 'Service' and 'Installation' pipework, with service pipework being discussed firstly.

Service pipework is defined by its pressure regime i.e. low pressure (up to and including 75mbar), medium pressure (above 75mbar and up to 2bar) and high pressure (above 2bar).

Medium pressure (MP) and high pressure (HP) service pipework is not discussed within this manual, given that most domestic service pipework installations will and are recommended to be regulated to 75mbar and below (low pressure) before the ECV at the dwelling. However, detailed guidance on MP & HP testing of service pipework can be found in UKLPG CoP 22. 2011.

Tightness testing low pressure service pipework – new and existing installations – using air/inert gas or LPG

The procedures discussed here involves the tightness testing of low pressure service pipework alone, which includes the short connection from the bulk storage vessel outlet/cylinder control valve to the regulator(s) at the vessel/cylinder (see Note) and pipework from there up to the ECV at the premises (see Figures 18.3).

Note: Service pipework which does not exceed 3m in length, which is above ground and which all joints are accessible can be tested using non-corrosive LDF only, with the system at working pressure – termed a 'connection leak detection test' – or via the applicable tightness testing procedure discussed here.

Procedure for tightness testing new or existing low pressure service pipework using either air or LPG – operating pressure after final regulator 37mbar Propane or 28mm Butane

Preparation

Cap or plug all open points in the system except for the one which will be used to admit air, inert gas or LPG to the system.

Procedure

1. Close the downstream control valve – typically the ECV at the premises.
2. Close the upstream bulk storage vessel/cylinder supply control valve
3. Connect an inline test tee. Attach a suitable pressure gauge to the test tee.
4. Open upstream supply control valve and activate UPSO reset. Admit the test medium – air or LPG – in to the system. Allow the regulator to lock-up.
5. Close the supply control valve and reduce the pressure to that shown in Table 18.4.
6. Leave for at least 5 minutes to allow the temperature in the system to stabilise.
7. Note the pressure gauge reading.
8. Leave for a further 2 minutes.
9. Note the pressure gauge reading. The pipework is considered gas tight if there is no discernible pressure drop (movement) from that noted in Step 7. If the pressure has fallen, examine the entire installation and check each joint with non-corrosive LDF to locate the source of escape.
10. Repair the escape and repeat the procedure from Steps 4 to 9 until the installation is proven gas tight.

After successfully completing a gas tightness test of the low pressure service pipework, the upstream supply control valve needs to be proved as being leak tight i.e. not letting-by during the preceding tightness test which would mask any gas escape.

Table 18.4 Tightness test pressures for new or existing low pressure service pipework

Test medium	Low pressure service pipework operating at:	
	≤37mbar	>37mbar ≤75mbar
Butane	20mbar [1& 2]	N/A
Propane	30mbar [1 & 2]	60mbar [1]
Air	45mbar [2]	350mbar [2*]

[1] Existing service pipework
[2] New service pipework
* This test is a combined strength and tightness test
≤ Less than or equal to
> greater than

Let-by test of upstream supply control valve

Procedure

1. Ensure downstream control valve remains closed.
2. Confirm upstream supply control valve is closed.
3. The pressure gauge should still be attached to a suitable test point – if not, attached a suitable pressure gauge.
4. Release gas slowly to the open air via the test tee connected to a gas torch at the end of a length of LPG hose or other safe method. When the pressure has dropped to between 5 and 10mbar, shut off the release of gas (see Note 1).

Note 1: If there is an UPSO downstream of the supply control valve, this will close as the pressure falls. In this case, once the pressure has been reduced to between 5 and 10mbar, as in Step 4, operate the UPSO re-set to release the trapped upstream pressure and then allow it to re-shut. There may be a small sudden rise in the pressure gauge reading as the upstream pressure is released in to the downstream pipework.

5. Leave for 2 minutes.
6. After 2 minutes, observe the pressure gauge reading. If there is no pressure rise greater than 0.5mbar, 'no let-by' can be recorded for the supply control valve (see Note 2).

Note 2: If the procedure in Note 1 has been necessary, the UPSO re-set will need to be operated again after the 2 minute period and before recording the second pressure gauge reading.

A rise in pressure will indicate either:

- that the gas temperature has risen during the test; or
- that the supply control valve is letting by (see '**What happens if a supply control valve is letting by?**' in this Part) – confirm let-by, by disconnecting the outlet of the supply control valve and applying non-corrosive LDF to the barrel/ball of the valve.

If you suspect a gas temperature rise, repeat the procedure from Steps 4 to 6.

7. If the supply control valve has passed the test, remove the pressure gauge and reseal the test point. Turn on the supply control valve and test all joints including the test point with non-corrosive LDF.

8. Record your results on suitable documentation and, where required, purge the installation (see **Purging LPG gas installations – precautions you need to take** in this Part and, **Part 15 – Re-establishing gas supply/re-lighting appliances** for further guidance).

What happens if a supply control valve is letting by?

You need to rectify the fault or change the valve and repeat the let-by test to prove 'no let-by' of the replaced supply control valve:

- if this is a cylinder outlet valve, change the cylinder and notify the fault to the gas supplier
- if this is a bulk storage vessel-fed installation, then notify the fault to the gas supplier

Important: On no account should unauthorised persons attempt to remove or dismantle a cylinder/bulk storage vessel supply control valve.

Note: Installations fed by two or more cylinders may be tested for let-by through the cylinder outlets simultaneously, i.e. with all valves closed. If you detect let-by, repeat the test for each individual valve with the other cylinders disconnected and the free end of the pigtails capped, until you locate the source of let-by.

Tightness testing new and existing Installation pipework

The procedures discussed in this Part involves the tightness testing of new and existing Installation pipework – downstream from the outlet of the ECV/AECV/supply control valve/final stage regulator, as appropriate, including a gas meter (where installed) up to and including gas appliances.

Procedure for tightness testing new installation pipework using air or inert gas only

New pipework should be tightness tested with air where it is not connected to a live gas supply, before being connected to a live supply and tightness tested and purged with fuel gas.

Preparation

Cap or plug all open points in the system except for the one which will be used to admit air or inert gas to the system. Visually inspect the installation to ensure all joints are correctly made and ensure that all appliance isolation valves are open, that all appliance control taps (including any pilot taps) are shut.

On cookers with a fold-down lid, raise the lid to a fully open position during the tightness test so that any Safety Shut-Off Valve (SSOV) on the supply to the hotplate taps is open.

Procedure

1. Connect an inline test tee. Attach a suitable pressure gauge to the test tee.
2. Slowly inject air or inert gas in to the system until the pressure gauge reads:
   - for LPG (Propane and Butane) installations – 45–46mbar
   - for LPG/air mix installations operating at 21mbar – 20–21mbar
   - for LPG/air mix installations operating at 14mbar – 13–14mbar

3. Isolate the pressure source.
4. Leave for one minute to allow temperature to stabilise. If after the stabilisation period the pressure has changed, re-adjust the pressure to the relevant figure stated in Step 2.
5. Note the pressure gauge reading.
6. Test over the next 2 minutes. The installation pipework is considered to be gas tight if there is no discernible pressure drop (movement) from that recorded in Step 5. If the pressure has fallen, examine the entire installation and check each joint with non-corrosive LDF to locate the source of escape.
7. Repair escape and repeat the test procedure from Step 2 to 6 until the installation passes the test. If the installation cannot be repaired, make safe by disconnecting all appliances and capping/sealing any open ends.
8. Where the installation is not to be immediately connected to a live gas supply and retested using fuel gas, remove the pressure gauge and re-seal the test point.
9. Complete all necessary documentation (i.e. recording the test results) and advise the responsible person for the installation.

Note: Where the installation has passed the tightness test and you are not connecting the installation to a live gas supply, cap or plug the pipework with an appropriate fitting. It would also be considered 'good practice' if you label the installation, stating that it has passed a tightness test using air only.

Procedure for tightness testing new and existing LPG installations – designed to operate at 37mbar Propane and 28mbar Butane

Preparation

Cap or plug all open points in the system. Visually inspect the installation to ensure that all appliance isolation valves are open, that all appliance control taps (including any pilot taps) are shut.

On cookers with a fold-down lid, raise the lid to a fully open position during the tightness test so that any SSOV on the supply to the hotplate taps is open.

Procedure

1. Turn off the ECV/supply control valve.
2. Attach a suitable pressure gauge to a suitable pressure test point on the installation, on the outlet of the ECV/supply control valve and final stage regulator.
3. Test the ECV/supply control valve for let-by, by safely reducing the pressure within the installation to between 7–10mbar.

- ensure that any regulator upstream of the ECV/supply control valve is active

Note 1: If there is an UPSO downstream of the supply control valve, this will close as the pressure falls. In this case, once the pressure has been reduced to between 7–10mbar, as in Step 3, operate the UPSO reset – to release trapped upstream pressure and then allow it to re-shut. There may be a small sudden rise in the pressure gauge reading as the upstream pressure is released in to the downstream pipework.

4. Leave for 1 minute.
5. After 1 minute, observe the pressure gauge reading. If there is no discernible pressure rise 'no let-by' can be recorded for the ECV/supply control valve (see Note 2).

Note 2: If the procedure in Note 1 has been necessary, the UPSO reset will need to be operated again after the 1 minute period and before recording the second pressure gauge reading.

A rise in pressure will indicate either:

- that the gas temperature has risen during the test; or
- that the supply control valve is letting by (see '**What happens if a supply control valve is letting by?**' in this Part) – confirm let-by, by disconnecting the outlet of the ECV/supply control valve and applying non-corrosive LDF to the barrel/ball of the valve.

If you suspect a gas temperature rise, repeat the procedure from Steps 3 to 5.

6. Where the let-by test is satisfactory – slowly raise the test pressure to:

- for installations where the regulator is included in the test –
  - Butane installation – 20–21mbar
  - Butane and Propane installation operating at 30mbar – caravan installation* – 28–29mbar
  - Propane installation operating at 37mbar – 30–31mbar
- for installations where the regulator is not included in the test –
  - Butane – 27–28mbar
  - Butane and Propane installation operating at 30mbar – caravan installation* – 29–30mbar
  - Propane installation operating at 37mbar – 36–37mbar

*For caravan installations that are designed for use on the road and comply with BS EN 1949, BS EN 12864 or BS EN 13785.

Note: Care needs to be exercised when testing installation pipework that includes the regulator – to high a pressure may cause the regulator to lock-up. Hence the differing test pressures mention in Step 6.

7. Turn off the supply control valve.
8. Leave for 1 minute to allow the temperature to stabilise. Where necessary and after the 1 minute stabilisation period, re-adjust the pressure to that in Step 6.

9a. Test for a further 2 minutes:

- for new installations, with or without appliance connected and existing installations without appliances connected – there should be no discernible pressure drop (movement) of the pressure gauge and no smell of gas (see 9b).

9b. Where a pressure drop is recorded in 9a and the installation is existing with appliances connected, then the Installation Volume (IV) needs to be calculated in order to determine if the drop recorded in 9a is acceptable (see **Part 6 – Testing for gas tightness (Natural gas)** for guidance on calculating IV; applies regardless of fuel type).

- for existing installations with appliances connected and no smell of gas, a pressure drop of the following is acceptable –
  - 2mbar (for IV no greater than $0.0025m^3$)
  - 1mbar (for IV between $0.0025m^3$ – $0.005m^3$)
  - 0.5mbar (for IV between $0.005m^3$ – $0.01m^3$); and
  - no pressure drop allowed (for IV between $0.01m^3$ – $0.035m^3$).

10. If a drop in pressure is recorded that exceeds those values provided in Step 9 and/or there is a smell of gas, trace, repair and re-test the installation. If the source of escape cannot be traced/repaired, make the installation safe and cap/plug all open points – implement the current GIUSP.
11. Upon completion of a satisfactory tightness test, remove the pressure gauge and seal the test point.
12. Open the supply control valve and test the test point, control valve, outlet and regulator connections with non-corrosive LDF.

- Where air has entered the installation pipework or the pipework is new, purge the installation (see **Purging LPG gas installations – precautions you need to take** in this Part and **Part 15 – Re-establishing gas supply/re-lighting appliances** for further guidance), and repeat the test procedure from Steps 1-11

Note: The tightness test procedure needs to be repeated where air has entered the system, as it resists flow differently to that of LPG and therefore, certain levels of escape may not be detected on the initial test.

What happens if a supply control valve is letting by?

You need to rectify the fault or change the valve and repeat the let-by test to prove 'no let-by' of the replaced supply control valve:

- if this is a cylinder outlet valve, change the cylinder and notify the fault to the gas supplier
- if this is a bulk storage vessel-fed installation, then notify the fault to the gas supplier

Important: On no account should unauthorised persons attempt to remove or dismantle a cylinder/bulk storage vessel supply control valve.

Note: Installations fed by two or more cylinders may be tested for let-by through the cylinder outlets simultaneously, i.e. with all valves closed. If you detect let-by, repeat the test for each individual valve with the other cylinders disconnected and the free end of the pigtails capped, until you locate the source of let-by.

Procedure for tightness testing new and existing LPG/air mixture installations – designed to operate at 14mbar and 21mbar

Preparation

Cap or plug all open points in the system. Visually inspect the installation to ensure that all appliance isolation valves are open, that all appliance control taps (including any pilot taps) are shut.

On cookers with a fold-down lid, raise the lid to a fully open position during the tightness test so that any SSOV on the supply to the hotplate taps is open.

Procedure

1. Turn off the ECV/AECV for installations with a maximum operating pressure not exceeding 75mbar (low pressure), or Meter Inlet Valve (MIV) where the operating pressure exceeds 75mbar (medium pressure), ensuring the ECV is open.

Note: For installations with a maximum operating pressure above 75mbar, where work has been undertaken on any part of the meter installation ensure every joint within that installation is tested with LDF as soon as the ECV is opened.

2. Connect a suitable pressure gauge to a suitable test point on the installation (this would typically be at the gas meter, where installed).
3. Test the ECV or MIV for let-by, by adjusting the pressure to between 7–10mbar.

Note 1: For medium pressure installations, ensure the regulator is active. Where there is an UPSO downstream of the valve (ECV or MIV), this will close as the pressure falls. In this case, once the pressure has been reduced to between 7–10mbar, as in Step 3, operate the UPSO reset – to release trapped upstream pressure and then allow it to re-shut.

There may be a small sudden rise in the pressure gauge reading as the upstream pressure is released in to the downstream pipework. Should the pressure remain higher than 10mbar, repeat the process until a stable pressure between 7 and 10mbar is achieved.

4. Leave for 1 minute.
5. After 1 minute, observe the pressure gauge reading. If there is no discernible pressure rise 'no let-by' can be recorded for the ECV or MIV (see Note 2).

Note 2: If the procedure in Note 1 has been necessary, the UPSO reset will need to be operated again after the 1 minute period and before recording the second pressure gauge reading.

A rise in pressure will indicate either:

- that the gas temperature has risen during the test; or
- that the supply control valve (ECV or MIV) is letting by (see '**What happens if a supply control valve is letting by?**' in this Part) – confirm let-by, by disconnecting the outlet of the ECV or MIV and applying non-corrosive LDF to the barrel/ball of the valve.

If you suspect a gas temperature rise, repeat the procedure from Steps 3 to 5.

6. Let-by test satisfactory – slowly raise the test pressure to:

- for installations operating at 14mbar – 13–14mbar
- for installations operating at 21mbar – 20–21mbar

Note: When adjusting to the required test pressure you inadvertently take the pressure higher than either 14mbar or 21mbar, it will be necessary to repeat the process by dropping the pressure back down to 7–10mbar, before again slowly raising the pressure to the required test pressure – this ensure the regulator isn't locked-up.

7. Turn off the control valve.
8. Leave for 1 minute to allow the temperature to stabilise. Where necessary and after the 1 minute stabilisation period, re-adjust the pressure to that in Step 6.

Note: The pressure must be stable before proceeding with the tightness test, which may involve leaving additional time for temperature stabilisation.

9. Test for a further 2 minutes:

- for new and existing installations without appliance connected – there should be no discernible pressure drop (movement) of the pressure gauge and no smell of gas.
- for existing installations with appliances connected –
  - with or without a gas meter (≤6m³/h – U6, E6 or G4) and pipework – a maximum pressure drop of 1.5mbar with no smell of gas
  - with a gas meter (>6m³/h ≤16m³/h – U16/G10) and pipework – a maximum pressure drop of 0.5mbar with no smell of gas

10. If a drop in pressure is recorded that exceeds those values provided in Step 9, trace, repair and re-test the installation. If the source of escape cannot be traced/repaired, make the installation safe and cap/plug or all open points – implement the current GIUSP.
11. Upon completion of a satisfactory tightness test, remove the pressure gauge and seal the test point.

12. Open the supply control valve (ECV or MIV) and test the test point, control valve, outlet, regulator and meter connections with non-corrosive LDF.
13. Complete all necessary documentation (i.e. recording the test results) and advise the responsible person for the installation

What happens if a supply control valve is letting by?

You need to rectify the fault or change the valve and repeat the let-by test to prove 'no let-by' of the replaced supply control valve:

- notify the fault to the gas supplier and make the installation safe – suspend the tightness test until the control valve is rectified.

Once the valve has been replaced, repeat the complete tightness test procedure and where satisfactory, purge the installation (see **Purging LPG gas installations – precautions you need to take** in this Part and **Part 15 – Re-establishing gas supply/re-lighting appliances** for further guidance).

Important: On no account should unauthorised persons attempt to remove or dismantle a supply control valve.

Purging LPG gas installations – precautions you need to take

When you carry out work in relation to any meter or pipework, you must ensure that all installation pipework and fittings are purged of air.

During purging, take precautions to ensure gas is not allowed to accumulate in any confined space and:

- do not turn on an electrical switch or operate an appliance
- ensure there is no smoking or naked lights
- ensure adequate ventilation is provided within the vicinity of the purge point

An additional precaution needed with LPG:

When LPG is used, there is a danger during purging that gas may fail to disperse – due to the fact that LPG is heavier than air and may accumulate at low level. Therefore, in addition to the above precautions, carry out the purge:

- through the appliance where a source of ignition (e.g. lighted match) can be applied to the burner head until the purge is complete and the flame(s) stable

For further guidance on purging gas installations see **Part 15 – Re-establishing the gas supply/re-lighting appliances**.

Checking and/or setting LPG regulators – to maintain correct pressure

How pressure is reduced before reaching premises

LPG is generally supplied in bulk storage vessels or cylinders at pressures of up to 9 bar. There must be a reduction in pressure before reaching the premises.

This is achieved by using a pressure regulator to maintain a constant outlet pressure, irrespective of demand.

For LPG bulk storage systems, the pressure is broken down in two stages (occasionally three):

1. 1st stage regulator – the storage vessel pressure is reduced to a medium pressure, i.e. below 2 bar. The regulator is mounted on the bulk storage vessel.
2. 2nd stage regulator – medium pressure is further reduced to low pressure (operating pressure of 37mbar). The regulator can be mounted either on the bulk storage vessel downstream of the 1st stage regulator, or at the building externally on the outside wall.

1st and 2nd stage regulators for bulk storage vessels should comply with BS EN 13785: 2005 + A1: 2008 (or BS EN 16129: 2013 that supersedes BS EN 13785).

Note: The 2nd stage regulator will normally have an integral under pressure/over pressure shut off (UPSO/OPSO).

Typical bulk storage LPG installations are shown in Figures 18.3, 18.4 and 1811.

For cylinder installations, a single stage regulator is used to reduce the cylinder pressure to operating pressure and is mounted on/at the cylinder(s).

This regulator should comply with BS EN 12864: 2001 + A3: 2009 (or BS EN 16129: 2013 that supersedes BS EN 12864). See also **Operation and checking of gas safety devices and controls (LPG)** in this Part.

Where an automatic change over valve is fitted for multiple cylinder installations, this should comply with BS EN 13786: 2004 + A1: 2008 (or BS EN 16129: 2013 that supersedes BS EN 13786).

Maintaining the correct pressure

Most gas appliances used in domestic applications and in some commercial applications employ atmospheric burners.

A good example of an atmospheric burner would be a bunsen burner where a stream of gas issuing from the injector enters the mixing tube and in the process, entrains primary air through the primary air ports. This gas/air mixture is then ignited at the burner head.

To ensure satisfactory combustion at a gas burner, it is important that a constant pressure is maintained at the appliance burner injector.

Gas should therefore be supplied at the outlet connection of the regulator at a specified nominal pressure, e.g. 37mbar for Propane. All regulators will have set tolerances for minimum and maximum outlet pressures and these will vary dependent on what standard the regulator was manufactured to (see Table 18.5).

Note: It is your responsibility to check and ensure that the regulator pressure is correct.

LPG operating pressure – higher than that for NG

Because of the different characteristics of LPG, the operating pressure is significantly higher than that for NG. However, the principles for checking and/or setting the operating pressure, remain the same.

Note that the installation pipework should be of sufficient size to prevent a pressure absorption greater than 2 mbar across the installation (see **Sizing gas supplies (LPG) – important considerations** and Figures 18.5, 18.6 and 18.7 in this Part).

Method of checking the operating pressure

1. Attach a suitable pressure gauge to a test point immediately after the regulator (where possible).
2. Ensure that the pressure gauge is vertical and the water levels are zeroed i.e. water levels have the same reading in each limb, or the electronic gauge is calibrated in accordance with the manufacturer's instructions.

Note: With no gas in use, the pressure recorded will be the standing pressure. Do not confuse this with the operating pressure.

To establish the operating pressure, gas must be passing through the regulator at a sufficient flow rate. Where appliances have already been installed and commissioned, this can be achieved by operating all appliances at full gas flow rate. Where no appliances are installed and commissioned, this can be achieved by using a device that allows a flow rate of at least 0.5 m^3/h.

Figure 18.11 Typical bulk storage vessel LPG installations

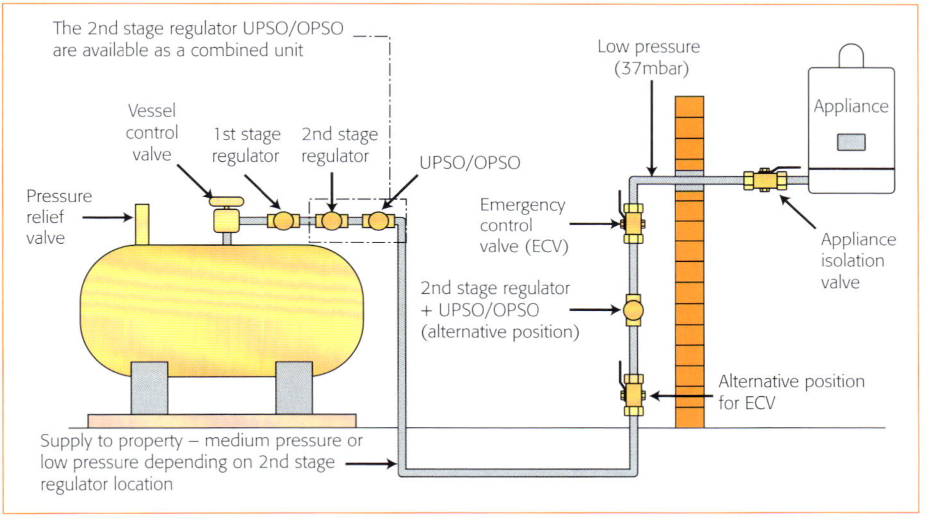

LPG – Operation/positioning of Emergency Control Valves (ECVs) and isolation valves

Bulk storage vessels

Bulk storage vessels are fitted with a service valve, which controls the high pressure stage. The gas supplier is responsible for the repair and maintenance of this valve.

During a gas emergency, or in adverse weather conditions, gaining access to the bulk storage vessel service valve (which will normally be fitted under a locked cover) could result in an unnecessary delay before the gas is turned off.

To overcome this potential problem:

- an ECV must be installed at the time of connecting the vessel to the gas supply pipework
- the valve should be fitted on an external wall with all relevant identification labels as near as possible to the point where the gas supply enters the dwelling (see **Part 10 – Emergency notices, warning labels and forms**)

The owner of the property will normally be responsible for the repair and maintenance of this ECV.

Cylinder(s)

Multi-cylinder installations:

- connect the 'in service' and 'reserve' cylinders to a manifold fitted with non-return valves, allowing one cylinder to be removed without shutting down the whole system
- for a continuous supply, the cylinders may be inter-connected via an automatic change-over device, which switches the gas supply from the 'in service' cylinders to the 'reserve' cylinders as the 'in service' cylinders empty (see Figure 18.12)

- fit multi-cylinders with an ECV located as near as practicable to the point where the gas supply enters the property and use the appropriate identification labels (see **Part 10 – Emergency notices, warning labels and forms**)

The property owner will normally be responsible for the repair and maintenance of this ECV.

On single cylinder installations:

- the vapour (on/off) valve on the cylinder regulator typically acts as the ECV

LPG meter installations

Every installation supplied through a meter should have an ECV as part of the associated meter controls.

It should be as follows:

- in a readily accessible position
- so that the lever (handle) of the valve is parallel to the pipe to which it is connected when it is in the open position
- so that the lever only moves downwards one quarter of a turn to the off position (see Figure 18.13)
- the lever should be securely attached to the valve spindle
- with a test point fitted downstream of the ECV
- labelled with the on/off position of the valve and the gas emergency procedure (see **Part 10 – Emergency notices, warning labels and forms**)

Operation and checking of gas safety devices and controls (LPG)

General

LPG regulators should conform to BS EN 12864 for cylinders, BS EN 13785 for bulk storage vessels and BS EN 13786 for change over valves for multiple cylinder installations.

Note: All three Standards are declared (by BSi) as being current and superseded by BS EN 16129: 2013.

They should have a throughput suitable for the rated gas heat input connected to the supply. The regulator is normally set by the manufacturer to provide the correct operating pressure for the type of LPG supply chosen. However, check this and if necessary or possible, adjust it to ensure that the correct operating pressure is achieved at the maximum demand (see Note).

Note: Many LPG regulators and associated controls are factory preset and should not require altering on site during their service life. Only the gas supplier or their authorised agent should break regulator seals that prevent interference with the pressure settings.

Propane

Propane appliances generally operate with a nominal supply pressure of 37mbar. This is lower than the vapour pressure within the storage vessel, which is approximately 7 bar. However, taking climatic conditions into consideration, the stored pressure can vary between 2 and 9 bar.

The vapour pressure needs to be reduced and this is normally achieved by using two pressure regulators, known as 1st stage and 2nd stage regulators.

A safety pressure relief valve protects a bulk storage vessel against over pressure. For cylinders, this safety device is part of the cylinder vapour valve.

Figure 18.12 LPG multi-cylinder installation

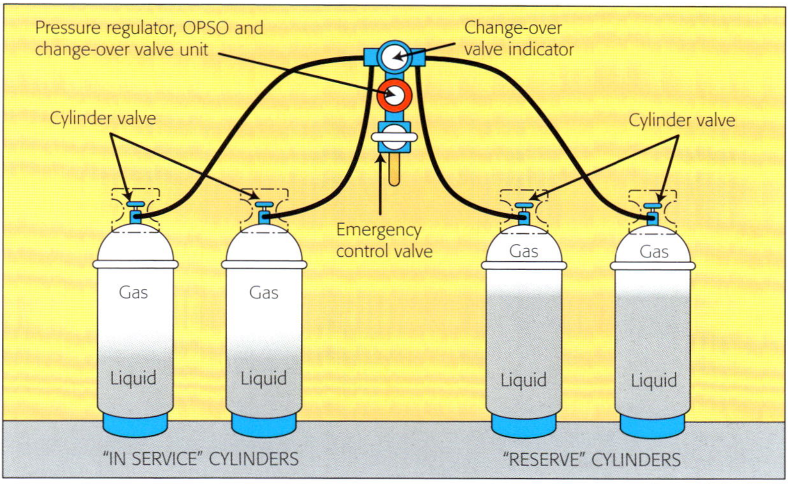

Figure 18.13 LPG meter installation

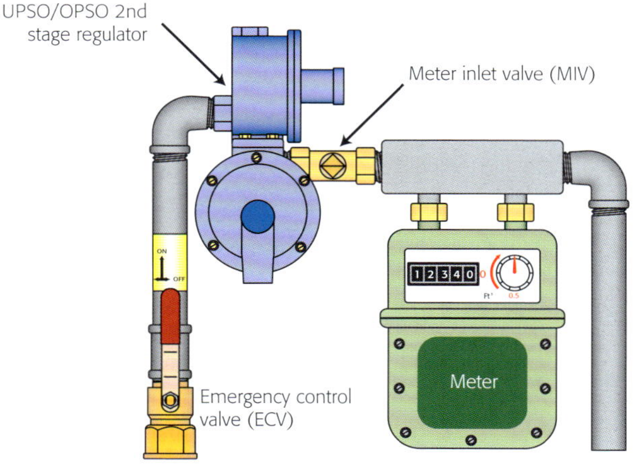

For single cylinder installations, a single stage regulator is normally used to reduce the pressure in the cylinder down to the nominal operating pressure (37mbar).

First (1st) stage regulator – bulk storage vessel (Propane)

The 1st stage regulator, which is normally fitted at the bulk storage vessel, is sited under the control cover hood and adjacent to the vapour take-off. It accepts the varying high pressure within the vessel and reduces it down to a constant medium pressure below 2 bar.

It must be capable of maintaining a constant nominal outlet pressure, not fluctuating when the pressure within the vessel varies due to weather conditions.

If the downstream pipework is not capable of withstanding the bulk storage vessel pressure should the 1st stage regulator fail, a high pressure OPSO is fitted. This is either incorporated into the 1st stage regulator or a separate device. The OPSO set pressure is normally between 2 and 4 bar, dependent on the application.

These devices should always be used where polyethylene (PE) pipework is installed after the 1st stage and before the 2nd stage regulators.

Second (2nd) stage regulator – bulk storage vessel (Propane)

The 2nd stage regulator on domestic installations normally forms part of a single unit, which incorporates both 2nd stage regulation together with under and over pressure protection as illustrated in Figure 18.14.
(see **Under pressure/over pressure shut off (UPSO/OPSO) safety device** in this Part).

Where separate, fit the 2nd stage regulator down stream of the 1st stage regulator. You may also fit it onto a gas meter inlet, if a metered supply is required.

For installations involving long pipe runs, you may fit the 2nd stage regulator remote from the storage vessel, immediately adjacent to the point of entry into the property.

The 2nd stage regulator's function is to reduce the medium pressure and provide a constant nominal operating pressure (low pressure) of 37 mbar. The regulator rating should be capable of passing the sufficient capacity of gas required by the installation.

Cylinder pressure regulation

Cylinder systems operate on a similar principle to bulk storage vessel installations, except that they are fitted with a small pre-set single stage regulator, which is connected directly onto the cylinder valve outlet connection generally known as a POL valve (see **Part 19 – Definitions**) Alternatively, you may fit an automatic cylinder changeover valve which also acts as a pressure regulator.

Under pressure/over pressure shut off (UPSO/OPSO) safety device

GSIUR require that where gas:

- is supplied from a bulk gas storage vessel; or
- an installation is supplied with gas from 4 or more cylinders

there must be:

- a means of preventing the installation pipework and gas fittings downstream of the 2nd stage regulator from being subjected to a pressure different to that for which they are designed
- an adequate alternative automatic means for preventing the gas fittings from being subjected to a greater pressure than that for which they are designed

The UPSO/OPSO valve carries out all of these statutory requirements.

The current ACoP and Guidance to the GSIUR and BS 6891: 2015 require all fixed pipework installations to be protected from over pressurisation by the fitting of an OPSO and a limited relief valve. This is irrespective of the size of the instalation or the size and number of cylinders connected to it. A typical single stage regulator designed for direct connection to the vapour valve on a cylinder is pictured in Figure 18.1.

The UPSO/OPSO valve (see Figure 18.14) is designed to protect an LPG installation and appliances from under and over pressure situations.

It is normally constructed as a single unit and can be mounted on the bulk storage vessel, or wall-mounted externally adjacent to the entry into the dwelling. Each device operates by stopping gas flow when the internal diaphragms sense abnormal pressures.

The device can be re-set manually – but it is essential that the ECV for the installation and all appliances are in the 'off' position before this is done.

The UPSO/OPSO valve may also be incorporated in a 2nd stage regulator, set to nominally 37 mbar outlet operating pressure. You do not then need a further 2nd stage regulator.

2nd stage regulators with an OPSO valve also include a limited relief valve. This may vent small amounts of gas atmosphere to protect the installation from over pressurisation whilst preventing nuisance tripping of the OPSO.

The settings for the UPSO, Limited Relief and OPSO are as follows:

- **UPSO** – shuts off the gas flow, should the outlet operating pressure be interrupted or should a malfunction cause the regulator to fully open (low pressure). The UPSO will also operate should the maximum capacity be exceeded. The UPSO will close between 25 mbar and 32 mbar (see Table 18.5).
- **OPSO** – senses the pressure downstream of the 2nd stage regulator, but isolates the line upstream of the regulator. The standard the OPSO was manufactured to will determine the pressure range in which it will close (see Table 18.5).
- **Limited relief valve** – The standard the relief valve was manufactured to will determine the pressure range in which it operates (see Table 18.5).

For many years regulators with slightly different settings will be installed and encountered, but there is no requirement to replace older regulators that may operate at different pressures to the current BS EN 16129 regulators.

Some regulators have warning indicators to indicate the operation of the UPSO and OPSO.

The UPSO can be re-set by the gas user/gas operative after shutting off all controls downstream of the valve. Under pressure is usually caused by an empty storage vessel/cylinder(s).

The OPSO is sealed and if installed by the gas supplier can only be re-set by them. Over pressure is usually caused by the failure of a pressure regulator.

The limited relief valve is self-acting and does not require re-setting.

Butane

Butane regulators are generally mounted on the cylinder and operate at a nominal outlet operating pressure of 28 mbar or 29 mbar dependent on the standard the regulator was manufactured to (see Table 18.5).

Note: Only the gas supplier or their authorised agent should break regulator seals that prevent interference with the pressure settings.

Figure 18.14 Under pressure/over pressure shut off valve (UPSO/OPSO) safety device

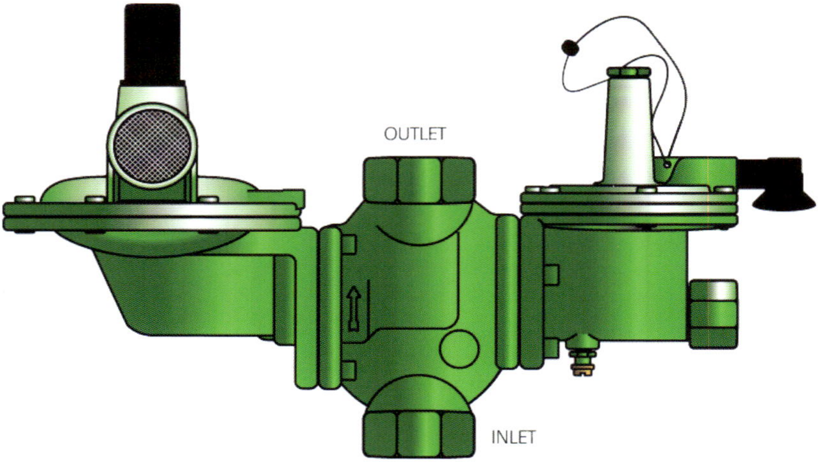

Figure 18.15 Single stage regulator incorporating an OPSO

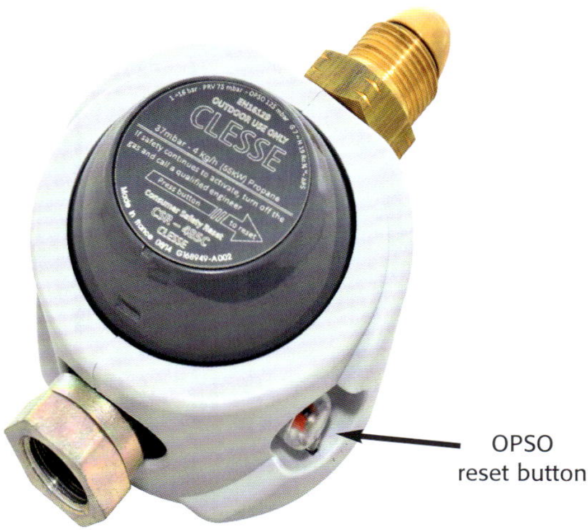

LPG cylinder valve

Operation and use:

A cylinder valve is fitted into the top of every refillable storage cylinder that weighs 4.5kg and above, via a screwed bung welded into the cylinder. The valve serves two purposes:

1. To allow re-filling of the cylinder with liquid phase gas by the supplier.
2. To allow the user to draw off gas from the cylinder and close the gas supply off.

The valves are an integral part of the cylinder: you must not remove them, nor must gas users.

Many different makes will be found with the outlet connection being either a 'clip-on' or 'screw type' fitting.

The most popular outlet sizes for Butane cylinders are 20mm or 21mm 'clip-on' valves, with a regulator fitting directly onto the cylinder valve. Some butane cylinders though have 'left hand male threads'.

Valves for Propane cylinders normally have 'left hand female threads'; These are frequently refered to as POL connections. A number of Propane cylinders are on the market for use with leisure appliance, such as BBQ's and patio heaters, that have 27mm 'clip-on' valves.

Do not alter threads and adapters or force them onto dissimilar connections.

Cylinder valves are protected from accidental damage by a metal guard welded to the cylinder.

Inspect connections and valves

1. Check they are not damaged and are kept free of debris.
2. Only use the valve for the gas pressure and/or gas type for which it has been designed.

LPG cylinder change-over valve

Operation and use:

When a single LPG cylinder is unable to provide an adequate capacity of gas for an installation, it is possible to increase the capacity by linking together sets of cylinders.

These cylinders are joined via a manifold, a manual changeover valve or an automatic change-over valve.

In a bank of four cylinders, such as that illustrated in Figure 18.12, fitted with an automatic changeover valve, two would be 'in service' and two would be in 'reserve'.

Change-over from 'in service' to 'reserve' is achieved automatically by a regulator that senses the pressure difference between the near-empty and full cylinders.

An indicator on the valve assembly warns the user of the 'change-over' from 'reserve' to 'service'.

Checks

Ensure the whole assembly is accessible for both the gas user and gas operative – and for cylinder changing by the gas supplier / gas consumer.

The change-over valve should be firmly attached to the fabric of the building.

Table 18.5 Pressure characteristics of various regulators (LPG)

Fuel	Nominal outlet pressure mbar	Minimum outlet pressure mbar	Maximum outlet pressure mbar	Maximum lock-up pressure (above set pressure) mbar	Relief valve operating range mbar[A)]	UPSO operating range mbar	OPSO operating range mbar[A)]
BS 3016 - cylinder single stage regulator set pressures:							
Butane	28	23	33	+10	-	-	-
Propane	37	32	42	+15	-	-	-
BS 3016 - Final (second) stage regulator set pressures (low pressure):							
Propane	37	32	42	+10	50 - 62[B)]	25 - 32	70 - 80
BS 3016 - Cylinder automatic changeover device set pressure (low pressure):							
Propane	37	32	42	+15	50 - 62[B)]	-	70 - 80
BS EN 12864 or BS EN 16129 - Cylinder single stage regulator set pressure:							
Butane	29	22	35	40	48 - 150	-	48 - 150
Propane	37	27[C)]	45	50	60 - 150	-	60 - 150
BS EN 13785 or BS EN 16129 - Final stage regulator set pressures (low pressure):							
Propane	37	32[D)]	45	50	60 - 150	25 - 32[E)]	60 - 150
BS EN 13786 or BS EN 16129 - Cylinder automatic changeover device set pressures (low pressure):							
Butane	29	22	35	40	48 - 150	-	48 - 150
Propane	37	27[C)]	45	50	60 - 150	-	60 - 150

A) For BS EN 13785 and BS EN 16129 regulators this operating range is the maximum and minimum values within which the device needs to activate. BS EN 16129 specifies a tolerance on the manufacture's declared set pressure of ±20% for a relief valve and ±15% for an OPSO

B) Normal setting 55 mbar.

C) Value is the minimum outlet pressure specified in the regulator standard to align with BS EN 437. However, it is recommended that, for the UK market, the minimum operating pressure for propane regulators is 32 mbar.

D) Due to the UK-agreed UPSO range the minimum outlet pressure for regulators manufactured for the UK market is taken as 32 mbar NOT the 27 mbar declared in BS EN 13785 or BS EN 16129.

E) Industry - agreed figures for regulators manufactured for the UK market (not formally documented in the standards

Definitions – 19

Definitions

'R' type adaptor: see **Ridge terminal adaptor ('R' type adaptor) in this Part**.

1st family gas: at present normally only LPG-air.

2nd family gas: Natural gases.

3rd family gas: Liquefied Petroleum Gases (LPG).

Access plate: a removable plate giving access into a combustion chamber and/or heat exchanger.

Additional emergency control valve (AECV): a valve, which is in addition to the emergency control valve (ECV). The AECV is not located at the end of the network and may not isolate all of the consumer's pipework (see **Emergency control valve (ECV) in this Part**).

Adventitious ventilation: ventilation obtained through gaps around doors, floors and windows, for example.

Aerated burner: a burner in which some or all of the required air has been mixed with the gas before it leaves the burner port(s).

Air guide plate: an integral part of a gas fire designed to guide secondary air to the burner flames – may be part of the radiant support plate.

Air heater: appliance for heating air to be used for space heating.

Air vent free area: total area of the unobstructed openings of an air vent.

Air vent: non-adjustable grille or duct, which allows the passage of air at all times.

Annular space: space between a flue liner and chimneybreast.

Appliance (gas): appliance designed for heating, lighting, cooking or other purposes.

Appliance compartment: an enclosure (not being a habitable space) specifically designed or adapted to house a gas appliance(s) (see also **Balanced compartment in this Part**).

Appliance manufacturer's installation instructions: the instructions prepared by the appliance manufacturer giving particular information and requirements on how the appliance should be installed.

Appliance ventilation duct: provided to convey combustion or cooling air for an appliance or component.

Aspect ratio: means the ratio of width to depth when measured in a horizontal plane.

Atmosphere sensing device: (also known as an oxygen depletion system) shuts off the gas supply to an appliance burner before there is a build up of a dangerous quantity of combustion in the room concerned.

Back boiler: water heating appliance designed to fit behind a fire, in a fireplace recess or builder's opening to provide central heating and/or hot water.

Back circulator: appliance with a rated heat input (net) not exceeding 6kW, designed to fit behind a fire, in a fireplace recess or a builder's opening, primarily for the supply of hot water.

Balanced compartment: method of installing an open-flued appliance as room-sealed, so that flueing and ventilation provide a balanced flued effect.

19 – Definitions

Balanced flued appliance: a room-sealed appliance which draws its combustion air from a point adjacent to that at which the combustion products are discharged, the inlet and outlet being so disposed that wind effects are substantially balanced.

Basement (LPG appliances): a room, which is completely or partly below ground level on all or some sides.

Bayonet connector: the plug end on a flexible pipe, which is inserted into a wall-mounted valve on the gas supply. Insertion of the plug opens the valve and its withdrawal closes it.

Bedsitting room: any room or space used for living and sleeping purposes.

Benchmark scheme: a voluntary code of practice, designed to ensure that central heating installers fit boilers in accordance with the manufacturers' installation instructions. Manufacturers provide a 'Log Book' as part of the literature pack, which is supplied with the boiler and this remains with the appliance throughout its life.

Boiler: appliance designed to heat water for space heating and/or water supply.

Bottled gas: normally Butane or Propane stored as a liquid under pressure in refillable portable containers.

Bourdon gauge: a dial type pressure gauge.

Branched flue system: a shared open-flued system serving appliances situated on two or more floors.

Builder's opening: a recess made in a chimney or wall into which an appliance may be placed and from which a flue is run.

Bulk storage vessel: permanently installed vessel of approved design for the storage of LPG under pressure, which is filled on site.

"Calorific Value" (or CV): the Calorific Value is the quantity of heat (energy) produced when a unit volume of the fuel, measured under standard conditions of temperature and pressure, is burned completely in excess air.

A distinction is made between:

a) Gross Calorific Value – (also known as the Higher Calorific Value or HCV) – in the determination of which the water vapour produced by the combustion of the fuel is deemed to have been condensed into a liquid phase at the standard temperature and its latent heat released.

b) Net Calorific Value – (also known as the Lower Calorific Value or LCV) – in the determination of which water vapour produced by the combustion of the fuel is assumed to remain in the vapour phase. The net Calorific Value is therefore the gross Calorific Value minus the latent heat of the water vapour contained in the combustion products.

Canopy: facilitates the passage of the products of combustion into the flue from an open-flued/flueless appliance.

Capacity (of a gas meter): maximum rate that gas will flow through the meter, measured in m^3/hr or ft^3/hr.

Catalytic space heater: a flueless space-heating appliance in which neat fuel gas is passed through a catalyst bed and is flamelessly oxidised by air that has diffused into the bed from the surrounding atmosphere at a temperature below that at which combustion would usually occur.

Catchment pit: specially constructed pit, sited in a safe place, in which LPG can collect to enable it to evaporate in a controlled manner. Used after a controlled or uncontrolled release of LPG to atmosphere.

Catchment space: correctly sized void below the flue spigot connection for the collection of soot, rubble and other debris (see **Void in this Part**).

Central heating system: a fixed system for warming a building from a single source of heat, with manual or automatic control of the operation of the whole system and of the temperatures in the heated space.

Chimney: a structure consisting of a wall or walls enclosing the flue or flues.

Chimney bird guard: a device which when fitted to a chimney pot prevents the entry of nesting birds, but allows the safe dispersal of the products of combustion.

Chimney breast: projection beyond the thickness of a wall that contains the flue and builder's opening.

Chimney component: any part of a chimney.

Chimney flue blocks: pre-formed sections of flue for building into a wall.

Chimney liner: pipe inside a chimney to form a flue. May be rigid or flexible.

Chimney plate: a permanent plate or label giving details of the chimney or flue installation.

Chimney pot: prefabricated unit fitted at the outlet of a chimney.

Chimney system: a complete assembly of chimney components from one or more appliances to a single terminal, including primary flue(s) and draught diverter(s), if any.

Circulator: boiler with a rated heat input not exceeding 8kW (gross), designed primarily for the supply of domestic hot water in conjunction with a separate storage vessel.

Closed flue system: an obsolete term for a flue system that does not have a draught diverter, flue break or any draught break on the appliance, so that the system is closed to the room or internal space.

Closure plate: non-combustible plate to close off a fireplace opening when installing a gas fire.

Combination (combi) boiler: a wet central heating boiler capable of providing hot water directly (some models contain an internal hot water store).

Combination hot water storage unit:

a) a hot water supply apparatus comprising of a hot water storage vessel with a cold water feed cistern immediately above it, the two being fabricated together as a compact unit.

b) a hot water supply apparatus comprising a hot water storage vessel with a cold water feed cistern beside it or inside it.

Commissioning: initial start-up of an installation to check and adjust for safe and reliable operation.

Common flue system: a shared open-flued system that serves two or more appliances installed in the same room or space.

Compartment (caravan): separate habitable living space, which includes any area that can be divided by the use of a curtain or similar.

Compartment: housing specifically designed or adapted to house a gas appliance(s).

Competence: competence in safe gas installation requires gas operatives to have enough knowledge, practical skill and experience to carry out the job in hand safely, with due regard to good working practice. Knowledge must be kept up-to-date with awareness of changes in law, technology and safe working practice.

19 – Definitions

Concealed connection: pipe connection to an appliance installed so that the pipe or fittings are not visible.

Condensate drain: a device in a flue where condensate can be removed.

Condensate-free length: the length of individual open-flue, which can be calculated to maintain the temperature of the flue gases above the dew point.

Condensing appliance: designed to use latent heat from water vapour in the combustion products by condensing the water vapour within the appliance.

Convector heater: discharges heated air only into the space in which it is installed.

Cooker: a composite appliance comprising an oven or ovens, a hotplate and normally a grill, designed for domestic use.

Cooler plate/device: fitted at the spigot outlet of a gas fire to prevent overheating of a flue-block chimney.

Credit meter: a meter in which the volume registered by the index is the basis of a periodic account rendered to the customer.

Damper: a device used to vary the volume of air passing through a confined cross-section by varying the effective cross sectional area.

Data plate: a durable, permanently fixed plate bearing specified information relative to the appliance.

Decorative Fuel Effect (DFE) gas appliance: primarily for decorative purposes but designed to simulate a solid fuel appliance and intended to be installed so that the products of combustion pass unrestricted from the firebed to the flue.

Dew point: the temperature of a mixture of combustion products and water vapour at which further cooling results in condensation of the water vapour.

Diaphragm meter: a positive displacement meter in which the measuring chambers have deformable walls, e.g. a U6/G4 meter.

Direct flue connection: when an appliance is connected directly, or via an adaptor, to an open-flued chimney system. Products of combustion pass directly from the appliance into the flue.

Direct gas-fired tumble dryer: an appliance in which textile material is dried by tumbling in a rotating drum through which heated air and products of combustion are forced or induced by mechanical means.

Downstream: that part of a gas installation after a certain point e.g. the gas meter is downstream of the meter regulator.

Draught brake: an opening into any part of an open-flued chimney system, including that part integral with the appliance.

Draught diverter: prevents interference to the combustion of an open-flued appliance; must be fitted to the manufacturer's instructions and in the same room, space or compartment as the appliance, with at least 600mm of vertical flue above it.

Dress-guard: fitted to the front of a gas fire to prevent contact between a heat source and clothes. Should conform to BS 1945.

Drop-out time: time taken for the flame supervision system to respond to a loss of flame.

Dry run: verification of the correct operation of components and interlocks without fuel being available to the combustion space.

Duct: purpose-designed enclosure used to contain gas pipes.

Ducted warm air heater: flued appliance, which uses ducts to distribute the heated air.

Electrical bonding: see **Equipotential bonding** in this Part.

Electronic meter: a meter that measures indirectly the volume of gas passing through it, e.g. an E6 meter.

Electronic portable gas analyser: apparatus that will detect and measure the presence of combustion gases and clearly display the results

Electronic token meter (ETM): see **Prepayment meter** in this Part.

Emergency control valve (ECV): valve for shutting off the supply of gas in an emergency; not a service isolation valve see also **Additional emergency control valve (AECV)** in this Part.

Enforcing authority: an authority with a responsibility for enforcing the Health and Safety at Work Etc. Act 1974 and other relevant statutory provisions; normally Health and Safety Executive (HSE) or the local authority for the area as determined by the Health and Safety (Enforcing Authority) Regulations 1977.

Equipotential bonding: electrical conductor between a point close to the outlet of a gas meter and the earth terminal (such bonding does NOT involve connecting electrical power to gas pipework).

Equivalent flue height: the height of the straight vertical flue pipe, measured from the flue spigot to the terminal, which will produce the same flue flow rate as the flue under consideration.

Fan diluted flue system: permits dilution with air of the combustion products from an open-flued system, providing safe termination in otherwise unacceptable locations.

Fanned draught flue system: flue system in which the draught to remove the products of combustion is produced by a fan.

Fanned draught room-sealed appliance: an appliance that, when in operation, has the combustion system including the air inlet and the products of combustion outlet, isolated from the room or space in which the appliance is installed. The draught to operate the flue is created by an integral fan.

Fire compartment: room or space constructed to prevent the spread of fire.

Fire stop: a barrier or seal of non-combustible material that is designed to prevent or retard the passage of smoke or flames.

Fire surround: purpose-designed setting for a gas fire, fitted against a wall at the base of a flue and usually incorporating a hearth.

Fire wall: requirement on bulk LPG storage vessel; a wall, screen or partition in the open air to reduce radiated heat on an LPG vessel and ensure adequate distance for dispersion of any leaking LPG.

Firebed: that part of a fuel effect gas appliance on which the gas burns.

Fireplace opening: recess formed by the inclusion of fireplace components in the builder's opening. Also known as a fireplace recess.

Fireplace recess: a recess formed by the addition of fireplace components in the builder's opening.

19 – Definitions

Flame failure: the loss of a flame from the normally detected position.

Flame retention: prevention of flame-lift off.

Flame supervision device: control, which detects the presence of a flame and in the absence of that flame, prevents the uncontrolled release of gas to the burner.

Flat roof: a section of a roof, which is horizontal or virtually horizontal such that any airflow over it will not affect the operation of flue terminating above the roof, i.e. within 20° of the horizontal. This includes a "bow string truss" roof as may be found in a residential park home.

Flexible connector: pipe connector to a bayonet valve that allows a moveable flueless appliance to be safely disconnected; allows an appliance to be moved a short distance without the need for disconnection.

Flue: this is the passage or space for conveying the products of combustion to the outside atmosphere.

Flue box: non-combustible substitute for a builder's opening or fireplace recess.

Flue break: an opening in the secondary flue in the same room as and in addition to, the opening at the draught diverter.

Flue liner: this is the wall of a chimney consisting of components, the surface of which is in contact with products of combustion. This is not just a flexible flue liner but also any suitable material to convey the products of combustion.

Flue outlet: the part of the appliance that allows the exit of products of combustion from the appliance.

Flue pipe: a pipe enclosing a flue; for a double-walled type it is the inner pipe.

Flue safety device: a device designed to detect adverse flue conditions (down draught) at the appliance draught diverter; also known as a TTB.

Flue terminal guard: fitted to prevent human contact (especially that of children) with a terminal and to prevent interference with the terminal or damage to it.

Flue terminal: device fitted at the flue outlet to:

- assist the escape of products of combustion.
- minimise downdraught.
- prevent flue blockages.

Flue termination: the outlet of a flue system where products of combustion discharge into external air.

Flueless appliance: designed for use without a flue under conditions in which the products of combustion do not present a safety risk.

Free area: total area of the individual unobstructed openings of an air vent.

Freestanding cooker: cooking appliance resting directly on the floor comprising:

- a hotplate, including one or more burners.
- one or more ovens, with or without a thermostat.
- possibly a grill and/or griddle.

Gas fittings: means gas pipework, valves (other than emergency controls), regulators and meters and fittings, apparatus and appliances, designed for use by consumers of gas for heating, lighting and other purposes, for which gas can be used (other than the purpose of an industrial process carried out on industrial premises), but it does not mean:

a) any part of a service pipe.

b) any part of a distribution main or other pipe upstream of the service pipe.

c) a gas storage vessel.

d) a gas cylinder or cartridge designed to be disposed of when empty.

Gas meter: an instrument for measuring and recording the volume of gas that passes through it without interrupting the flow of gas.

Gas Safety Regulations: legally binding requirements for safe gas work.

Gas transporter: a person conveying gas in a network as defined in GSMR.

Gather: a smooth transitional reduction in area above the builder's opening or fireplace recess into the flue.

Greenhouse heater: flueless convector heater of not more than 4.2kW specially designed to be able to tolerate a partially vitiated environment.

Grill: appliance for cooking food by radiant heat.

Grille: a fixed non-closeable mesh or latticework of holes designed and constructed to prevent vermin from entering, but allow the passage of air at all times.

Hearth: a floor level slab of fire-resistant material to prevent overheating of the floor beneath a heating appliance.

High level grill: a grill above the hotplate of a cooker.

High pressure cut-off: automatically shuts off the gas supply if a high pressure is registered on the low pressure side of the regulator; normally sealed and cannot be reset by the gas user. On LPG, this is known as an over pressure shut-off (OPSO).

High pressure stage (LPG): that part of the LPG installation between the take-off valve of the bulk storage vessel or cylinder and the inlet of the 1st stage regulator. For Propane, the pressure will be in the region of 6.9bar. For Butane, the pressure will be in the region of 1.93bar. These pressures may vary dependant upon ambient temperatures.

Hotplate (hob): comprises of one or more covered or uncovered burners and designed to support cooking vessels.

Identity badge: see **Data plate** in this Part.

Index: a series of dials or rows of figures indicating the volume of gas that has passed through a gas meter.

Infill panel: panel of fire-resisting material used in a fireplace opening that is too large for a closure plate alone.

Inset Live Fuel Effect (ILFE) gas fire: appliance designed to simulate a solid fuel appliance. Installed so that the passage of the products of combustion from the firebed to the chimney or flue is restricted; air enters the fireplace only through purpose-designed openings.

Instantaneous water heater: water heater with a nominal heat input up to 45kW intended to supply hot water to a sink, basin, bath or a comprehensive service to a number of draw-off points, the water being heated only as it passes through the appliance to the tap/outlet.

19 – Definitions

Insulated chimney – factory-made: complete assembly of all essential factory-made insulated sections, fittings and accessories to convey the products of combustion to the outside air.

Interlock device: monitor, which inhibits action if safe operating conditions are not met.

Intermediate pressure stage (LPG): the part of the LPG installation between the outlet of the 1st stage regulator and the inlet of the 2nd stage regulator. For Propane, the pressure will be in the region of 75mbar. Also known as medium pressure.

Internal space: an interior space such as a passageway, hall, stairway or landing, which is not itself a room.

Intrinsically safe: apparatus in which no spark or any thermal effect produced under prescribed test conditions (including normal operation and specified fault conditions) is capable of causing ignition of a given explosive atmosphere.

Intumescent air vent: special air vents, designed to prevent the spread of smoke in the event of a fire. They incorporate a latticework of holes to allow continuous ventilation, but in the event of extreme heat/fire, they will expand and close.

Latent heat of condensation: the quantity of heat removed, at constant temperature during the change from the gaseous to liquid state.

Latent heat: the quantity of heat (energy) added to or removed from, a substance at constant temperature during a change of state.

Leak detection fluid (LDF): used to identify leaks on pipework and appliances, should contain no greater than 30ppm of halogens when used on stainless steel, should not contain ammonia when applied to copper and brass and should have a ph value no greater than 7.0.

Leisure accommodation vehicle: a unit of living accommodation for temporary or seasonal occupation that may meet the requirement for the construction and use of road vehicles, e.g. a caravan.

Lint arrester: a fibre pad or metallic mesh designed to trap a mixture of dust, fluff, fibres and droplets of grease which would otherwise collect in the airways of a burner.

Liquefied Petroleum Gas (LPG): normally commercial Propane or Butane gas, stored in a vessel under pressure, which turns into a liquid state.

Lock-out: a safety shut-down condition of a control system such that restart cannot be accomplished without manual intervention.

Lock-up: is where at zero flow conditions, to limit the maximum pressure to an acceptable level at its outlet, the meter regulator closes off.

Louvre: an assembly of fixed angled blades or vanes contained within a framework, which is designed to resist light and rain penetration but allow the passage of air at all times.

Low pressure (LPG): the part of the LPG installation between the outlet of the 2nd stage regulator and the gas appliance(s). For Propane, the normal operating pressure is 37mbar. For Butane, the normal operating pressure is 28mbar.

Low pressure (Natural gas): gas inlet pressure to the meter regulator not exceeding 75mbar.

Low pressure regulator: apparatus for automatically maintaining a constant gas outlet pressure at the level recommended for an appliance.

Main flame: flame on the main burner.

Manometer: instrument for the measurement of gas pressure. Includes 'U' and electronic gauges.

Manual reset low pressure cut-off valve: automatically shuts off a gas supply if the pressure falls below a predetermined level; needs to be reset by hand. Also known as under pressure shut-off (UPSO) on LPG installations.

Manufacturer's instructions: documents supplied with the gas appliance/equipment by the manufacturer giving guidance on how to use, service, maintain and install the product.

Maximum Incidental Pressure (MIP): the maximum operating pressure that a gas installation is allowed to experience under fault conditions, limited by safety devices.

Maximum Operating Pressure (MOP): the maximum operating pressure that a gas installation is allowed to operate continuously under normal operating conditions.

MDPE pipe: medium density polyethylene pipe.

Mechanical ventilation: air supplied by a fan.

Medium pressure (Natural gas): gas inlet pressure to the meter regulator between 75mbar and 2bar.

Meter box: a receptacle or compartment designed and constructed to contain a meter with its associated gas fittings.

Meter bracket: purpose-made support including means of securing the meter union, from which a unit construction meter can be hung.

Meter compound: an area or room designed and constructed to contain one or more meters with their associated gas fittings.

Meter filter: a filter fitted between the ECV and the meter.

Meter housing: a meter box or meter compound.

Meter inlet valve (MIV): a valve, situated between the outlet of the meter regulator and the inlet connection of the meter on medium pressure fed meter installations. Used for gas tightness testing purposes of the downstream gas installation.

Meter rating: the maximum volumetric rate at which the performance of a meter is certified to be within statutory requirements or some other agreed requirements.

Meter regulator: a device located in close proximity and upstream of a primary meter which is used solely to control the pressure of the gas within the gas installation.

Meter union: fitting used to connect the meter to the pipework.

Mounded vessel: storage vessel for LPG sited above ground or partly buried, completely covered by a mound of earth or similar material.

Natural draught chimney system: a chimney system in which the draught is provided by the thermal force arising from the heat of the products of combustion.

Non-combustible material: that which has passed tests for non-combustibility in accordance with British Standards.

Non-ferrous: denoting any metal other than iron, i.e. any metal not containing iron.

Non-return valve: device to prevent the reversal of gas, water or air flow.

Ofgem approved meter installer (OAMI): a registered gas meter installer is an organisation or person which performs meter installation work and is registered in accordance with regulation 3(3) of the Gas Safety (Installation and Use) Regulations 1998 with Ofgem.

19 – Definitions

Ofgem: The Office of Gas and Electricity Markets (Ofgem) was formed early in 1999 by combining the functions of the former Office of Gas Supply (Ofgas) and the Office of Electricity Regulation (OFFER).

Open basket type: a freestanding DFE gas appliance not fitted into a chairbrick or sealed to a fireplace recess.

Open-flued appliance: an appliance designed to be connected to an open-flued chimney system which draws combustion air from the room or space in which it is installed.

Open-flued chimney system: system that is open to a room or internal space at each appliance.

Operating pressure: the pressure at which the gas installation operates under normal conditions.

Oven directly/internally heated: an oven in which the burner(s) is situated and products of combustion circulate, inside the oven space.

Oven forced convection/circulation: an oven within which an even temperature is achieved with the aid of a fan.

Oven indirectly/externally heated: an oven in which the burner(s) is situated outside the oven space and the products of combustion do not pass into it.

Oven semi-directly/semi-externally heated: an oven in which the burner(s) is situated outside the oven space, but the products of combustion pass into it.

Oven: an appliance for baking and roasting.

Oxygen depletion system: shuts off the gas supply to an appliance burner before there is a build up of a dangerous quantity of products of combustion in the room concerned, also known as an atmosphere sensing device.

Permanent dwelling: structure of a permanent nature, to which appropriate building regulations apply and which is used primarily for domestic purposes.

Pliable connector: a stainless steel tube formed with annular corrugations and having factory-fitted end connections.

Pluming: visible cloud of combustion products from an outside flue terminal, which are cooled to below dew point by mixing with external air.

POL valve: an abbreviation describing a Prestolite connection (originating from the USA), which is manufactured with two ground surfaces of a predetermined shape which are mated together by fittings using left hand threads. In the UK a slightly different design of POL connection is utilised.

Pre-aerated burner: a burner to which gas and air are supplied already mixed.

Prepayment meter: a meter (mechanical or electronic) fitted with a mechanism, which on the insertion of a coin or token, permits the passage of a predetermined quantity of gas.

Pressure gauge: instrument for indicating or recording pressure.

Pressure regulator: automatically maintains a constant outlet pressure.

Pressure test point: small plug type fitting on a meter, installation pipe or appliance, allowing attachment of a pressure gauge.

Prevailing wind: the direction of wind, which at a given place, occurs more frequently than any other during a specified period. Over all parts of the British Isles statistics show the prevailing wind to be south, southwest or west. There is however, an annual variation, e.g. in some places the prevailing wind in spring and early summer has an easterly component.

Primary meter: meter connected to the service pipe, which forms the basis of charge by the supplier of gas for that premises.

Protected area: an area such as a hall corridor or stairway, which is enclosed with fire resisting materials.

Protected shaft: a shaft which enables persons, air or objects to pass from one compartment to another, enclosed within a fire-resisting construction.

Protected stairway: a stairway including an exit passageway leading to its final exit, enclosed within a fire-resisting construction (other than any part that is an external wall of a building).

Radiant convector gas fire: gas fire designed to emit heat by both radiation and convection.

Radiant gas fire: gas fire designed to emit heat mainly by radiation.

Radiant support plate: an integral part of a gas fire designed to support and align gas fire radiants – design may direct secondary air to the burner flames
(see **Air guide plate in this Part**).

Radiant: component designed to become incandescent when heated by a gas flame.

Range: a heavy duty (cast iron) cooker.

Register plate: a fire-resistant plate used to seal the annular space around the base of a flexible flue liner and masonry chimney
(see **Annular space in this Part**).

Remotely operated emergency valve: shut-off valve capable of remote operation, which will close automatically on loss of the actuating power or fire engulfment and which, preferably, is fire safe in accordance with BS 5146, API607 or API6FA. Electrically driven valves need not close automatically on fire engulfment if the power supply is adequately fire protected.

Residential park home: a mobile home designed for permanent residential accommodation that does not meet the requirement for construction and use of road vehicles.

Responsible person: the occupier of the premises or, where there is no occupier or the occupier is away, the owner of the premises or any person with authority to take appropriate action in relation to any gas fitting therein.

RIDDOR: The Reporting of Injuries, Diseases and Dangerous Occurrences Regulations 1995.

Ridge terminal adaptor ('R' type adaptor): a fitting for connecting a circular cross-sectional pipe to the rectangular-section connection of a ridge flue terminal.

Ridge terminal: a terminal designed for fitting at the ridge of a building.

Room-sealed: an appliance that, when in operation, has the combustion system, including the air inlet and the products outlet, isolated from the room or space in which the appliance is installed.

Safety shut-off valve: actuated by the safety control so as to admit and stop gas flow automatically.

Sealed heating system: not open to the atmosphere; has a sealed diaphragm expansion vessel and other safety controls.

Secondary flue: the part of the open-flued chimney system connecting a draught diverter or draught break to the terminal.

Secondary meter: a subsidiary meter for measuring gas, used to separate parts of premises or for separate appliances, after the gas has passed through a primary meter.

Se-duct: a duct serving special room-sealed appliances; it is open at both ends and rises vertically in buildings to bring combustion air in and take combustion products out.

Service isolation valve (SIV): valve for controlling a supply of gas incorporated in a service pipe and not situated inside a building; intended for use by a supplier or transporter of gas.

Service pipe: a pipe for distributing gas to premises from a distribution main, being any pipe between the distribution main and the outlet of the first Emergency Control Valve (ECV) downstream from the distribution main (Natural gas system).

Service pipework: means a pipe for distributing gas to premises from a gas storage vessel, being any pipe between the gas storage vessel and the outlet of the ECV (LPG systems).

Shared open-flued system: serves two or more appliances e.g. Se-ducts and U-ducts flue systems.

Sleeve: duct, tube or pipe embedded in the building structure allowing the gas installation pipework to pass through a wall or floor; capable of containing the gas.

Solid-top hotplate: a hotplate in which the burner(s) is fitted below a solid plate(s) on which the cooking vessels stand.

Stoichiometric mixture: a mixture of gas and air in the proportions determined by the theoretical air requirement.

Storage water heater: storage system with an integral storage vessel that is intended to supply hot water to a bath or a comprehensive service to a number of draw-off points.

Temporary continuity bond: a means of providing electrical continuity on a gas supply for safety reasons.

Thermal cut-off device: safety device designed to stop the flow of gas when the surrounding air temperature exceeds the predetermined value.

Thermostat: a thermally actuated control device for maintaining a desired temperature.

Throat: contraction in the flue immediately above the fireplace recess to accelerate flue gases.

Through-room: any room formed by the removal of an intercommunicating wall between two rooms, or any large room formed by two open plan smaller rooms. The opening/archway present between two smaller rooms may have sliding or intercommunicating doors.

Tightness test: the testing of installation pipes and equipment for escapes from the system (also known as a soundness test).

Transducer: electrical device, which can convert an electrical signal into a mechanical signal, i.e. it can act as a transmitter and receiver.

Transfer grille: non-adjustable fitment in a wall, door, or partition, to transfer air between adjacent rooms and/or spaces.

U-duct: literally, a u-shaped flue system; combustion air is provided by one limb and special room-sealed appliances are connected to the other. The U-duct ends are open and adjacent.

Under pressure shut-off: automatically shuts off LPG supply if the pressure falls below a predetermined level; needs to be reset by hand. Also known as a manual reset low pressure cut-off valve (UPSO).

UPSO/OPSO (Under and Over pressure shut-off): automatically shuts off the LPG gas supply if the pressure increases, or decreases from predetermined levels. The safety device incorporates a manual reset for both functions.

Upstream: that part of a gas installation prior to, or before a certain point e.g. the meter is upstream of the appliance.

Valve: device to stop or regulate the flow of gas by the closure or partial closure of an orifice by means of a gate, flap or disc.

Ventilation opening: includes any means of ventilation, which opens directly to external air, such as the openable parts of a window, a louvre, airbrick, progressively openable ventilator or window trickle ventilator. It also includes any door, which opens directly to external air.

Ventilation: the process of supplying fresh air to and removing used air from, a room or internal space.

Void: a correctly sized catchment space below a flue spigot connection for the collection of soot, rubble and other debris.

Warning pipe: an overflow pipe so fixed that its outlet, whether inside or outside a building, is in a conspicuous position where the discharge of any water from it can be easily seen.

Water diaphragm: a flexible component within the water section of a water heater that moves when subjected to water pressure and thus controls the flow of gas.

Work: in relation to a gas fitting this includes any of the following activities carried out by any person, whether an employee or not:

a) Installing or reconnecting the fitting.

b) Maintaining, servicing, disconnecting, permanently adjusting, repairing, altering or renewing the fitting or purging it of air or gas.

c) Where the fitting is not readily movable, changing its position; and

d) Removing the fitting.

Note: work in this context does not include the connection or disconnection of a bayonet fitting or other self-sealing connector.

Useful data – 20

Symbols

A	ampere
Btu	British thermal unit
c	centi (prefix) (0.01)
CV	calorific value
c/s	cycle per second
cal	calorie
cc	used loosely for cm^2 (cubic centimetre)
cd	candela
cm	centimetre
cwt	hundredweight
d	deci (prefix) (0.1)
dg	decigram
dm	decimetre
eV	electronvolt
F	farad
fl oz	fluid ounce
ft	foot
ftH_2O	foot head of water
g	gram
gal	gallon
h or hr	hour
hp	horsepower
Hz	hertz
in Hg	inch of mercury
in	inch
in w.g.	inch water gauge
j	joule
K	kelvin
k	kilo (prefix) (1000)
kcal	kilocalorie
kg	kilogram
kgf	kilogram force
kj	kilojoule
km	kilometre
kPa	kilopascal
kW	kilowatt
l	litre
l/s	litres per second
lb	pound
lbf	pound force
lm	lumen
M	mega (prefix) (1000000)
MJ	megajoule
MJ/m^3	megajoule per cubic metre
m	metre
m	milli (prefix) (0.001)
m/s	metres per second
mb	sometimes used for mbar
mbar	millibar
mg	milligram
mile	mile
min	minute
ml	millilitre
mm	millimetre
MN	meganewton

20 – Useful data

mol	mole
MPa	megapascal
N	newton
n	nano (prefix) (0.000 000 001)
nm	nanometre
Pa	pascal
pt	pint
qt	quart
s	second
stp	standard temperature and pressure
t	tonne
tf	tonne force
ton	ton
tonf	ton force
USgal	US gallon
V	volt
W	watt
yd	yard
°C	degree Celsius
°F	degree Fahrenheit
$\geq$	equal to or more than
$<$	less than
$>$	greater than
$\leq$	equal to or less than

Conversion – Metric/Imperial

To convert to metric, multiply by the factor shown.
To convert to imperial, divide by the factor shown.

Imperial	Factor	Metric
	Length	
ft	0.3048	m
ins	25.4	mm
ins	2.54	cm
	Area	
sq. ft	0.0929	m^2
sq. ft	929.03	cm^2
sq. ins	645.16	mm^2
sq. ins	6.4516	cm^2
	Volume	
cu yds	0.7646	m^3
cu ft	0.0283	m^3
cu ft	28.3168	dm^3 or litre
cu ins	16.3871	cm^3
	Capacity	
gal	4.546	dm^3 or litre
pt	0.568	dm^3 or litre
	Velocity	
mph	1.0693	km/h
ft/sec	0.03048	m/s
	Mass	
tons	1.016	tonnes
lbs	0.4536	kg
	Fuel	
mpg	0.354	km/litre
Btu/ft^3	0.0373	MJ/m^3
Btu	0.000293	kW
	Pressure	
in wg	2.491	mbar
lbf/ft^2	0.4788	mbar
lbf/in^2	0.06895	bar

Conversion – kW-Btu/hr and Btu/hr-kW

kW	Btu	Btu	kW
0.5	1706	1000	0.29
1.0	3412	5000	1.47
1.5	5118	10000	2.93
2.0	6824	15000	4.40
2.5	8530	20000	5.86
3	10236	25000	7.33
4	13649	30000	8.79
5	17061	35000	10.26
6	20473	40000	11.72
7	23885	45000	13.19
8	27297	50000	14.65
9	30709	55000	16.12
10	34121	60000	17.58
11	37534	70000	20.51
12	40946	75000	21.98
13	44358	80000	23.45
14	47770	90000	26.38
15	51182	100000	29.31
20	68243	110000	32.24
25	85304	120000	35.17
30	102364	130000	38.10
35	119425	140000	41.03
40	136486	150000	43.96
45	153546	160000	46.89
50	170607	170000	49.82
55	187668	180000	52.75
60	204729	190000	55.68
70	238850	200000	58.61
80	272971	250000	73.27
90	307093	300000	87.92
100	341214	350000	102.57
110	375336	400000	117.23
120	409457	450000	131.88
130	443579	500000	146.54

Conversion – Temperatures

°C	°F
-40	-40
-20	-4
-17.5	0
-10	14
-5	23
0	32
5	41
10	50
15	59
20	68
25	77
30	86
35	95
40	104
45	113
50	122
55	131
60	140
65	149
70	158
75	167
80	176
85	185
90	194
95	203
100	212

References – 21

Manufacturers/Suppliers

Aga-Rayburn

Tel: 0845 712 5207
Website: www.agaliving.com

Adey Professional Heating Solutions

Tel: 01242 546 700
Fax: 01242 546 777
Website: www.adey.co.uk

Airflow (Nicoll Ventilators) Ltd

Tel: 01425 611 547
Fax: 01425 638 912
E-mail: sales@airflow-vent.co.uk
Website: www.airflow-vent.co.uk

Alpha

Tel: 0844 871 8764
Website: www.alpha-innovation.co.uk

Arctic Products Ltd

Tel: 0113 271 5245
Fax: 0113 271 5779
E-mail: sales@arctic-products.co.uk
Website: www.arctic-products.co.uk

Ariston

Tel: 0333 240 7777 (technical)
Website: www.ariston.com/uk

Anton-group

Tel: 0870 428 0075
E-mail: sales@anton-group.com
Website: www.anton-group.com

Baxi

Tel: 0344 871 1545
E-mail: info@baxi.co.uk
Website: www.baxi.co.uk

B.E.S Ltd

Freefone: 0800 80 10 90
Website: www.bes.co.uk

Biasi

Tel: 01922 714 600
E-mail: technicalandservice-service@biasi.uk.com
Website: www.biasi.co.uk

Brewer Cowls

Tel: 01243 539639
Fax: 01243 533184
E-mail: sales@brewercowls.co.uk
Website: www.brewercowls.co.uk

Mantec Technical Ceramics (formerly Ceramic Gas Products)

Tel: 01782 377 550
Fax: 01782 377 599
E-mail: info@mantectc.com
Website: www.mantectechnicalceramics.com

Comark Instruments

Tel: 0207 942 0712
Fax: 0207 942 0714
E-mail: technical@comarkinstruments.com
Website: www.comarkinstruments.net

Crosslee PLC

Tel: 01422 203 555
Fax: 01422 206 304
E-mail: info@crosslee.co.uk
Website: www.crosslee.co.uk

Danfoss Randall Ltd

Tel: 01234 320 257 (technical)
Fax: 01234 320 297 (technical)
E-mail: ukheating@danfoss.com
Website: heating.danfoss.co.uk

Drager Safety UK Ltd

Tel: 01670 352 891
Fax: 01670 356 266
Website: www.draeger.com

Drayton Controls

Tel: 0333 7000 622 (technical)
Fax: 0333 1230 522
E-mail: customer.care@invensys.com
Website: www.draytoncontrols.co.uk

Electrolux plc

Tel: 08445 613 613
Fax: 08445 618 920
Website: www.electrolux.co.uk

Elster Metering Ltd

Website: www.elstermetering.co.uk

Evinox Energy Ltd

Tel: 01372 722 277
Fax: 01372 744 477
Website: www.evinoxenergy.co.uk

Fernox

Tel: 0330 100 7750 (Technical)
E-mail: technical@fernox.com
Website: www.fernox.com

Ferroli

Tel: 0330 205 0005 (Technical)
Website: www.ferroli.co.uk

Flueboost Ltd (flue boosters)

Tel: 01565 755 599
Fax: 01565 755 055
E-mail: flueboost@yahoo.co.uk
Website: www.flueboost.co.uk

Focal Point Fires Plc

Tel: 01202 499 330
Fax: 01202 499 326
E-mail: sales@focalpointfires.co.uk
Website: www.focalpointfires.co.uk

F&P Wholesale

Tel: 0116 256 7380 (Head Office)
Fax: 0116 256 7473
E-mail: info@fpwholesale.co.uk
Website: www.fpwholesale.co.uk

Gas & Environmental Services

Tel: 01707 373 751
Fax: 01707 373 752
E-mail: info@gesuk.com
Website: www.gesuk.com

Gas Appliance Spares Limited

Tel: 01772 702 755
Fax: 0845 305 4281
E-mail: info@gas-spares.co.uk
Website: www.gas-spares.co.uk

Kiwa Gastec

Tel: 01242 677 877
Fax: 01242 676 506
E-mail: gastecenquiries@kiwa.co.uk
Website: www.kiwa.co.uk

Glow-Worm

Tel: 01773 824 639
Tel: 0330 100 7679 (technical)
Website: www.glow-worm.co.uk

Grundfos Europump

Tel: 01525 850 000 (South)
Tel: 01942 263 628 (North)
Tel: 01506 461 666 (Scotland)
Website: uk.grundfos.com

Glen Dimplex Boilers

Tel: 0844 371 1111
E-mail: sales@glendimplexboilers.com
Website: www.glendimplex.com

Harton Services

Tel: 020 8310 0421
E-mail: office@hartonservices.com
Website: www.hartons.co.uk

Heatline

Tel: 01773 596 611
Fax: 01773 828 123
E-mail: info@heatline.co.uk
Website: www.heatline.co.uk

Hepworth Building Products

Tel: 01709 856 300
Fax: 01709 856 301
Website: www.wavin.co.uk

Hepworth Heating (Please see Glow-worm)

Honeywell For Your Home

Website: www.yourhome.honeywell.com

Horstmann

Tel: 01179 788 700
Fax: 01179 788 701
E-mail: reception@horstmann.co.uk
Website: www.horstmann.co.uk

Ideal Boilers

Tel: 01482 498 663 (Technical)
Tel: 01482 498 660 (Customer services)
Website: idealboilers.com

Indesit Company UK

Tel: 01733 568 989
Fax: 01733 341 783
Website: www.gda.uk.com

Intergas Heating Ltd

Tel: 01527 888 000
Fax: 01527 888 001
E-mail: info@intergasheating.co.uk
Website: www.intergasheating.co.uk

Invensys Controls

See **Drayton**

Johnson & Starley Ltd

Tel: 01604 762 881
Fax: 01604 767 408
Website: www.johnsonandstarley.co.uk

Kane International Ltd (Analysers)

Tel: 01707 375 550 (switchboard)
Fax: 01707 393 277
Website: www.kane.co.uk

Keston Boilers

Tel: 01482 443 005
Fax: 01482 467 133
E-mail: technical@keston.co.uk
Website: www.keston.co.uk

Kitchen Economy

Tel: 02920 451 222
E-mail: spares@kitchen-economy.co.uk
Website: www.kitchen-economy.co.uk

Leisure Consumer Products

Tel: 0845 600 4916
Tel: 0345 600 4916 (Moble callers)
Website: www.leisurecooker.co.uk

Malvern Boilers Limited

Tel: 01684 893 777
Fax: 01684 893 776
Website: www.malvernboilers.co.uk

Mecserflex Manufacturing Company Ltd

Tel: 01793 773 320
Fax: 01793 773 330
Website: www.mecserflex.co.uk

Midtherm Flue Systems Ltd

Tel: 01384 458 800
Fax: 01384 458 811
E-mail: sales@midtherm.co.uk
Website: www.midtherm.co.uk

Midtherm Fans Limited

Tel: 01384 253 475
Fax: 01384 253 229
E-mail: info@midthermfans.co.uk
Website: www.midthermfans.co.uk

Myson Radiators

Tel: 0845 402 3434
Fax: 0191 491 7568
E-mail: sales@myson.co.uk
Website: www.myson.co.uk

Myson Underfloor

Tel: 0845 402 3434
Fax: 0191 491 7465
E-mail: underfloor@myson.co.uk
Website: www.myson.co.uk

New World Domestic Appliances Ltd

see **Stoves**

OH Ltd

Tel: 0800 644 4222 (steel cowls)
Tel: 0800 644 4400 (clay pots, boat cowl & general technical)
Website: www.ohlimited.co.uk

Pactrol Controls Ltd

Tel: 01942 529 240
Fax: 01942 529 241
Website: www.pactrol.com

Parts Center

Tel: 0845 270 9800
Website: www.partscenter.co.uk

Pegler Yorkshire

Tel: 0800 156 0050
Website: www.pegleryorkshire.co.uk

PH Products Ltd (Artic)

See **Artic Products Ltd**

Plumbase

Tel: 0800 169 9862
Fax: 01293 552 100
Website: www.plumbase.com

Polygroup

Tel: 01772 632 850
E-mail: sales@polytank.co.uk
Website: www.polytank.co.uk

Potterton

Tel: 0344 871 1560 (technical)
E-mail: info@potterton.co.uk
Website: www.potterton.co.uk

Purmo–UK

Tel: 0845 070 1090
Fax: 0845 070 1080
E-mail: uk@purmo.co.uk
Website: www.purmo.com

Purmo–Ireland

Tel: 090 9673 006
E-mail: info@matrixbathrooms.ie
Website: www.matrix

Quinn Radiators UK

Tel: 01633 657 000
Fax: 01633 657 151
E-mail: enquiries@qrl-radiators.com
Website: www.qrl-radiators.com

QMI Europe Ltd

Tel: 0844 482 4497
Fax: 0844 247 121
E-mail: enquireies@qmieurope.com
Website: qmieurope.com

Teddington Appliance Controls Ltd

Tel: 01726 222 530 (technical)
Fax: 01726 67953 (technical)
Website: www.tedcon.com

Telegan Gas Monitoring Ltd

see **Anton Group**

Telford Copper Cylinders Ltd

Tel: 01952 257 961
Fax: 01952 253 452
E-mail: sales@telford-group.com
Website: www.telford-group.com

Testo Ltd

Tel: 01420 544 433
Website: www.testolimited.com

TFC Group LLP

Tel: 01732 351 680
Fax: 01732 354 445
Website: www.tfc-group.co.uk

TracPipe (OmegaFlex Ltd)

Tel: 01295 676 670
Fax: 01295 267 302
E-mail: tech-@omegaflex.net
Website: www.tracpipe.co.uk

Vaillant Ltd

Tel: 0344 693 3133 (technical)
Website: www.vaillant.co.uk

Valor Ltd

Tel: 0844 879 3588
Website: www.valor.co.uk

Viessmann Ltd

Tel: 01952 675 070 (technical)
Fax: 01952 675 040
E-mail: technical-uk@viessman.com
Website: www.viessmann.co.uk

Vogue (UK) Ltd

Tel: 01902 387 000
Fax: 01902 387 001
E-mail: info@vogueuk.co.uk
Website: www.vogueuk.co.uk

Vokera Ltd

Tel: 0844 391 0999
Fax: 0844 391 0998
E-mail: technical@vokera.co.uk
Website: www.vokera.co.uk

White Rodgers (brand of Emerson Climate Technologies)

Tel: 01525 375 655
Fax: 01525 378 075
Website: www.emersonclimate.com

Wilo (UK) Ltd

Tel: 01283 523 000
Fax: 01283 523 099
Website: www.wilo.co.uk

Worcester Bosch Group

Tel: 0330 123 3366 (technical)
E-mail: technical.enquiries@uk.bosch.com
Website: www.worcester-bosch.co.uk

Yorkshire Copper Tube Ltd

Tel: 01902 499 766 (technical)
Fax: 0151 545 5018
E-mail: technical@yorkshirecopper.com
Website: www.yorkshirecopper.com

Zehnder Retail Division (Radiators)

Tel: 01276 408 401
E-mail: radiator.technical@zehnder.co.uk
Website: www.zehnder.co.uk

Ravenheat Manufacturing Ltd

Tel: 0113 252 7007 (sales & technical)
E-mail: enquiries@ravenheat.co.uk
Website: www.ravenheat.co.uk

Remeha

Tel: 0118 978 3434 (switchboard)
Fax: 0118 978 6977
E-mail: technical@remeha.co.uk
Website: www.uk.remeha.com

Robinson Willey

Tel: 01282 686 791
Fax: 01282 686 799
E-mail: info@robinsonwilley.com
Website: www.robinsonwilley.com

Sangamo (control switches)

Tel: 01475 745131
E-mail: enquiries@sangamo.co.uk
Website: www.sangamo.co.uk

Saunier Duval

Website: www.saunierduval.com

Selkirk (distributed by Deks Distribution UK)

Tel: 01275 858 866
Fax: 01275 855 887
E-mail: sales@deks.org.uk
Website: www.selkirkchimney.co.uk

Servowarm

Tel: 0800 731 0321
Website: www.servowarm.org

Siemens plc

Tel: 01276 696 000 (HQ)
Website: www.siemens.co.uk

Siemens Metering Ltd

Tel: 01159 906 6000
Website: www.siemens.co.uk

SIT Controls UK

Tel: 01302 836 340
Fax: 01302 836 341
E-mail: sit.uk@sitgroup.it
Website: www.sitgroup.it

Sperryn Gas Controls

Tel: 01462 443 225
Fax: 01462 443 311
E-mail: sperrynsales@cranebsu.com
Website: www.sperryn.co.uk

Spirotech UK Ltd

Tel: 0208 451 3344
Fax: 0208 451 3366
E-mail: info@spirotech.co.uk
Website: www.spirotech.co.uk

Stadium

Tel: 01843 854 006
Fax: 01843 854 010
Website: www.flambeau.co.uk

Stoves

Tel: 0344 815 3745 (spares)
Website: www.stoves.co.uk

Strebel Ltd

Tel: 01276 685 422
Fax: 01276 685 405
E-mail: info@strebel.co.uk
Website: www.strebel.co.uk

Sugg Lighting Ltd

Tel: 01293 540 111
E-mail: sales@sugglighting.co.uk
Website: www.sugglighting.co.uk

Sunvic Controls Ltd

Tel: 01698 810 945 (technical)
Fax: 01698 813 637
E-mail: technical@sunvic.co.uk
Website: www.sunvic.co.uk

Organisations

Association for the Conservation of Energy (ACE)

Website: www.ukace.org

Association of Independent Gas Transporters (AIGT)

Tel: 07790 877148
E-mail: john.barrett@aigt.org.uk

Association of Plumbing and Heating Contractors (APHC)

Tel: 0121 711 5030
Fax: 0121 705 7871
E-mail: info@aphc.co.uk
Website: www.aphc.co.uk

Blue Flame Certification

Tel: 0845 194 90 31
E-mail: cps@blueflamecertification.com
Website: www.blueflamecertification.com

British Marine Federation (BMF)

Tel: 01784 473 377
Fax: 01784 439 678
E-mail: info@britishmarine.co.uk
Website: www.britishmarine.co.uk

British Standards Institution (BSI)

Tel: 0208 996 9000
Website: www.bsigroup.com

Builders Merchants Federation (BMF)

Tel: 02476 854980
Fax: 02476 854981
E-mail: info@bmf.org.uk
Website: www.bmf.org.uk

Building Research Establishment (BRE)

Tel: 01923 664 000
Fax: enquiries@bre.co.uk
Website: www.bre.co.uk

Building Services Research & Information Association (BSRIA)

Tel: 01344 465 600
Fax: 01344 465 626
E-mail: bsria@bsria.co.uk
Website: www.bsria.co.uk

Chartered Institution of Building Services Engineers (CIBSE)

Tel: 020 8675 5211
Fax: 020 8675 5449
Website: www.cibse.org

Chartered Institute of Plumbing and Heating Engineering (CIPHE)

Tel: 01708 472 791
Fax: 01708 448 987
E-mail: info@ciphe.org.uk
Website: www.ciphe.org.uk

City and Guilds of London Institute

Tel: 0207 294 2468
Fax: 0207 294 2413
E-mail: centresupport@cityandguilds.com
Website: www.cityandguilds.com

Confederation of British Industrys (CBI)

Tel: 020 7379 7400
Fax: 020 7379 7200
Website: www.cbi.org.uk

Construction Industry Training Board (CITB)

Tel: 0344 994 4400
Fax: 01485 577 793
Website: www.citb.co.uk

UK Government Departments (centralised resource for accessing numerous governmental departments)

Website: www.gov.uk/government/organisations

Energy Savings Trust (EST)

Tel: 020 7222 0101
Fax: 020 7654 2444
Website: www.energysavingtrust.org.uk

Energy & Utilities Alliance (EUA) has four divisions: Utility Networks, HHIC (see HHIC in this Part), ICOM (see ICOM in this Part) & Hot Water Association (HWA)

Tel: 01926 513 777
Fax: 01926 511 923
E-mail: mail@eua.org.uk
Website: www.eua.org.uk

Energy & Utility Skills

Tel: 0845 077 99 22
E-mail: enquiries@euskills.co.uk
Website: www.euskills.co.uk

Federation of Master Builders (FMB)

Tel: 0330 333 7777
Website: www.fmb.org.uk

Gas Industry Safety Group (GISG)

Tel: 020 7706 5155
E-mail: info@gisg.org.uk
Website: www.gisg.org.uk

Gas Safe Register®

Tel: 0800 408 5500
E-mail: enquiries@gassaferegister.co.uk
Website: www.gassaferegister.co.uk

HSE (Health and Safety Executive) Books

Tel: 01787 881 165
Fax: 01787 313 995
E-mail: hsebooks@prolog.uk.com
Website: www.books.hse.gov.uk

Heating & Hotwater Industry Council (HHIC)

Tel: 01926 513 777
Website: www.centralheating.co.uk

HVCA - See The Building & Engineering Services Association

ICOM Energy Association (Formally BCEMA)

Tel: 01926 513 748
Fax: 01926 855 017
Website: www.icomenergyassociation.org.uk

Institute of Domestic Heating & Environmental Engineers

Tel: 01895 820 954
Fax: 02380 66 08 88
E-mail: admin@idhee.org.uk
Website: www.idhee.org.uk

Institution of Gas Engineers and Managers (IGEM)

Tel: 01509 678 182
Fax: 01509 678 198
E-mail: general@igem.org.uk
Website: www.igem.org.uk

Manufacturers' Association of Radiators and Convectors (MARC) - a subsidiary of BARMA (Boiler and Radiator Manufacturers Association Ltd)

Tel: 01926 513 742
Website: www.marcuk.com

National Association of Chimney Sweeps (NACS)

Tel: 01785 811 732
Fax: 01785 811 712
Website: www.nacs.org.uk

National Caravan Council (NCC)

Tel: 01252 318 251
Fax: 01252 322 596
E-mail: info@thencc.org.uk
Website: www.thencc.org.uk

National Federation of Builders (NFB)

Tel: 0345 057 8160
Fax: 08450 578 161
Website: www.builders.org.uk

National Housing Federation

Tel: 020 7067 1010
Fax: 020 7067 1011
Website: www.housing.org.uk

National Inspection Council for Electrical Installations Contracting (NICEIC)

Tel: 0870 013 0382
Website: www.niceic.com

Ofgem
(Office of the Gas and Electricity Markets) – see also Consumer Direct

Tel: 020 7901 7295
Fax: 020 7901 7066
Website: www.ofgem.gov.uk

Ofwat
(the Water Services Regulation Authority)

Tel: 0121 644 7500
Fax: 0121 644 7559
E-mail: mailbox@ofwat.gsi.gov.uk
Website: www.ofwat.gov.uk

Royal Society for the Prevention of Accidents (RoSPA)

Tel: 0121 248 2000
Website: www.rospa.com

Scottish and N. I. Plumbing Employers' Federation (SNIPEF)

Tel: 0131 225 2255
Fax: 0131 226 7638
E-mail: info@snipef.org
Website: www.snipef.org.uk

The Association for Decentralised Energy (ADE)

Tel: 020 3031 8740
E-mail: info@theade.co.uk
Website: www.theade.co.uk

The British Electrotechnical and Allied Manufacturers Association

Tel: 0207 793 3000
E-mail: technical@beama.org.uk
Website: www.beama.org.uk

The Building & Engineering Services Association (BESA) – formerly the HVCA

Tel: 020 7313 4900
Website: www.besa.com

The Energy Institute

Tel: 0207 467 7100
Website: www.energyinst.org

The Stationary Office (TSO)

Tel: 0870 600 5522
Website: www.tso.co.uk

United Kingdom Accreditation Service (UKAS)

Tel: 020 8917 8400
E-mail: info@ukas.com
Website: www.ukas.com

UK Government Departments
(centralised resource for accessing numerous governmental departments)

Website: www.gov.uk/government/organisations

UKLPG

Website: www.uklpg.org

Nationally Accredited Certification Scheme (ACS) for Individual Gas Fitting Operatives

UKAS ACCREDITED CERTIFICATION BODIES

Blue Flame Certification
Chatterley Whitfield Enterprise Centre, Chatterley Whitfield, Stoke on Trent, Staffordshire, ST6 8UW

Tel: 0845 194 90 31
Website: www.blueflamecertification.com

BPEC Certification
2 Mallard Way, Pride Park, Derby, DE24 8GX

Tel: 0845 644 6558
E-mail: info@bpec.org.uk
Website: www.bpec.org.uk

CITB
CITB-ConstructionSkills, Bircham Newton, Kings Lynn, Norfolk, PE31 6RH

Tel: 0344 994 4020
E-mail: cskillsawards@cskills.org
Website: www.cskills.org

Cert-ain Certification Ltd
58a High Sgtreet, Melbourne, Derbyshire, DE73 8GJ

Tel: 01332 987220
E-mail: info@cert-ain.com
Website: www.cert-ain.com

ERS
Lincoln Court, Bryggen Road, Kings Lynn, Norfolk, PE30 2HZ

Tel: 01553 771 147
E-mail: enquiry@erscert.co.uk
Website: www.erscert.co.uk

LOGIC Certification Limited
Unit 2, 1 Rowdell Road, Northolt, Middlesex, UB5 5QR

Tel: 020 8839 2439
E-mail: enquiries@logic-cert.com
Website: www.logiccertification.com

NICEIC Certification
Warwick House, Houghton Hall Park, Houghton Regis, Dunstable, LU5 5ZX

Tel: 0800 519 5190
E-mail: traininginfo@niceic.com

UK Certification
Unit 5, Station Close, Westover Trading Estate, Langport, Somerset, TA10 9RB

Tel: 01458 252 757
E-mail: info@ukcertification.co.uk
Website: www.ukcertification.org

Health and Safety Executive (HSE) Area Offices

London

Rose Court

2 Southwark Bridge, London, SE1 9HS.
Fax: 020 7556 2102

Covers: London only.

Westminster Office

Caxton House, Tothill Street, London, SW1H 9NA.
Fax: 020 7227 3802

Wales & South West

Cardiff

Government Buildings, Phase 1, Ty Glas, Llanishen, Cardiff CF14 5SH.
Fax: 029 2026 3120

Covers: Merthyr Tydfil, Rhondda Cynon Taff, Vale of Glamorgan, Bridgend, Neath Port Talbot, Powys, Blaenau Gwent, Caerphilly, Cardiff, Monmouthshire, Newport, Torfaen and part of Powys.

Bristol

2 Rivergate, Bristol, BS1 6EW.
Fax: 01179 262 998

Covers: Bristol, Somerset, Bath and North East Somerset, North Somerset, Gloucestershire, South Gloucestershire, Dorset, Swindon and Wiltshire.

Plymouth

North Quay House, Sutton Harbour, Plymouth, PL4 0RA.
Fax: 01752 226 024

Covers: Devon and Cornwall.

Clwyd

Regent House,Regent Street, Wrexham, Clwyd, LL11 1PR.
Fax: 01978 355 669

Covers: Anglesey, Conwy, Denbighshire, Flintshire, Wrexham, Gwynedd and part of Powys.

Dyfed

Tŷ Myrddin, Old Station Road, Carmarthen, Carmarthenshire, SA31 1LP.
Fax: 01267 223 267

Covers: Carmarthenshire, Pembrokshire, Ceredigion and Swansea.

Dorset

14 New Fields, Stinsford Road, Nuffield Industrial Estate, Poole, Dorset, BH17 0NF.
Fax: 01202 667 224

Covers: Dorset

East & South East

Basingstoke

Priestley House, Priestley Road, Basingstoke, RG24 9NW.
Fax: 01256 404 100

Covers: Berkshire, Hampshire, Oxfordshire and Isle of Wight.

Bedford

Woodlands, Manton Lane, Manton Lane Industrial Estate, Bedford.
Fax: 01234 220 633

Covers: Hertfordshire, Cambridgeshire, Bedfordshire and Milton Keynes and Buckinghamshire.

Chelmsford

Wren House, Hedgerows Business Park, Colchester Road, Springfield, Chelmsford, CM2 5PF.
Fax: 01245 706 222

Covers: Essex (except Barking, Havering, Redbridge and Waltham Forest, these are covered by London), Norfolk and Suffolk.

Kent

International House, Dover Place, Ashford, Kent, TN23 1HU.
Fax: 01233 634 827

Covers: Kent.

Norwich

Rosebery Court, 2nd Floor, St Andrew's Business Park, Norwich, Norfolk, NR7 0HS.

Covers: Norfolk, Suffolk and Essex (see also Chelmsford).

Oxted

The council Offices, Station Road east, Oxted, Surrey, RH8 0BT

Covers: East & West Sussex, Surrey

Midlands

Birmingham

1 Hagley Road, Birmingham, B16 8HS.
Fax: 0121 607 6349.

Covers: West Midlands.

Cheltenham

St James House, St KJames Square, Cheltenham, Gloucestershire, GL50 3PR
Fax: 01242 587133

Chelterham office only covers Nuclear Directorate

Crewe

Hornbeam House, Electra Way, Crewe, CW1 6GJ

Covers: Staffordshire and Shropshire

Northampton

900 Pavilion Drive, Northampton Business Park, Northampton, NN4 7RG.
Fax: 01604 738333.

Covers: Leicestershire, Northampton, Rutland and Warwickshire.

Nottingham

City gate West, Level 6 (First Floor), Toll House Hill, Nottingham, NG1 5AT.
Fax: 0115 971 2802

Covers: Nottinghamshire, Derbyshire, Lincolnshire (North Lincolnshire covered by Sheffield Office).

Worcester

Haswell House, St Nicholas Street, Worcester, WR1 1UW.
Fax: 01905 723 045

Covers: Worcestershire and Herefordshire.

North West

Bootle Headquarters

Redgrave Court, Merton Road, Bootle, Merseyside, L20 7HS.

Covers: Chesire, Cumbria, Greater Manchester, Lancashire & Merseyshire

Carlisle

2 Victoria Place, Carlisle, CA1 1ER.
Fax: 01228 548 482

Covers: Cumbria.

Yorkshire and North East

Leeds

The Lateral, 8 City Walk, Leeds, LS11 9AT.
Fax: 0113 283 4382.

Covers: West and North Yorkshire.

Sheffield

Health and Safety Executive, Foundry House, 3 Millsands, Riverside Exchange, Sheffield, S3 8NH.
Fax: 0114 291 2379.

Covers: East Yorkshire, North Lincolnshire, NE Lincolnshire, South Yorkshire and Kingston-upon-Hull.

Newcastle

Alnwick House, Benton Park View, Newcastle-Upon-Tyne, NE98 1YX.
Fax: 0191 202 6300

Covers: Northumberland, Tyne and Wear, Durham and Cleveland.

Scotland

Edinburgh

Belford House, 59 Belford Road, Edinburgh, EH4 3UE.
Fax: 0131 247 2121

Covers: Borders, Lothian, Central Perth, Kinross, Fife and Dundee.

Glasgow

1st Floor, Mercantile Chambers, 53 Bothwell Street, Glasgow, G2 6TS.
Fax: 0141 275 3100

Covers: West Scotland.

Aberdeen

Field Operations Division, Lord Cullen House, Fraser Place, Aberdeen, AB25 3UB.
Fax: 01224 252 525

Covers: Angus, Aberdeenshire, Moray and Shetland.

Inverness

Longman House, 28 Longman Road, Longman Industrial Estate, Inverness, IV1 1SF.
Fax: 01463 713 459.

Covers: Highlands and Orkney.

Statutory and Normative Documents Relevant to Gas Work

Please refer to Legislative, Normative & Informative Document List published by Gas Safe Register available from their website: https://engineers.gassaferegister.co.uk login.

CORGI*direct* Publications – 22

Gas – Domestic

Manual Series

GID1 Essential Gas Safety

GID2 Gas Cookers and Ranges

GID3 Gas Fires and Space Heaters

GID5 Water Heaters

GID6 Gas Meters (now included in GID1)

GID7 Central Heating – Wet and Dry

GID8 Gas Installations in Timber/Light Steel Frame Buildings

GID9 LPG – Including Permanent Dwellings, Leisure Accommodation Vehicles, Residential Park Homes and Boats

GID11 Using Portable Electronic Combustion Gas Analysers for Investigating Reports of Fumes

GID12 Using Portable Electronic Combustion Gas analyser – Servicing and Maintenance

FFG2 Fault Finding – wet central heating systems Domestic

Pocket Guides

USP1 The Gas Industry Unsafe Situations Procedure

CPA1 Combustion performance testing – Domestic

SRB1 Ventilation Slide Rule

GRB1 Gas Rating Slide Rule Natural Gas – Domestic

GRB2 Gas Rating Slide Rule LPG – Domestic

TTP1 Tightness Testing and Purging

TTG1 Terminals and Terminations

Design Guides

DCH1 Domestic Central Heating Installation Specification

DHG Domestic Heatuing Design Guide

UHDI1 Underfloor Heating Design and Installation Guide

Forms

All CORGI*direct's* gas forms carry Gas Safe Register® logo under licence from the HSE.

CP1	Gas Safety Record
CP2	Leisure Industry Landlord's Gas Safety Record
CP3FORM	Chimney/Flue/Fireplace and Hearth Commissioning Record
CP4	Gas Safety Inspection
CP6	Service/Maintenance Checklist
CP9	Visual Risk Assessment of Gas Appliances
CP12	Landlord/Home Owner Gas Safety Record
CP14	Warning Notice
CP26	Fumes Investigation Report
CP32	Gas Testing and Purging – Domestic (NG)
CP43	Risk Assessment for Existing Chimney Systems in Voids Where Inadequate Access for Inspection is Provided

Labels

WL1	'Danger Do Not Use' labels
WL5	Gas Emergency Control Valve labels
WL8	Compartment/Ventilation labels
WL9	Electrical Bonding labels
WL13	Serviced By Label
TG5	Tie on Uncommissioned Appliance labels
TG8	Void Property Tag
CP3PLATE	Chimney/Hearth Notice Plate

Gas – Non Domestic

Manual Series

ND1	Essential Gas Safety Non-domestic
ND2	Commercial Catering and Laundry Non-domestic
ND3	Commercial Heating Non-domestic

Pocket Guides

USP1	The Gas Industry Unsafe Situations Procedure
CPA2	Combustion performance testing – Non Domestic
VENT1	Boiler Ventilation – Non Domestic

Forms

CP15	Plant Commissioning/Servicing Record (Non-domestic)
CP16	Gas Testing and Purging (Non-domestic)
CP17	Gas Installation Safety Report (Non-domestic)
CP42	Gas Safety Inspection (Commercial Catering Appliances)
CP44	Mobile Catering Vehicle/Trailer Safety Check

Labels

WL1	'Danger 'Do Not Use' labels
WL10	Emergency Control Valve labels
WL35	Manual Gas Isolation Valve
WL36	Automatic Gas Isolation Valve

Electrical

Form

CP22 Minor Electrical Installation Works Certificate

Plumbing

Pocket Guides

CWCB Cleansing of Wet Central Heating Systems

Design Guides

DCH1 Domestic Central Heating Installation Specification

DHG Domestic Heatuing Design Guide

UHDI1 Underfloor Heating Design and Installation Guide

Forms

CP8 Domestic Unvented Hot Water Storage Vessel Commissioning/Inspection Record

CP20 Central Heating Commissioning/Inspection Record

CP33 Commissioning of Water Pipework

CP34 Central Heating Cleansing Record

CP40 Bathroom Quality Check Sheet

CP41 Combined Pressure Test Record Sheet

Label

TG9 Cleansing service label

Renewables

Manual Series

EEM1 Ground Source Heat Pumps

EEM2 Domestic Solar Hot Water Systems

EEM3 Domestic Biomass Systems

Form

CP11 Solar Thermal Commissioning Record

Labels

WL33 Solar Thermal Installation

WL34 Solar Thermal Pressure Relief Valve

Business

Form

CP19 Invoice form

